MULTIPURPOSE TREE SPECIES
FOR SMALL-FARM USE

W9-CKB-784

Proceedings of an international workshop
held November 2-5, 1987
in Pattaya, Thailand

Editors: **Dale Withington**
Kenneth G. MacDicken
Cherla B. Sastry
Norma R. Adams

Co-sponsors: Winrock International Institute for Agricultural Development
International Development Research Centre of Canada
Food and Agriculture Organization of the United Nations,
Regional Office for Asia and the Pacific

Co-publishers: Winrock International Institute for Agricultural Development, USA
International Development Research Centre of Canada

Contents

Preface

One of the most serious problems facing developing countries today is the volume and rate of deforestation. Destruction of tropical forests has been especially rapid in the last 50 years due to logging; transmigration; systematic clearing and conversion to agriculture; and unauthorized felling and nomadic farming for food, shelter, fodder, fuelwood, and pasture. Each year, some 11 million hectares are removed for these purposes and less than 10 percent are returned to forest vegetation. The loss of forest resources can only damage a nation's economy and its environment. The trend, therefore, has to be reversed, and all available land must contribute to rural income by alleviating shortages of fodder, fuelwood, and small timber, especially among resource-poor rural communities of Asia.

Fortunately for mankind, some interesting developments have resulted from national reawakening and donor agency support. One such movement is intensified planting of multipurpose tree species (MPTS) on smallholdings and farms in the developing world. These trees not only provide shade, shelter, food, fodder, fuelwood, and timber, but in some cases may improve soil fertility. Although the practice of planting trees around homesteads and small farms is age-old, scientific development and improvement of these tree species are recent. It is therefore not surprising that available published information on how MPTS are presently used is scarce.

The papers contained in this proceedings help fill this information gap. They were presented at the workshop *Multipurpose Tree Species for Small-Farm Use*, held November 2-5, 1987 in Pattaya, Thailand. This workshop, the first of its kind in Asia, focused on advancing biological and social-science research on MPTS, with the ultimate goal of improving the livelihood of small farmers in Asia. Specifically, the workshop was designed to:

o identify and describe what is known of small-farm use of MPTS in Asia,

o identify knowledge gaps and constraints to improving production and use of MPTS on small farms, and

o produce a strategy for future MPTS research efforts to improve the livelihood of small farmers in Asia.

Over 40 participants from eight Asian countries participated in this multidisciplinary and multidonor effort. Six major sessions, consisting of papers and posters, focused on the following:

o small-farm use of multipurpose tree species,

o role of Eucalyptus on small farms,

o nitrogen-fixing trees as MPTS for small-farm use,

o fruit trees and other woody perennials,

o socioeconomic considerations for MPTS research, and

o research strategies to fill information gaps.

We hope this publication serves as a useful guide to anyone involved in research targeted to uplifting the resource-poor.

Cherla B. Sastry
Senior Program Officer, Forestry
IDRC Singapore

Kenneth G. MacDicken
F/FRED Team Leader and MPTS
Network Specialist
Winrock International

Acknowledgments

We would like to thank the Kasetsart University Faculty of Forestry for organizing and conducting the workshop field trip and Dean Sathit Wacharakitti for the welcome address. We would also like to thank each participant for preparing an abstract for review by the Participant Selection Committee and for preparing and presenting papers or posters at this workshop. We also acknowledge the efforts of the three co-sponsoring organizations: Food and Agriculture Organization of the United Nations, Regional Office for Asia and the Pacific; International Development Research Centre (IDRC) of Canada; and Winrock International Institute for Agricultural Development, Forestry/Fuelwood Research and Development (F/FRED) Project. Finally, we would like to thank the publications staff of Winrock International and IDRC who helped make this publication possible.

Session I: Small-Farm Uses of Multipurpose Trees

Chairman: George F. Taylor II
Discussants: Pradeepmani Dixit
Cor Veer

Session I Summary

George F. Taylor II

U.S. Agency for International Development
Kathmandu, Nepal

Session I introduces the workshop's theme by examining small-farm uses of multipurpose tree species (MPTS) across the tropics of Asia. K.G. Tejwani reviews in detail research on MPTS used by small farmers in India. While he notes substantial research on several species, most has not been devoted to understanding MPTS uses within the integrated production system context of small farms. Tejwani presents a typology of agrosilvicultural, silvoagricultural (with the tree component predominant), pastoral/silvicultural (with grazing as the major component), and agrosilvopastoral systems, noting research results where available. On the whole, his review indicates that virtually none of the many species grown by farmers on agricultural fields, farm boundaries, field bunds, and as woodlots has been researched. In fact, these species have been studiously ignored. While the species grown by farmers have been catalogued and MPTS practices have been described, researchers cannot yet claim real progress in making these traditional systems more productive. Important challenges and opportunities thus exist in an essentially uncharted area of research.

Evans examines tree planting on small farms in the tropics and reviews why farmers grow trees, what they grow, and what they want to grow. He notes that research on a tree crop must not be an end in itself. The primary focus of research should be farmers, their overall production systems, and their priorities. As a start, farmers need to determine what needs to be researched. The challenge is not doing more research to fill knowledge gaps, he says, but principally in finding ways to fill gaps in farmers' information. He cautions against excessive scientific formality as a possible source of conflict and misunderstanding between scientific culture and local needs and perceptions. He stresses the importance of effectively communicating ideas to farmers by meeting their needs at their point of reference. Among other things, this requires that the technologies developed must have low input, low risk, and provide high returns. In addition, these technologies should build on traditional practices, be developed in close cooperation with the local people who determine how resources are used and managed, and include an element of ongoing evaluation. As to which species should be chosen for priority research, Evans suggests that, provided perceived needs are met, native species should be chosen before exotics, MPTS before single-purpose species, and nitrogen-fixing species before others.

Session I continues with four papers on small-farm uses of MPTS in parts of Bangladesh, India, Nepal, and Thailand. The paper by Abedin et al. focuses on MPTS of the Ganges Floodplain soils of Bangladesh. It provides results of a survey carried out by the On-Farm Research Division of the Bangladesh Agricultural Research Institute. Using five representative locations, a pre-tested questionnaire was used to collect information from marginal, small, medium, and large farmers. The purpose was to understand species distribution on farms, tree-crop interactions, and the various uses of predominant species. Survey results indicate wide variability in the species grown and the density of on-farm planting. None of the trees are grown for a single purpose. Although there has been increased tree planting in recent years, many trees have been cut before attaining the farmer-defined optimum age. Trees are important as insurance during difficult periods, particularly for marginal and small farmers. On tree/crop associations, the survey found virtually all trees grown with all crops. Predominant associations with different crops and crop sequences are grouped on the basis of land type. The authors conclude that as this is, to their knowledge, the first organized study on trees grown on crop fields in Bangladesh, there are many agronomic and socioeconomic aspects of agroforestry practices that could not be covered adequately here. They identify an impressive array of issues that deserve careful consideration in future studies, including detailed analysis of tree-crop interactions in relation to several factors, identification of improved management practices for both trees and crops, the role of trees in food security for small farmers, the economics of various agroforestry production systems, and a better understanding of the historical and social processes involved in species distribution and use.

Sapkota's paper on MPTS for small-farm use in Nepal briefly describes the demand for multipurpose species in Nepal and presents results from the Tribhuvan University/IDRC Farm Forestry Project. As these results are also covered in Sessions IV and V (Karki and Dixit, respectively), they will not be discussed at length here. The Farm Forestry Project is a pioneering research effort in Nepal, involving faculty from the Institute of Agriculture and Animal Science and the Institute of Forestry. They are working together on an action research program designed to better understand and encourage increased use of multipurpose trees on farms in selected communities.

Hegde's paper complements the overview presented by Tejwani. After introducing major proportions (86 million small-farm families with a substantial and increasing livestock population requiring large amounts of fuel and fodder currently unavailable on a sustainable basis), Hegde outlines the scope for MPTS, notes several programs currently underway, and presents in some detail various species most suitable for cultivation by small farmers in different climatic zones. Profitability, presented as a key to the successful expansion of tree planting, is directly linked to marketing. Hegde notes that programs will not succeed without setting up suitable infrastructure for marketing and, if possible, post-harvest processing of the products.

Suree Bhumibhamon's paper on MPTS for small-farm use in the Central Plain of Thailand presents results of recent surveys. Like Abedin et al., Suree found considerable diversity in the types of trees planted and the variety of their uses. He highlights numerous important socio-cultural factors that bear directly on species selection in central Thailand. For example, Borassus is planted extensively in the western portion of the Central Plain in areas of Burmese cultural influence, and is completely absent in the eastern region under Thai cultural influence. Socio-cultural factors, including strict restrictions on the planting of certain species (e.g., *Ficus religiosa*) and cultural mandates to plant selected species at various points around homesteads, are shown as fundamental determinants of species selection and planting practices.

The final paper in Session I is a case study presented by Zhu Zhaohua on the extensive use of a single species, *Paulownia elongata*, intercropped with food and cash crops in Northern China. Zhu, whose presentation includes both a paper and videotape, describes the work he has led at the Chinese Academy of Forestry since 1973. The story is an impressive one of careful, systematic, and extensive research and extension efforts on this indigenous, multipurpose species. Some 230 scientists were involved in the first national Paulownia research coordination meeting in 1976, which has been followed by the planting of some 500 million trees over 1.5 million hectares of farmland on the plains of Northern China.

In his paper, Zhu presents the main research results to date. These have concentrated on determining optimum spacing and management methods through careful observation of the ecological, biological, and socioeconomic effects in various cropping patterns. Paulownia intercropping has been developed on the basis of farmers' experience, and the rapid, large-scale extension of this species with continual improvements in genetic stock and cultural practices has been based on close cooperation between research scientists, extension agents, and farmers. Extension methods used have included the news media, several large training courses, the production of three films shown throughout China, and extensive use of demonstration farms.

As noted earlier, the seven papers presented in Session I outline the enormous diversity of small-farm uses of multipurpose species across Asia and highlight a wide range of issues, challenges, and opportunities for future research. In attempting to summarize the most salient points of the papers and the discussion they generated, I propose that the overriding challenge facing this workshop's participants is to "Identify Feasible Solutions for Multipurpose Tree Species (MPTS) Use," as outlined below:

Identification and
Description of
Existing or
New
Types of production systems followed by
Interdisciplinary
Farming system based
analYsis carried out by

Foresters,
Economists,
Agronomists,
Social scientists
Innovative plant/animal breeders,
Botanists,
Livestock experts, and
Extension specialists.

Seeking
Optimal
Low input/low risk land
Use
Technologies designed to
Increase
On farm
iNcome and well-being with particular focus on
Small and marginal farmers, through

Farmer (including female)
Oriented
Research to increase/improve use of

Mix of
Preferred
Tree
Species that are

Understood by farmers, are
Socially acceptable and technically sound, and
Enhance both the productivity and sustainability
     of farming systems.

Small Farmers, Multipurpose Trees, and Research in India

K. G. Tejwani

International Centre For Integrated Mountain Development
Kathmandu, Nepal

This paper focuses on the role of multipurpose trees in the socioeconomic life of small farmers in India and research efforts to improve current practices. Indian farms are usually small, and farmers have traditionally integrated trees in their production systems. This fact has been recognized by researchers and development agencies in the recent past. Agroforestry systems contribute much to the use of tree products by small farmers. Many specific examples of tree species used under various agro-ecological conditions and available research data are described. Current and future directions of research are discussed. This author concludes that efforts to date have mostly described the practices and have not greatly increased their productivity. This is where challenges and opportunities await scientists.

India is a country of small, subsistence farmers, most of whom do not have access to irrigation. These farmers have to produce their own food, fodder, and fuel. Already they have integrated trees with crop production more intensively than many researchers and extension agencies realize. Indian farmers not only make use of the trees grown on their own agricultural land, but they also use trees grown in government forest and community grazing lands.

Traditional agroforestry practices are common in India. Some have been researched extensively-- e.g., growing tea under shade trees (since the late 19th century) and fuelwood plantations (since the 1950s). Many institutes conduct research on agroforestry systems. Realizing their importance, an All-India Coordinated Research Project on Agroforestry was begun in 1983 to operate initially at 8 institutes of the Indian Council of Agricultural Research and 12 universities. This project was expanded considerably in the seventh Five-Year Plan (1985-90).

Agrosilviculture Systems

Trees in Agricultural Fields

Farmers grow trees in agricultural fields for many reasons. *Prosopis cineraria* and *Zizyphus nummularia* are grown extensively in arid parts of Rajasthan and Gujarat states, and in areas of Punjab and Haryana that border on the Rajasthan Desert. Annual rainfall varies between 200 to 500 mm. Soils are low in organic matter and have little water-holding capacity. Unstable crop production leads to low and unprofitable yields of rainfed crops (*Pennisetum typhoides*, *Sorghum vulgare*, *Sesamum indicum*, *Phaseolus radiatus*, *P. aconitifolius*, and *Cyamopsis tetragonoloba*).

Livestock husbandry plays an important role in the economy. Cattle, sheep, goats, and camels are the most important livestock, although their productivity is generally low (Acharya, Patnayak, and Ahuja 1977). By arid zone standards, the Rajasthan Desert is one of the most populated deserts of the world, having an average density of 48 people/km^2 compared to an average density of 3 people/km^2 in most other desert regions (Malhotra 1977). Size of individual holdings varies from 3.27 to 24.6 ha, increasing with aridity (Jodha 1977).

Prosopis cineraria grows slowly in the early stages. It takes 10-15 years to be useful in the 200-350 mm rainfall zone, and 20 years in the 100-200 mm rainfall zone (Mann and Saxena 1980, Saxena 1980). The planting density of *P. cineraria* varies from 5 to 80 trees per ha, depending upon soil and rainfall. Its density increases from western to eastern Rajasthan as rainfall increases and soil regeneration conditions improve (Shankar 1980b). The tree has a high regeneration capacity.

Every part of the tree is used. Leaves and tender twigs are used as fodder (green and dried), branches as fuelwood, thorny twigs as fencing material, green and dry pods as vegetables, ripe pods as fruit, and flowers and bark as medicine. The bark and galls formed on branches are used for tanning; gum is used in making sweets; and the roots are used for making cot frames, handles for agricultural implements, rakes, bullock cart frames, and butter churning sticks.

The tree has many indirect benefits. Apart from improving soil fertility (Aggarwal 1980), it binds soil, decreases the velocity of hot summer winds, and provides shade to humans and animals during the summer and greenery in acute dry periods. As a source of income during droughts, it helps farmers maintain economic self-sufficiency. No wonder the tree is called the "king of the desert" and is an integral part of cultural and socio-religious festivities, including weddings, funerals, and other rites (Purohit and Khan 1980, Shankar 1980a).

The system of growing agricultural crops and *Prosopis cineraria* has been described extensively (Central Arid Zone Res. Inst. 1981, Mann and Saxena 1980, and Shankarnarayan 1984). Although intensive efforts have been made to improve agricultural crop yields, none has been made to improve the tree component or the system as a whole. Although the trees are well-maintained by good farmers, lopping for fodder and other uses is often indiscriminate. The methods of propagation, growth, and utilization of *P. cineraria* are based on traditional knowledge. Research needs are increasing production by joint management of the trees and crops; selection and evaluation of fast growing trees with high foliage and fodder production (commensurate with high protein content, high digestibility, and low tannin content of leaves); development of management practices with respect to the lopping cycle, lopping intensity, plant spacing, control of gall formation and insect pests, etc.; and economic evaluation of the system with respect to direct and indirect benefits (Mann and Saxena 1980).

Zizyphus nummularia is an important woody shrub maintained extensively in cultivated fields in the arid zone. There is often a high density of this shrub (250-500 per ha) in cultivated fields with a well-defined calcium carbonate pan at 100-150 cm soil depth. Exposed, gravelly plains support a density of 120-150 shrubs per ha (Saxena 1984).

The air-dried foliage is used as fodder. Yearly cut shrubs on cultivated fields do not bear fruits. In dry years, farmers still can harvest the dry fodder. *Z. nummularia* has remarkable regenerative powers. Its branches and twigs are used as fencing material and fuel.

Other indigenous trees and shrubs in the arid zone of Rajasthan identified as providing valuable, palatable forage are *Acacia nilotica*, *A. senegal*, *Albizia lebbek*, *Anogeissus rotundifolia*, *Azadirachta indica*, *Grewia tanax*, and *G. spinosa* (Ganguli, Kaul, and Nambiar 1964). Most of these tree and shrub species are multipurpose (Table 1).

Other species are grown in agricultural fields in the Himalayan regions in Himachal Pradesh and Uttar Pradesh at elevations of 500-2,300 m. The climate is subtropical to temperate, according to elevation, with distinct rainy, winter, and summer seasons. Mean annual rainfall is 1,000-2,500 mm. Topography and soils vary. Agriculture is generally subsistence with some cash crops. Individual land holdings are usually less than 2 ha and fragmented (Ghildyal 1981, Shah 1981). Major agricultural crops are maize, paddy rice, wheat, barley, and potato. Off-season crops are tomato, cabbage, and ginger. Main fruit trees are apple, plum, peach, apricot, and mango (at lower elevations). Fruit trees also are grown in agricultural fields in an agro-horticulture system. Only 10% of the agricultural fields are irrigated. Farmers grow *Grewia optiva* (300-2,500 m), *Morus serrata* (1,000-2,700 m) and *Celtis australis* (500-2,500 m) for fodder, fiber, fuel, and timber.

Grewia optiva is a small to medium-sized deciduous tree. It is multipurpose and pollards and coppices well, especially during winter when no other green fodder is available. Its bark yields fiber; its leaves provide nutritious fodder; the fruit is edible; and its wood is used for axe handles, shoulder poles, and cot frames. It is seldom used as fuelwood because of its unpleasant odor (Gupta 1956). It is most often propagated by planting seedlings on terrace risers.

Celtis australis is a moderate to fairly large deciduous tree that grows over a wide altitudinal range of 500-2,500 m. It pollards and coppices well. It is raised primarily for fodder, but also is used for firewood, tool handles, and plows. Summer-lopped fodder has higher nutritive value than fodder lopped in autumn.

Table 1.   Important multipurpose tree species grown by farmers in India.

Botanical Name	Climatic Zone	Important Uses
Acacia albida (A. leucophloea)	Dry to moist tropical	Timber turning, fodder, and fuel
A. nilotica (spp. indica)	Dry to moist tropical	Fodder, soil conservation, agri-cultural tools, tannin, and fuel
A. planifrons		
A. senegal	Dry tropical	Gum, fodder, and fuel
Achras zapota		
Ailanthus excelsa	Dry to moist tropical	Fodder and soil conservation
Albizia falcataria	Tropical moist	Soil conservation and fuel
A. lebbek	Very dry to wet tropical	Fodder, oilseed, soil conservation, timber, and fuel
Anacardium occidentale	Dry to moist tropical	Fruit, fuel, and soil conservation
Annona reticulate	Moist tropical	Fruit, timber, and fuel
A. squamosa		
Areca catechu	Moist tropical	
Artocarpus altilis		
A. hirsuita		
Azadirachta indica	Very dry to moist tropical	Fodder, soil conservation, timber, medicinal oil, and fuel
Bambusa arundinaea		
Bassia latifolia		
Bauhinia purpurea	Moist tropical	Tannin, fodder, and agricultural tools
Borassus flabellifer		
Calophyllum inophyllum		
Calotropis procera		
Cassia siamea	Dry and moist tropical	Soil conservation, fodder, and fuel
Casuarina equisetifolia	Moist tropical	Timber and fuel
Ceiba pentandra		Floss and fuel
Celtis australis	Moist tropical	Soil conservation and fuel
Citrus sinensis (orange)		
Cocos nucifera	Moist tropical	Fruit, oil, fiber, and fuel
Commiphora caudata		
C. berryii		
Dalbergia sissoo	Dry to moist tropical	Timber, soil conservation, fodder, and fuel
Delonix delta		
Emblica officinalis	Very dry to wet tropical	Fruit and fuel
Eriobotrya japonica (loquat)		
Erythrina indica		
Eucalyptus camaldulensis	Dry to moist tropical	Industrial timber, oil, gum, tannin, ornamental, and soil conservation

Table 1. Continued.

Botanical Name	Climatic Zone	Important Uses
E. globulus	Moist tropical and subtropical	Timber, pulpwood, fuel, and oil
E. tereticornis	Dry to moist tropical	Pulpwood and fuel
Ficus benegalensis	Moist tropical	Fruit, fodder, fuel, and along roads
F. religiosa		
Garcinia indica (mangosteen)		
Gliricidia sepium		
Grewia spp.	Subtropical to temperate	Soil conservation, fodder, fuel, and timber
G. optiva		
G. spinosa		
G. tanax		
Hardwickia binata	Dry tropical	Fodder, fuel, and agricultural tools
Holoptelea integrifolia	Moist tropical	Fodder, timber, and fuel
Leucaena leucocephala	Dry to moist tropical	Fodder and soil conservation
Lichi chinensis (lichi)		
Madhuca longifolia var. *latifolia*	Dry to moist tropical	Oilseed, fruit, timber, and fuel
Malus sylvespris (apple)		
Mangifera indica	Moist tropical to subtropical	Fruit, fodder, timber, packing cases, plywood, and fuel
Morinda tinctoria		
Moringa oleifera	Moist tropical	Fruit and fodder
M. spp.		
Morus serrata	Subtropical	Industrial timber, small timber poles, fodder, ornamental, and soil conservation
Phoenix dactylifera (date palm)		
Pithecellobium dulce	Tropical	Tannin, dye, medicine, and timber
Pongamia pinnata	Dry to moist tropical	Oilseed, soil conservation, and fuel
Populus spp.	Subtropical to temperate	Soil conservation, fodder, packing cases, pulpwood, and fuel
Prosopis chilensis	Very dry to moist tropical	Soil conservation, fodder, and fuel
P. cineraria	Dry tropical	Fodder, fuel, fruit, cultural uses
P. juliflora		
Prunus domestica (plum)		
P. amygdalus (almond)		
P. armeniaca (apricot)	Temperate	Small timber poles, fuel, fruit, ornamental, and soil conservation
P. persica (peach)	Temperate	Small timber poles, fuel, edible fruits, oil, gum, tannin, ornamental, soil conservation, and medicine

Table 1. Continued.

Botanical Name	Climatic Zone	Important Uses
Psidium guajava (guava)	Tropical and subtropical	Edible fruit and fuel
Pterocarpus marsupium	Moist tropical	Fodder, timber, medicine, and fuel
P. santalinus	Dry tropical	Fodder, gum, dye, and timber
Quercus incana poles, fuel, fodder, tannin,	Temperate	Industrial timber, small timber ornamental, and soil conservation
Q. spp.	Temperate	Fodder and fuel
Salix spp.	Temperate	Soil conservation, match wood, sporting goods, fodder, and fuel
Santalum album	Dry tropical	Industrial timber, oil, ornamental, soil conservation, and medicine
Sesbania grandiflora	Tropical moist to subtropical	Fence posts, tannin, gum, medicine, fodder, fuel, and pulp (fiber)
Swietenia mahagoni	Subtropical, tropical moist	Industrial timber, small timber poles, gum, tannin, ornamental, and soil conservation
S. macrophylla		
Tamarindus indica	Dry and moist tropical	Fruit, timber, and fuel
Tectona grandis	Dry and moist tropical	Industrial timber, small timber poles, fuel, fodder, tannin, ornamental, and medicine
Thespesia populenea		Timber and fruit
Wrightia tincotoria		
Zizyphus jujuba	Dry tropical	Fodder, edible fruit, tannin, ornamental, soil conservation, and medicine
Z. nummularia	Dry tropical	Fodder, fruit, fences, and medicine

Morus serrata is a moderate to large deciduous tree found in mixed lower western Himalayan forest at 1,200-2,700 m elevation. It is found in subtropical as well as temperate climates and is raised for fodder, timber, and fuelwood. Its leaves make a valuable base for sericulture. Its wood is excellent for furniture, carving, cabinetmaking, sporting goods, toys, and agricultural tools. Its fruit is sweet and edible. It is lopped in summer (Gupta 1956).

Grewia optiva, Celtis australis, and *Morus serrata* can be lopped at 10 years and continue to be lopped for 30 years. They are raised on field bunds/terrace risers at a 5-8 m linear spacing. The average number of trees per farm holding is 30. There is no specific mixture of trees and crops, and animals are part of the system. *Moringa oleifera,* Quercus spp., Bauhinia spp., and other minor species are also grown (Table 1).

Eucalyptus

Many Eucalyptus species are grown in and around agricultural fields. *Eucalyptus globulus* was introduced in 1843 (Samraj 1981) from Tasmania into the temperate, high-rainfall Nilgiri Hills in Tamil Nadu. It is grown by farmers on the outer edges or risers of bench terraces. The terrace crops are potato, cabbage, and cauliflower. The trees are grown primarily for harvesting of leaves to extract oil. No economic analyses of this practice are available. It has been reported that the yield of potato is adversely affected by growing *E. globulus* (Samraj, Haldorai, and Henry 1982, Tejwani 1981). Farmers apparently are willing to sacrifice partial yield of cash crops to be able to harvest Eucalyptus leaves and fuelwood.

A Eucalyptus hybrid is being grown extensively in many parts of India, mostly on field bunds. The economics of this practice are unknown. It is reported that this hybrid reduces the yield of rainfed wheat (Khybri and Ram 1984) and that *E. citriodora* reduces yield of many rainfed monsoon crops (e.g., *Phaseolus radiatus* and *Sorghum vulgare*).

Acacias

Acacia nilotica subspecies *indica*, found in dry to moist tropical climates, is allowed to grow extensively in and around agricultural fields throughout India. Almost every part of the tree is used. Leaves, young twigs, and pods are used as fodder, branches are used as fuelwood and fencing material, bark for tannin extraction, and timber for making agricultural tools. It also yields gum.

Jumbulingam and Fernandes (1986) reported that *Acacia leucophloea* is grown in over 100,000 ha in the dry tracts of Coimbatore and Periyar districts of Tamil Nadu. Rainfall is erratic from year to year. *A. leucophloea* regenerates profusely when the land is plowed after the first rain. Farmers encourage the growth of young seedlings and also sow crops like *Pennisetum glaucum* and *Dolichros biflorus* (horse gram). Depending upon the rainfall, the crops are harvested for grain or fodder.

At the end of the first year, about 1,000 tree seedlings remain. This number is further reduced by natural mortality to around 500-800 per ha at the end of the third year. Farmers continue to till the soil up to the base of the trees to sow crops. After about 10 years, the trees, which are about 10 m tall

and 20 cm diameter at breast height, are thinned to 60-100 trees per ha. It is reported that sorghum yields 20-23% more dry fodder when it is grown in association with *Acacia leucophloea* (Seshadri 1976). The trees develop a large spreading canopy when they are 15-20 years old and are thinned out to 25-60 trees per ha. Such trees yield up to 100 kg of pods annually, which provide excellent high-protein fodder supplement in the dry season. Wood is used for timber and fuel. Due to erratic rainfall and socioeconomic reasons, farmers increasingly plant *Cenchrus ciliaris* grass instead of cereals and pulses under the trees. Thus, agrosilvicultural practices are changing to pastoral silviculture.

Jambulingam and Fernandes (1986) reported that *Acacia planifrons* grows extensively along the coast in Tirunelveli and in localized pockets of Ramanathapuram and Coimbatore districts. Species regeneration is achieved by penning goats that have been fed the pods. To ensure successful regeneration (up to 80%), the land is plowed prior to penning. At the end of one year, the population is thinned to 1,250-1,500 seedlings per ha at an approximate spacing of 3 x 3 m or 3 x 4 m. Between the trees, sorghum, pulses, peanut, or chili are grown for 4-5 years until the tree canopy closes. The trees are felled at 7-10 years, depending on the socioeconomic conditions of farmers. Trees provide wood for fuel and valuable fodder pods during the dry season.

Other Tree Species

Leucaena leucocephala, which has attracted much attention recently, is being attributed many benefits. While it is popularized for its reported good qualities, little information is available about its interaction with agricultural crops or its economics. When it is intercropped, its fodder and wood yield are reduced and crops grown with it yield less. The types of crops grown and their yields are determined by the amount of rainfall. However, it is premature to draw valid conclusions at this early stage (Khybri, Ram, and Bhardwaj 1984; Mittal and Singh 1983; Prasad and Verma 1983).

Jambulingam and Fernandes (1986) estimated that about 40 million trees of *Borassus flabellifer* grow in the plains of Tamil Nadu. Usually the tree is grown randomly on croplands, but it also is planted on the bunds of paddy fields. It grows on a wide range of soil types from littoral sands to fine clays and tolerates water logging and salinity well. It is surmised that the effect of its shade on crops is negligible because of the tree's small crown and usual spacing of 10-15 m. Sale of various products obtained from the palm probably compensate for reduced crop yields due to root competition. Moreover, the tree provides cash income during the dry season when no crops are grown.

The major product of the palm is nera, a sugary exudate from the cut end of a flowering spathe. Tapping of nera usually begins when the tree is 14-16 years old and is limited to about 5 months each year. Other products from the palm include leaves and the fiber from petioles, widely used for thatching and a wide variety of handicrafts.

Jambulingam and Fernandes (1986) have reported that *Tamarindus indica* grows on a wide variety of soils in dry to moist tropical climates. Spacings commonly adopted are 8 x 8 m, 8 x 12 m, and 12 x 12 m. The trees begin producing fruit pods in about 7 years and the yield stabilizes at about 15 years. A mature tree (20 years) can yield 100-500 kg of pods a year, giving 40-250 kg of edible pulp. Usually, pod yield is cyclical with bumper yields every third year. The pulp is used as a condiment; the seed is used to extract starch for textile, paper, and jute production; and wood is used as fuel and timber. Depending upon site conditions, the tree can mature to 20-25 m with a dense crown. Reduced yields of interplanted crops caused by shading is reportedly severe. Once pod yields decline (after 50 years or more), the trees are cut down.

Jambulingam and Fernandes (1986) have reported that the Kapok tree (*Ceiba pentandra*) is grown by the farmers in the Madusai, Ramanathapuram, Timnelvi, Coimbatore, Periyar, and Salem districts of Tamil Nadu. The trees are planted in agricultural fields and field bunds at a spacing of 5 x 5 m or 8 x 8 m, and intercropped with cereals or pulses until their canopies close at about 7-8 years. The trees begin producing fruit at 3-4 years. Stabilized pod yield, usually obtained in the seventh or eighth year, can vary from 1,000-1,200 pods per tree depending upon management. Well-managed 15-year-old trees yield 2,000-3,000 pods annually. The small-pod variety yields up to 6 kg of floss per 1,000 pods, while the large-pod variety yields up to 7-9 kg. The wood is soft and used by the match industry. Yields from interplanted crops are expected to decline from the fourth to seventh year when the tree canopy closes, especially with the Singapore variety, which has a dense canopy. If a farmer wishes to prolong intercropping, the local variety, with its sparser canopy, should be used. The Singapore variety can be used on field bunds and boundaries.

Prosopis juliflora (mesquite) is used by farmers in Tamil Nadu to reclaim wastelands and saline soils for agricultural crops (Jambulingam and Fernandes 1986). In Ramanathapuram district of Tamil Nadu, farmers use *P. juliflora* as a fallow species for four years. This improves the soil sufficiently so that farmers can grow annual food crops for at least two years, after which the trees are allowed to grow for another four years. The tree is a major source of fuelwood and income for many rural people in Andhra Pradesh and Tamil Nadu. The wood yields good quality charcoal.

Trees on Farm Boundaries

Trees grown in the above setting also are grown on farm boundaries in India. Apart from their specific uses, these trees also help demarcate farm and field boundaries and serve as windbreaks and shelterbelts. *Leucaena leucocephala* planted along field boundaries in Bundelkhand is reported to yield 3.21 tons of fuelwood and 1.44 tons of forage per ha three years after planting (Patil, Deb Roy, and Pathak 1981). Willows (*Salix* spp.) and poplars (*Populus* spp.) are grown on field boundaries in Lahaul in Himachal Pradesh. Willows provide fodder and wood useful as fuel and for making sporting goods. Poplars yield fodder and their wood is used for making packing cases, pulpwood, matches, and fuel. *Delonix delta*, a leguminous tree, is grown in Tamil Nadu on field bunds and boundaries to provide green manure for rice fields. The trees are pollarded at a height of about 2 m to yield 20-50 kg green leaves for incorporation into soil. Branches are used as fuelwood (Jambulingam and Fernandes 1986).

Eugenia jambolana and *Gliricidia* spp. also are grown as green manure trees. *Tectona grandis* is popular with the farmers of Tanjavur district in Tamil Nadu. *Prosopis juliflora* is grown extensively.

Information is unavailable on the interaction of these species with agricultural crops and the economics of their production. Like *Prosopis cineraria*, some leguminous trees (e.g., *Acacia nilotica* or *Tamarindus indica*) and other species (e.g., *Ceiba pentandra*) shed leaves that may benefit the soil. *P. juliflora*, on the other hand, affects agricultural crop yields adversely when grown on farm boundaries in semi-arid areas. Reports indicate sorghum yield was adversely affected 30 m from the tree or 4-5 times the tree height (Prajapati 1971).

Wood Lots

In many parts of India, farmers grow trees in separate wood lots. This practice is expanding rapidly as fuelwood shortages become more acute. *Casuarina equisetifolia* is grown extensively in Andhra Pradesh, Tamil Nadu, and Karnataka states on lands too poor or unsuitable for profitable agriculture. This method was practiced by farmers in Tamil Nadu long before India began to popularize farm forestry. The trees are planted densely and harvested in 5-7 years when a yield of about 120 tons of fuelwood per ha may be obtained (Reddy 1981). Casuarina is clear-felled, and stumps, which also secure a good price, are uprooted. Often the land is sequentially brought under agriculture for a year or two and then allowed to revert back to Casuarina.

Farmers in the Punjab have started diversifying their farming practices due to market pressures. They now perceive that an acute shortage of fuelwood makes Eucalyptus plantations economically more viable than annual crops. Farmers in Assam grow wood lots of bamboo along with paddy fields. Some tree species grown only under specific conditions include *Bambusa arundinacea* in depressed and waterlogged areas in Andhra Pradesh and *Pterocarpus santalinus* in the arid/semi-arid Cuddapah District of Andhra Pradesh (Reddy 1981).

Silvoagricultural Systems

Under silvoagricultural systems, trees are the major component and agricultural crops are grown either sequentially, as in shifting cultivation, or as intercrops, such as in taungya systems, intercropping under coconut, areca nut and horticultural trees, and growing cardamom under forest trees (Tejwani 1987a and 1987b).

Taungya systems are usually practiced by the Forest Department, which leases government land to small farmers who have no claim on the trees.

Recently, large farmers have started growing poplars under a taungya system. Since the system concerns the government or large farms, the tree components are not described here.

Growing Agricultural Crops with Commercial Trees

Since commercial nut and fruit trees require much time to mature and yield cash returns, growers seek ways to generate interim returns. The most outstanding examples are intercropping with coconut (Nair 1979), areca nut (Nelliat 1979), many horticultural trees, and cardamom.

Coconut (*Cocos nucifera*) is essentially a small-growers crop in India. Over 90% of the coconut holdings in southern India are less than one ha and the average is only 0.22 ha. Hardly 2% of the holdings have an area of two or more ha.

Coconut grows in eastern and southern India, with temperature determining the boundaries with respect to altitude and latitude. Kerala, Tamil Nadu, Karnataka, and Andhra Pradesh in south India account for over 90% of coconut production area. A mean annual temperature of 25°C with a diurnal variation of 6-7°C is optimal. Coconut does well if the annual rainfall is 1,500-2,250 mm and is well distributed (Menon and Pandalai 1960).

Within 8 years of planting the coconut palm, the crown size increases gradually and the interspaces can be used for intercropping with annuals or short-duration crops that do not compete with the developing palm. From 8-25 years, when the canopy covers 80% of the ground and the trunk is short, there is little or no scope for growing other crops. The most important period for intercropping occurs after 25 years (Nelliat 1979). A wide variety of crops (tuber crops, tapioca, elephant foot yam, rice, banana, pulses, oilseed, cacao) are grown under the coconut canopy. Data on yield of coconut and the crops growing with it in farmers' fields are not available.

Coconut is truly a multipurpose tree. All its parts are used by the farmer and for industry. The Central Plantations Crops Research Institute, Kasargod, Kerala, has been engaged in research on coconut intercropping. Many important aspects, such as selection of compatible crops, spacing, water management, fertilizer application, soil fertility, pest and disease management, long-term production, protective benefits and interactions, and socioeconomic concerns need to be researched.

Areca nut

India produces 75 % (about 191,400 tons from 184,500 ha) of the world's areca nut (*Areca catechu*) (Vellapan and George 1982). It grows in Kerala, Karnataka, Tamil Nadu, Assam, Meghalaya, and West Bengal. For good growth, a temperature range of 18-38°C without extreme variation and a well-distributed annual rainfall of 2,250 mm appear necessary. Because of its height, crown shape, and wide spacing, areca can be grown with annual, biennial, and perennial crops such as banana, cacao, black pepper, pineapple, betel vine, elephant foot yam, tapioca, turmeric, ginger, and Guinea grass. Reports indicate that the yield and condition of areca nut are not affected adversely when intercropped with elephant foot yam, yam, sweet potato, tapioca, banana, pineapple, ginger, turmeric, betel vine, black pepper, and small cardamom. Increased yields of areca nut have been reported when cacao (Bhat 1978), cinnamon, clove, nutmeg, and coffee are grown with it (Sanamarappa and Murlidharan 1982).

Farmers use every part of the areca nut tree (nut, husk, leaf, and stem). This crop is studied extensively by the Central Plantation Crops Research Institute, Kasargod, Kerala. With respect to intercropping, areas requiring research emphasis are similar to those of coconut.

Horticultural Trees

Some fruit trees that take longer to bear fruit and are spaced widely are amenable to intercropping. Examples are mango (*Mangifera indica*), chiku (*Achras zapota*), loquat (*Eriobotrya japonica*), orange (*Citrus sinensis*), guava (*Psidium guajava*, apple (*Malus sylvespris*), apricot (*Prunus armeniaca*), peach (*P. persica*), almond (*P. amygdalus*), plum (*P. saliciana*), nectarine, lichi (*Litchi chinensis*), and date palm (*Phoenix dactylifera*). Most orchards are irrigated. Although information is incomplete, different species may require different management practices (Gangolly et al. 1957, Singh, Krishnamurthi, and Katyal 1963). About 2.1 and 2.5 million ha were under fruit trees in India in 1976 and 1982, respectively. This represents a 19% increase in area over 6 years, indicating that many young orchards are fit for intercropping. Mango is the most dominant fruit, occupying over 40% of the total area (Ministry of Agriculture 1983).

Though intercropping is generally practiced for economic reasons, neither detailed descriptions nor research results are available. Mango is a multipurpose tree species. Most other fruit tree species are valuable sources of fodder and fuel. The Indian Horticultural Research Institute in Bangalore is responsible for fruit tree research. However, as mentioned above, intercropping with horticultural trees has received no attention.

Pastoral/Silvicultural Systems

Indian farmers have integrated crop farming with large numbers of livestock (Ministry of Agriculture 1985). Most of these livestock graze freely on village lands and adjoining forest lands, where villagers have grazing rights. Livestock also graze freely on agricultural lands where no crops grow. This is well regulated by traditional and social practices. The agro-ecological conditions in India lead to tree growth even in grasslands (except in alpine pastures). The practices that include grazing as the major component and a scattering of trees are referred to as pastoral/silvicultural systems. The practices that have trees as the major component with associated grazing are referred to as silvopastoral systems (Tejwani 1987a and 1987b).

Pastoral/silvicultural systems are practiced extensively by most farmers in the country. Types of grasses and trees vary with local conditions. Some practices are well developed, particularly in the semi-arid and arid zones.

In the semi-arid zones, farmers leave the land fallow with existing trees. Natural succession of grasses follows. The grass crop is protected for 1-3 years, depending on farmers' economic conditions. The grasses (usually the Sehemia-Dicanthium grassland type) grow fast and can peak in 4-10 years, depending on location and site conditions.[*] Farmers determine which of the many trees represented in this climate should be encouraged to be grown or planted. Commonly planted trees are Eucalyptus hybrid and *Casuarina equisetifolia*. Palms like *Borassus flabellifer* and

_____

[*] S. Chinnamani 1983: personal communication.

21

Phoenix sylvestris also are common. Trees are lopped for fodder and fuel. Custard apple (*Annona squamosa*), mango (*Mangifera indica*), Zizyphus, and tamarind (*Tamarindus indica*) fruits are collected for home consumption. Neem (*Azadirachta indica*) fruits are collected for sale. Mahua (*Madhuca latifolia*) fruits yield edible oil and the tree's flowers are used to brew an alcoholic drink.

Animals are an integral part of the system, commonly termed *kanchas*. The system is practiced in the semi-arid parts of Andhra Pradesh, Karnataka, Tamil Nadu, Maharasthra, and Madhya Pradesh in the Deccan Plateau. The elevation is 300-1,300 mm with 45-70 rainy days occurring annually mainly during the southwest monsoon. Farmers own the kanchas although some are managed under the government forest departments. Resource input is restricted to enclosing the area, planting/protecting the grass and/or trees, and managing them. Kanchas are ideal production and protective systems in the semi-arid lands.

In the hot arid zone, *Prosopis cineraria* and *Zizyphus nummularia* are vital in the pastoral/silvicultural system. Community and village grazing lands in Rajasthan are termed *Oran* or *Bir*. Both species are encouraged and managed in orans. When *P. cineraria* is cut, it produces numerous new buds that result in several (5-12) new shoots. Within a year, these assume a bushy structure. Coppiced shoots are grazed continuously in an oran, leading to a "cushion" form of crown. Under severe grazing and trampling, the newly sprouted branches spread horizontally and provide feed to small animals like sheep (Saxena 1984). *P. cineraria* is compatible with grass (Shankar 1984). The shrubs of *Z. nummularia* have remarkable regenerative powers through root suckers. Apart from leaf fodder, mature Zizyphus yields about 3.0-5.0 kg of fruit per bush in an oran with 150-250 mm rainfall (Saxena 1984).

Experimental results have been reported on closure to biotic interference and planting of grasses and trees in pastoral/silvicultural systems. These include *Dalbergia sissoo*, *Albizia lebbek*, *Grewia optiva*, *Banhinia purpurea* and *Leucaena leucocephala* in high rainfall, subtropical humid climates and *Acacia nilotica* in semi-arid to medium rainfall areas (Tejwani 1987a and 1987b).

Agrosilvopastoral Systems

Since Indian farmers usually integrate animals with their farming operations, many of the agroforestry systems are, in fact, agrosilvopastoral systems. Some examples include trees in agricultural fields and on farm boundaries and intercropping with coconut and horticultural trees if grasses also are included.

Home gardens support a variety of livestock (cows, buffaloes, bullocks, goats, sheep, and sometimes pigs), birds (chickens and ducks) and fish. These animals are combined with crops and trees. In Kerala and Tamil Nadu states, which have humid tropical climates, coconut is the main crop. Edaphic conditions are similar to those for coconut and areca nut. It is only recently that home garden practices in India have been described (Nair and Sreedharan 1986).

The wide variety of food crops includes arrow root (*Maranta arundinacea*), cassava (*Manihot esculenta*), Chinese potato (*Coleus parviflorus*), dioscorea (*Dioscorea* spp.), rice (*Oryza sativa*), sweet potato (*Ipomoea batatus*), taro (*Colocasia* spp.), and elephant foot yam (*Amorphophallus campanulatus*). Pulses include cowpea (*Vigna unguiculata*), horse gram (*Dolichos uniflorus*), mung bean (*Vigna radiata*), and pigeon pea (*Cajanus cajan*). Fruits include banana (*Musa* spp.), passion fruit (*Passiflora edulis*), pineapple (*Ananas comosus*), papaya (*Carica papaya*), and pomegranate (*Punica granatum*). Oilseeds are groundnut (*Arachis hypogaca*) and sesame (*Sesamum indicum*). Spices and condiments include cardamom (*Eletteria cardamomum*), ginger (*Zingiber officinale*), pepper (*Peper nigrum*), turmeric (*Curcuma longa*), and betel vine (*Piper betel*). Vegetables are lady's finger (*Abelmoschus esculentus*), bitter gourd (*Momordica charantia*), eggplant (*Solanum melongena*), cucumber (*Cucumis sativus*), snake gourd (*Trichosanthes cucumerina*), water melon (*Citrullus lanatus*), and winged bean (*Psophocarpus tetragonolobus*). Beverages include coffee (*Coffea* spp.).

Fodder grasses like Brazilian lucerne (*Stylosanthes guianensis*), Guatemala grass (*Panicum maximum*), Napier grass (*Pennisetum purpureum*), lemon grass (*Cymbopogon citratus*), and Vetiver (*Veliveria zizanoides*) are grown in combination with a large variety of trees, most of which are multipurpose. Many are fruit trees such as custard apple (*Annona* spp.), bread fruit (*Artocarpus altilis*), garcinia (*Garcinia indica*), amla (*Emblica officinalis*), guava (*Psidium guajava*), jack fruit (*Artocarpus heterophyllus*), sapota (*Achras zapota*), and tamarind (*Tamarindus indica*). Other trees produce beverages (*Theobroma cacao*); spices and condiments like cinnamon (*Cinnamomum zeylanicum*) and nutmeg (*Myristica fragrans*); and vegetables such as moringa (*Moringa* spp.).

Other tree species often present include bamboo (*Bambusa arundinaea*), erythrina (*Erythrina indica*), gliricidia (*Gliricidia sepium*), subabul (*Leucaena leucocephala*), mahogany (*Swietenia macrophylla*), morinda (*Morinda tinctoria*), portia (*Thespesia populenea*), teak (*Tectona grandis*), wild jack (*Artocarpus hirsuita*), areca nut (*Areca catechu*), cashew (*Anacardium occidentale*), palmyrah palm (*Borassus flabellifer*), and rubber (*Hevea braziliensis*). It is important to restate that the practice revolves around coconut. Mangroves also are essential in home gardens of backwater areas in lowlands and along coastal tracts.

Wide variation in the intensity of tree cropping is noticeable among home gardens located in the same agroclimatic zone. Reduction in the size of the holding intensifies cultivation (Nair and Krishnankutti 1984). As the intensity of tree cropping increases, miscellaneous trees having no immediate benefit are replaced with multipurpose species. Tree cropping intensity also is influenced by farmers' socioeconomic conditions and their response to external changes, particularly those related to input and product prices and land tenure systems (Nair and Krishnankutti 1984, Nair and Sreedharan 1986).

Information is unavailable on relative yield of agricultural and tree crops and animals in individual home gardens. Although it appears that trees and crops are not grown in specific patterns, it seems unlikely a system that has evolved over a long period could casually consider location, spacing, and site conditions of important crops. It may be assumed that those who practice home gardening have a practical knowledge of what, where, and when to plant and remove crops.

Economic yields have been reported in one case study of a 0.12 ha home garden (Kerala Gandhi Samarak Nidhi 1984). Although the values are

speculative, these results and others obtained outside India indicate that the system is remunerative and provides good subsistence. The system's major constraint is that it is the least understood scientifically.

Research

A brief reference to research efforts was made in the introduction. Some results, though limited, are cited in the text. This review indicates clearly that the many tree species grown by farmers in agricultural fields, farm boundaries, field bunds, and wood lots have not been researched. In fact, these species have been studiously ignored. The reasons for this neglect are obvious. Agricultural scientists only see the crops and refuse to see the trees. Foresters do not acknowledge the role of trees except in forest lands. Current socioeconomic factors compel scientists to acknowledge these tree species. Most of those grown by farmers are cataloged or the practices described. Though these activities are important, substantive progress and discoveries in production and productivity require more from researchers.

Basic and applied research must be conducted side by side. The reasons why certain trees and crops are compatible while others are not need to be understood clearly. India's wide range of agro-ecological conditions and the phytoelasticity of many tree species make it imperative to record and evaluate the biological diversity available in each species. Many basic issues need to be understood before production and productivity can be improved. These include interactions between trees and crops/grasses with respect to shade, rooting patterns, competition for plant nutrients and moisture, soil fertility, etc. Scientists need to determine whether it is possible to select fast-growing, high-yielding, more compatible native tree species. Questions that need to be addressed relate to use of resources; spatial and temporal arrangement of trees; inputs/outputs; and increasing sustainable production per unit of space, time, and water.

Apart from the Forest Research Institute and Colleges and the research institutes of the Indian Council of Agricultural Research, agricultural universities scattered throughout India's agro-ecological zones are the rightful places to study multipurpose tree species grown by farmers. Some beginnings have been made, but the pace needs to be accelerated and goals set. It is obvious that research cannot be conducted on all species simultaneously, at least in the near future. It will be necessary for each university/institute to focus on 2-3 priority species within their area. Many challenges and opportunities await scientists embarking on this uncharted research.

Acknowledgments

To write so briefly about India's multipurpose tree species is a formidable task. This has been made possible by the opportunities I have had to work with many institutions and individuals. Among the institutions, I am particularly grateful to the Indian Council of Agricultural Research, New Delhi; the International Council for Research in Agroforestry, Nairobi; and the East-West Center, Honolulu. Among the individuals, I thank Bjorn Lundgren, P.K.R. Nair, Napoleon Vergara, Henry Gholz, K.A. Shankarnarayan, K.V.A. Bavappa, S.Chinnamani, Ram Prakash, and S.S. Teaotia. I am also thankful to the International Centre for Integrated Mountain Development, Kathmandu, Nepal and its Director, Dr. Colin Rosser.

REFERENCES

Acharya, R.M., B.C. Patnayak, and L.D. Ahuja. 1977. Livestock production problems and prospects. In *Desertification and its control*, pp 275-280. New Delhi, Indian Council Agric. Res.

Aggarwal, R.K. 1980. Physiochemical status of soils under Khejri (*Prosopis cineraria* Linn). In *Khejri* (Prosopis cineraria) *in the Indian desert*, eds. H.S. Mann and S.K. Saxena, pp. 32-37. Jodhpur, India: Central Arid Zone Res. Inst.

Bavappa, K.V.A. 1980. Plantation crops--areca nut. In *Handbook of agriculture*, pp. 865-920. New Delhi, Indian Council Agric. Res.

Bhat, K.S. 1978. Agronomic research in areca nut: A review. *J. Plant. Crops* 6:67-80.

Central Arid Zone Res. Inst. 1981. *Proceedings of the summer institute on agroforestry in semi-arid zones*. Jodhpur, India.

Gangolly, S.R., R. Singh, S.L. Katyal, and D. Singh. 1957. *The mango*. New Delhi: Indian Council Agric. Res.

Ganguli, B.N., R.N. Kaul, and K.T.N. Nambiar. 1964. Preliminary studies on a few crop species. *Annals Arid Zone* 3:33-37.

Ghildyal, B.P. 1981. Soils of the Garhwal and Kumaon Himalayas. In *The Himalaya: Aspects of change*, ed. J.S. Lal, pp. 120-151. New Delhi: Oxford Univ. Press.

Gupta, B.L. 1956. *Forest flora of Chakrata, Dehradun and Saharanpur forest divisions*. 3rd ed. New Delhi: Government of India.

Jambulingam, R. and E.C.M. Fernandes. 1986. Multipurpose trees and shrubs on farmlands in Tamil Nadu State (India). *Agrofor. Syst.* 4:17-32.

Jodha, N.S. 1977. Land tenure problems and policies in the arid region of Rajasthan. In *Desertification and its control*, pp. 335-347. New Delhi: Indian Council Agric. Res.

Kaul, R.N. and B.N. Ganguli. 1963. Fodder potential of Zizyphus in the scrub grazing lands of arid zone. *Indian For.* 89:623-630.

Kerala Gandhi Samarak Nidhi. 1984. *Balanced (small) farms: An appropriate garden technology*. Trivandrum, India.

Khybri, M.L. and Sewa Ram. 1984. Effect of *Grewia optiva, Morus alba* and Eucalyptus hybrid on the yield of crops under rainfed conditions. In *Annual report*, pp. 51-53. Dehra Dun, India: Central Soil Water Cons. Res. Training Inst.

Khybri, M.L, Sewa Ram, and S.P. Bhardwaj. 1984. Study on intercropping of field crops with fodder crop of "subabul" (*Leucaena leucocephala*) under rainfed conditions. In *Annual report*, pp. 61-62. Dehra Dun, India: Central Soil Water Cons. Res. Training Inst.

Malhotra, S.P. 1977. Socio-demographic factors and nomadism in the arid zone. In *Desertification and its control*, pp. 310-323. New Delhi: Indian Council Agric. Res.

Mann, H.S. and S.K. Saxena. 1980. *Khejri* (Prosopis cineraria) *in the Indian desert*. Monograph, no. 11. Jodhpur, India: Central Arid Zone Res. Inst.

Menon, K.P.V. and K.M Pandalai. 1960. *The coconut palm--A monograph*. Ernakulam, India: Indian Central Coconut Committee.

Ministry of Agriculture. 1983. State wise area and production of important fruits in India from 1976-77 to 1978-79 and projections up to 1982-83. Unpublished paper. New Delhi.

Ministry of Agriculture. 1985. *Indian agriculture in brief*. 20th ed. New Delhi.

Mittal, S.P. and P. Singh. 1983. Study on intercropping of field crops with fodder crop of "subabul" (*Leucaena leucocephala*) under rainfed conditions. In *Annual report*. Dehra Dun, India: Central Soil Water Cons. Res. Training Inst.

Nair, C.T.S. and C.N. Krishnankutti. 1984. Socio-economic factors influencing farm forestry. A case study of tree cropping in homesteads in Kerala, India. In *Community forestry: Socio-economic aspects*, eds. Y.S. Rao, N.T. Vergara, and G.W. Lovelace, pp. 115-130. Bangkok: FAO Regional Office for Asia and the Pacific.

Nair, M.A. and C. Sreedharan. 1986. Agroforestry farming systems in the homesteads of Kerala, southern India. *Agrofor. Syst.* 4:339-363.

Nair, P.K.R. 1979. *Intensive multiple cropping with coconuts in India*. Berlin and Hamburg: Verlaug Paul Parey.

Nelliat, E.V. 1979. Multiple cropping (in areca nut). In *Multiple cropping in coconut and areca nut gardens*, eds. E.V. Nelliat and K.S. Bhat, pp. 3-5. Tech. Bull. no. 3. Kasargod, India: Central Plantation Crops Res. Inst.

Patil, B.D, R. Deb Roy, and P.S. Pathak. 1981. Agroforestry research and development with reference to Indo-Gangetic plains. In *Proc. agrofor. seminar*, pp 40-68. New Delhi: Indian Council Agric. Res.

Prajapati, M.C. 1971. Effect of lateral development of *Prosopis juliflora* D.C. roots on agricultural crops. *Annals Arid Zone* 10 (2&3): 186-193.

Prasad, S.N. and B. Verma. 1983. Intercropping of field crops with fodder crop of subabul (*Leucaena leucocephala*) under rainfed conditions. In *Annual report*, pp. 95-96. Dehra Dun, India: Central Soil Water Cons. Res. Training Inst.

Purohit, M.L. and Wajid Khan. 1980. Socio-economic dimensions of Khejri. In: *Khejri (Prosopis cineraria) in the Indian desert*, eds. H.S. Mann and S.K. Saxena, pp. 56-63. Jodhpur, India: Central Arid Zone Res. Inst.

Reddy, C.V.K. 1981. Agroforestry in coastal Andhra Pradesh. In *Proc. agrofor. seminar*, pp. 77-82. New Delhi: Indian Council Agric. Res.

Samraj, P. 1981. Useful alien trees of the Nilgiris. *Bull. Bot. Surv. India* 23:243-249.

Samraj, P., B. Haldorai, and C. Henry. 1982. Economic utilization by intercropping in afforested areas. In *Annual report*, pp. 47-49. Dehra Dun, India: Central Soil Water Cons. Res. Training Inst.

Sannamarappa, M. and A. Murlidharan. 1982. Multiple cropping (in areca nut). In *The areca nut palm*, eds. K.V.A. Bavappa, M.K. Nair, and Prem Kumar, pp. 133-149. Kasargod, India: Central Plantation Crops Res. Inst.

Saxena, S.K. 1980. Taxonomy, morphology, growth and reproduction of Khejri and its succession in northwest India. In *Khejri (Prosopis cineraria) in the Indian desert*, eds. H.S. Mann and S.K. Saxena, pp. 4-10. Jodhpur, India: Central Arid Zone Res. Inst.

_____. 1984. Khejri (*Prosopis cineraria*) and bordi (*Zizyphus nummularia*): Multipurpose plants of arid and semi-arid zones. In *Agroforestry in arid and semi-arid lands*, ed. K.A. Shankarnarayan, pp. 111-121. Pub. no 24. Jodhpur, India: Central Arid Zone Res. Inst.

Shah, S.L. 1981. Agricultural planning and development in the northwestern Himalayas, India. In *Nepal's experience in hill agricultural development*, pp. 160-168. Kathmandu: Min. Food Agric.

Shankar. V. 1980a. Khejri (*Prosopis cineraria*) in Indian scriptures. In *Khejri (Prosopis cineraria) in the Indian desert*, eds. H.S. Mann and S.K. Saxena, pp. 1-3. Jodhpur, India: Central Arid Zone Res. Inst.

_____. 1980b. Distribution of Khejri (*Prosopis cineraria* Macbride) in western Rajasthan. In *Khejri (Prosopis cineraria) in the Indian desert*, eds. H.S. Mann and S.K.

Saxena, pp. 11-19. Jodhpur, India: Central Arid Zone Res. Inst.

_____. 1984. Inter-relationships of tree overstorey and understorey vegetation in silvipastoral system. In *Agroforestry in arid and semi-arid lands*, ed. K.A. Shankarnarayan, pp. 143-149. Jodhpur, India: Central Arid Zone Res. Inst.

Shankarnarayan, K.A., ed. 1984. *Agroforestry in arid and semi-arid lands*. Jodhpur, India: Central Arid Zone Res. Inst.

Sheshadri, P. 1976. *Acacia leucophloea*. Paper presented at the Forestry Scientific Workers Conference, Tamil Nadu Agricultural University, Coimbatore, India

Singh, S., S. Krishnamurthi, and S.L. Katyal. 1963. *Fruit cultivation in India*. New Delhi: Indian Council Agric. Res.

Tejwani, K.G. 1981. Agroforestry for the wasted lands in western Himalayas. In *Proc. agrofor. seminar*, pp. 220-233. New Delhi: Indian Council Agric. Res.

_____. 1987a. Agroforestry practices and research in India. In *Agroforestry: realities, possibilities, and potentials*, ed. H. L. Gholz. Dordrecht and Boston: Martinus Nijhoff Pub.

_____. 1987b. Agroforestry in hill regions of India. In *Proc. international workshop on agroforestry for rural needs*. New Delhi: IUFRO-ISTS.

Vellapan, E. and G.K. George 1982. Development (of areca nut). In *Areca nut*, eds. K.S. Bhat and K.V.A. Bavappa, pp. 305-318. Monograph series no. 2. Kasargod, India: Central Plantation Crops Res. Inst.

Overview of Tree Planting on Small Farms in The Tropics

Julian Evans

International Institute for Environment and Development
London, United Kingdom

Agroforestry and multipurpose trees are fashionable research topics. Although much new work is in hand, one must beware of making tree crop research an end in itself. Scientists fail if farmers and villagers are not their primary focus. Throughout the tropics small farmers plant, protect, and care for whatever trees are available. Familiarity, not science, generally determines which ones they select, and most know much about how to use them. The challenge today is not principally to conduct more research to fill knowledge gaps about species, but to fill information gaps farmers have regarding seedling supply lines and post-planting care. Examples from Central America, central Africa, Southeast Asia, and the Pacific illustrate the discussion.

In the developed world, farming and forestry are often viewed as separate activities yielding different products. The divorce is recent, and like the separation of husband and wife, creates many more problems than it solves. However, a current trend, certainly discernible in my own country, indicates a rekindled interest in reuniting the two and recognizes the advantages of integrating the growing of food and trees. For guidance and example, Westerners must turn to small farmers in the tropics.

My introductory remarks stress the fact that I wish to learn and perhaps help, and do not want to impose fixed views of right and wrong. I acknowledge that the tropical farmer practices one of the most complex arts of land use or ecosystem management in the world. Thus, in titling my paper "Overview of Tree Planting on Small Farms in the Tropics," I wish only to set the scene and indicate some things scientists should carefully consider to aid the poor seriously.

Although new on-farm plantings must be a priority worldwide, this overview stresses the importance of considering on-farm trees in general, regardless of how they arose. Most small farmers do not distinguish between a planted tree and one that arises naturally. Neither should scientists. By "small farm" is meant the farming unit (rarely more than a few hectares) that is cultivated or tended by one family using mostly hand tools. In this context, most trees grown on the farm are for personal or local use.

Why Farmers Grow Trees

Of the numerous reasons for growing and using trees, it is helpful to recognize the following main types of benefits:

o Environment - shade, shelter, and soil stability

o Tree crops - fodder, fruit, and organic fertilizer

o Wood produts - firewood, fencing materials, poles, and timber

o Savings and security - possession of or rights to trees as a realizable asset to meet special contingencies.

Categories 1 and 2 refer to living trees. Category 3 usually refers to a tree that is cut down or whose branches are chopped. Category 4 includes all the farmer's tree resources, both living and recently felled. Of course, the farmer does not usually distinguish among categories, and one tree may be cared for to supply two, three, or all four of the above uses. The fourth category is worth stressing because it has been largely neglected (Chambers and Leach 1987). Provided farmers own or at least have full rights over use of trees, they frequently treat them as assets for contingencies such as drought, feasts, marriage expenses, funeral pyres, pledges, and even redemption of debts and mortgages. With ownership, farmers will plant and build resources; without it, much pleading is to no avail. Poor people are often thought incapable of saving. In fact, deprived people are much concerned about vulnerability, indebtedness, and assets. They make sacrifices to maintain land and trees and to save for future needs and security.

What Farmers Grow

Although there are local differences of emphasis or relative importance, most farmers and villagers grow trees for the same reasons wherever they live. The interesting point is that they use different tree species to fulfill the same role (Table 1).

In temperate countries, most timber comes from pines, spruces, and firs. The species used differ among regions, depending largely on what is native to each locality. In the tropics, a much wider range of species and genera (both native and exotic) occur on farms.

Table 1 is not exhaustive, and numerous other examples could be cited. What is interesting is that all the species are internationally available and well-known. This has implications for research agendas, and raises the following questions about future research:

o Should forestry follow the path of agriculture, providing the tropics a few highly adaptable, multipurpose trees, such as plantations of pine, eucalyptus, or teak?

o Should scientists develop germplasm based on what farmers already use and know, thereby denying opportunities to introduce other potentially helpful species?

o How do scientists decide what to do?

Although what follows does not answer these questions comprehensively, it provides some pointers.

Table 1 . Examples of tree species used by farmers in analogous situations in the humid tropics.

Moist Lowland Tropics		Cool Highland Tropics	
Country	Shading cocoa/coffee	Country	Fuel, poles, timber
Belem, N. Brazil	*Inga edulis*	Costa Rica	*Cupressus lusitanica*
Costa Rica	*Gliricidia sepium* *Erythrina poeppigiana*	Ecuador and Ethiopia	*Alnus acuminata* *Eucalyptus globulus*
Papua New Guinea	*Leucaena leucocephala* (unimproved)	Papua New Guinea	*Casuarina oligodon*
		Rwanda	*Cupressus lusitanica*
Sabah, E. Malaysia	*Albizia falcataria*		*E. maidenii* *Grevillea robusta*

What Farmers Want To Grow

It is now common knowledge that what the expert recommends as most suitable is frequently not what is viewed as most useful by the farmer or his family (one must never overlook the multiplicity of labor in the rural tropics). The "give 'em Eucalyptus" syndrome reflects the oversimplified approaches of the past. Equally, as we recoil from such patronage, Eucalyptus species are loved and used by the Ethiopian, Peruvian, and Indian farmers for countless domestic purposes. Indeed, remove Eucalyptus from Ethiopia and the highlands will begin to resemble ravaged, cut-over rain forest.

Nevertheless, when farmers are asked about their preferences, their replies often differ from the techno-fix solutions of outsiders. Usually, fruit and fodder trees are high on their list, followed by native, slower-growing species, such as Acacias and Combretum in West Africa, rather than Eucalyptus and Prosopis. In particular, women who gather firewood in arid areas prefer these trees. There is increasing evidence that change in management of native species, e.g., frequency of pollarding, can be as useful in increasing production as introduction of new species (Heermans and Minnick 1987). The reference to women and their special perceptions teaches the importance of dialogue with all concerned when assessing needs. When asked what trees they would like to see planted, Turkana headsmen in northern Kenya cited cypress, pine and eucalyptus, which a few had seen in the moist Kenyan highlands but which are wholly unsuited to the aridity of the Turkana desert. The women simply wanted more *Acacia tortilis* woodland near their villages so they would not have to walk so far for firewood and fencing materials for livestock pens.

Formality and Informality

A second, neglected aspect of where and how trees are planted on small farms centers on questions of spacing, regularity, woodlot shape, and lines and plans of farming systems. Research trials and demonstration plots also are carefully laid out according to convention and the requirements of experimental design. Textbooks and leaflets showing tree planting invariably display orderliness. Can this formality become a constraint to the farmer who would like to try a species he sees growing but does not know how to modify his land or cropping patterns to accommodate it?

This might appear absurd since there is no need to reproduce faithfully the dimensions of a trial plot. But does the farmer know? Does our formality inhibit adoption of an idea? In a project in southern Shoa, Ethiopia, spacing of terraces and bunds was carefully measured at 8 m regardless of slope angle, because the initial report (by this writer) had specified such a distance. By contrast, although the German-funded agroforestry trial at Nyabisindu, Rwanda has regular, straight lines of Grevillea, bananas, maize and beans, it was the model for an apparently haphazard but astonishingly successful cropping system begun in neighboring villages. The important point is to be aware of possible conflict or misunderstanding between our Western scientific culture and local needs and perceptions.

Encouraging Tree Planting

Most tropical countries have extension services that promote tree planting directly or in cooperation with non-government organizations. Although their success varies considerably, what Mbithi (1971) wrote about agricultural extension in Kenya generally remains true. After finding that four-fifths of all the programs he examined failed, he stated "that our extension agents tend to possess a package of technological practices but very little skill in the art of communicating ideas to farmers." This remains a central problem. The necessity or urgency to plant trees is often lacking because it is not recognized. This is quite separate from the mechanics of being able to do it. Following a 1980 energy sector review, the general effort to encourage village tree planting for fuel in Tanzania was stopped. Attention was concentrated on those villages requesting such assistance, i.e., where lack of fuelwood was perceived as a real need and initiatives would command support.

The problem of post-planting care of trees is well-known. Of the billions of seedlings raised and distributed, how many survive and grow into healthy trees? Every stage is fraught with risk, but, without doubt, inadequate or inappropriate post-planting care is critical. Protection from browsing and control of competing weeds are the principal requirements, but in the tropics little effort focuses on follow-up programs.

Successful programs that encourage farmers to plant do exist in all countries, but these usually regard the political, social, and cultural milieu as critical rather than the species.

Meeting Farmers' Needs[*]

The majority of rural dwellers in the tropics are small farmers or landless people faced with many physical and socioeconomic constraints that prevent them from overcoming the daily struggle for subsistence. Although the relative importance of these factors varies from one region to another, several common characteristics can be identified.

In many regions, physical factors pose a degree of environmental risk to which the farmers must adapt. Examples include erratic rainfall and drought in arid and semi-arid areas. In humid areas, steeply sloping or relatively infertile soils depend on organic matter input to maintain their productivity and they often cannot support continuous cropping. These physical constraints are frequently compounded by a lack of infrastructure in many rural areas. Existing roads are often in poor condition, limiting transport and distribution. As a result, farmers are not well-integrated into the market economy. They have low incomes, little if any cash reserves, and typically lack access to credit. Government extension services are often weak. Thus, small farmers and the landless generally have limited access to outside inputs and technology, and must rely on locally available resources to meet a wide range of subsistence and cash needs.

Because agriculture is still based largely on traditional practices of low productivity, farmers are unable to produce a surplus that can be stockpiled as protection against environmental risk. Under these conditions, farmers take the rational course of adopting low-resource farming strategies to minimize the risk of failure. Any outside intervention to increase farm productivity must fit into this framework. That is, technologies must be low input, low risk, and provide high returns if poor farmers are to benefit.

By their very nature, tree planting and agroforestry are small-scale and oriented to producing multiple outputs to meet local needs. By building on traditional agricultural practices with little or no reliance on outside inputs, agroforestry approaches are affordable, relatively easy to adopt, and widely replicable. Households and communities can determine their own priorities and grow the types and numbers of trees they choose. Trees are grown where they are accessible to those who need them, making households more self-sufficient. Finally, agroforestry can help farmers minimize risk by producing a more diversified and stable farming system. Agricultural production can be made more sustainable because of the physical benefits of combining trees with crops and livestock, and more flexibility exists to distribute the workload over the course of the year.

Conclusions

This short paper has deliberately focused on people rather than trees. Scientists must never forget for whom they work. Successful tree-planting programs on small farms are achieved when villagers' perceptions and hopes are met and not when another's ideas are imposed. Referring to agroforestry, Winterbottom and Hazelwood (1987) state, "it seems to be most successful when it builds on traditional practices, is developed in close cooperation with the local people who daily determine how resources will be used and managed, and when there is a sustained commitment to continue evaluation. This points to the need for greater involvement of non-government organizations which have a demonstrated ability to foster the participation of local communities and respond to their perceived needs through integrated, flexible, and long-term effects."

Scientists agree with the writer of Ecclesiastes, who wrote, "There is a time to plant." One might say now is the time to plant, but these efforts and advocacy will be fruitless if the enthusiasm for science ignores the planter. Should scientists fail, their epitaph will change the meaning of the acronym MPTS to Many Proposals, Trivial Successes.

REFERENCES

Chambers, R. and M. Leach. 1987. *Trees to meet contingencies: savings and security for the rural poor.* IDS Discussion Paper, no. 32. UK: University of Sussex.

Evans, J. 1982. *Plantation forestry in the tropics.* UK: Oxford University Press.

Heermans, J. and G. Minnick. 1987. *Guide to forest restoration and management in the Sahel. Based on case studies at the national forests of Gueselsdbodi and Goru-Bassounga, Niger.* Niamey, Niger: Ministry of Hydrology and Environment.

---

[*] Adapted from Winterbottom and Hazelwood, 1987.

Mbithi, P.M. 1971. *Issues on rural development in Kenya*.
Discussion Paper, no. 131. Institute of Development
Studies, University of Nairobi.

Winterbottom, R.W. and P.T. Hazelwood. 1987. Agroforestry
and sustainable development: making the connection.
Ambio 16: 100-110.

Uses of Multipurpose Trees on the Small Farms of the Low-Rainfall Ganges Floodplain Soils of Bangladesh

Zainul Abedin, Shafiqul Aktar, Fazlul Haque, and Shamsul Alam

Bangladesh Agricultural Research Institute
Joydebpur, Gazipur

A survey was conducted in the Ganges floodplain of Bangladesh to understand the distribution and uses of multipurpose trees, tree-crop interactions, and the crafts/cottage industries these trees support. A predesigned survey questionnaire was used. Results showed that Acacia nilotica, A. catechu, Artocarpus heterophyllus, Phoenix sylvestris, Borassus flabellifer and Mangifera indica are the major tree species grown on the croplands in the low-rainfall Ganges floodplain area for fruit, timber, fuel, and building material. The trees support different crafts/cottage industries. Fuel was a common, though not primary, use of all the tree species. Uses of particular trees varied from place to place and their order of importance changed over time. Species distribution differed among regions. Tree-crop combinations and their interactions depended more on land type, age of the trees, canopy structure, and plot location of trees rather than the type of species grown. Determination of optimum tree densities, optimum economic age for cutting, relative economic importance, and improved management practices are critical issues for future research.

In Bangladesh, systematic data on agroforestry practices in different ecological zones are unavailable. Therefore, in this paper, the authors attempt to document the multipurpose tree species grown on the low-rainfall areas of the Ganges floodplain. They also raise future research and development issues.

Most of Bangladesh's land was formed by the alluvial deposits of the Ganges, Brahmaputra, Meghna, and Tista rivers and estuaries. The land is divided and named after the deposits of these rivers. The Ganges floodplain is the largest soil tract, consisting of the districts of Jessore, Faridpur, Kushtia, Khulna, Pabna, and parts of Rajshahi, Dhaka, and Barisal.

The soils of the tract are physiographically subgrouped on the basis of relative age of deposits into (i) active Ganges meander floodplain, (ii) long Ganges meander floodplain, (iii) old Ganges meander floodplain, (iv) mixed young and old Ganges meander floodplain, and (v) mixed old Ganges and its tributaries floodplain. These lands are arranged in a catenary sequence and are mainly level to gently undulating with ridges and basins (Joshua and Rahman 1983). The highlands are flood-free, but other areas are shallow to deeply flooded (Table 1). The surface drainage is carried by the Ganges and its estuaries. Subsurface drainage is efficient in the highlands but less so at lower altitudes.

Soil texture varies from sandy loam to heavy clay. Light textured soils occur in the highlands and ridges. The texture gradually becomes heavier in the basin. Unlike the other tracts, the Ganges floodplain is characterized by calcareous soil. The older soils are decalcified up to a depth of 25-60 cm. Soil pH ranges from 7.0 to 8.5.

The area enjoys a tropical monsoon climate with the lowest and least reliable rainfall in Bangladesh (Manalo 1976). The monsoons usually start in May and end in late September. The annual rainfall within the tract varies from 1,400-1,800 mm. About 90% of the total annual rainfall occurs during June-September. The mean annual rainfall increases from west to east and from north to south. The average daily temperature ranges from 12° C during December-January to about 31° C during May-August.

Common crops in the area include direct seeded rice (Aus), transplanted rice (Aman), winter transplanted rice (Boro), wheat, jute, pulses, sugarcane, and vegetables.

Table 1 .   Percent of area flooded in the Ganges floodplain.

District	Seasonal	Flooding		
		Shallow	Moderate	Deep
Jessore	54	31	19	4
Kushtia	57	38	16	3
Pabna*	100			50
Charghat	40	15	25	-

* 50 % shallow to moderately deep flooding.

Table 2 .   Major and minor tree species in crop fields of Ganges floodplain by location.

Location	Major Species	Minor Species
Bagherpara	*Phoenix sylvestris, Borassus flabellifer, Artocarpus hetero- phyllus*	*Mangifera indica, Bombax malabaricum, Acacia catechu, Albizia procera, Azadirachta indica*
Narail	*P. sylvestris, B. flabellifer*	not available
Kushtia	*P. sylvestris, B. flabellifer, Acacia nilotica M. indica, A. hetero- phyllus*	*Tamarindus indica, A. catechu*
Ishurdi	*P. sylvestris, B. flabellifer, A. nilotica, A. catechu, A. hetero- phyllus*	*M. indica, Eugenia javanica*
Charghat	*P. sylvestris, A. nilotica, A. catechu, M. indica, A. hetero- phyllus*	*B. flabellifer, B. malabaricum, Cocos nucifera, Litchi chinensis*

Methods

Five locations with different rainfall, land, soil, and land-use characteristics were selected to represent the study area: Narail, Jessore/Bagherpara, Charghat, Ishurdi, and Kushtia. Earlier knowledge about the study area's land use and soil types helped in selecting the locations. A pre-tested questionnaire was used to collect information from farmers (marginal, small, medium, and large) who grew trees on their crop fields.

The following standard classification of the Bangladesh Bureau of Statistics was used to categorize land cultivated by farm size in hectares:

    Marginal - up to 0.5
    Small - 0.51-1.0
    Medium - 1.01-2.0
    Large - more than 2.0

For this study, multipurpose trees were defined as "trees and shrubs deliberately grown or kept and managed for more than one intended use, usually economically and/or ecologically motivated major products and/or services in any multipurpose land-use system, especially an agroforestry system" (Gupta 1978).

Tree Species Grown

Survey data revealed that as many as 13 different tree species are grown in the crop fields of the low-rainfall Ganges floodplain. But their occurrence is not similar at selected locations of the study area (Table 2).

Only *Phoenix sylvestris*, the wild date palm, occurs at all locations. Except for Narail, at least four major species are grown at all locations. Because the land of Narail is predominantly medium-high to medium-low, only *P. sylvestris* and *Borassus flabellifer* are grown there. The other species are absent mainly because of ecological conditions. At the other locations, where elevation is higher and floods fewer and less severe, *Acacia nilotica*, *A. catechu*, *Artocarpus heterophyllus* (jackfruit), and *Mangifera indica* (mango) are also grown.

Species considered major at one place are considered minor at another. For example, *Mangifera indica* is the major tree species in the crop fields of Charghat and Kushtia while it is a minor species at Bagherpara and Ishurdi. Though mango is usually a highland crop, its minor occurrence at Bagherpara is probably caused by the strong competition with *Phoenix sylvestris*, *Artocarpus heterophyllus*, and *Acacia arabica*. At Ishurdi, its minor status is caused mainly by competition with *A. heterophyllus* and *A. arabica*.

A clear pattern of density changes exists in the six tree species. In the Jessore District of Narail, *Borassus flabellifer* is the major species grown on high and medium lands. *Phoenix sylvestris* is grown mainly in the highlands. *Mangifera indica* and *Artocarpus heterophyllus* do not thrive in the crop fields because of the high water table and poor drainage conditions (both surface and subsurface). Further research is required to determine other possible reasons for the conspicuous absence of these trees.

As one moves northwest to Bagherpara, the wild date palm (*Phoenix sylvestris*) is found growing abundantly both on highland and mid-elevation sites. Densities of 102 trees/ha on highlands and 84 trees/ha on mid-elevation sites are found (Tables 3 and 4). Although *Artocarpus heterophyllus* and *Acacia nilotica* grow well on the highlands of Bagherpara, tradition supports the growing of *P. sylvestris*. But some farmers have begun growing jackfruit rather than date palms to increase their income. Further north, at Kushtia, the densities of *P. sylvestris* and *Borassus flabellifer* are reduced in high- and mid-elevation sites.

Further north across the Ganges at Ishurdi, the densities of *Phoenix sylvestris* and *Artocarpus heterophyllus* are found to increase, probably because this area has more well-drained highland areas than Kushtia. Although the density of *Acacia nilotica* is greater in the highlands of Ishurdi (38/ha) than in those of Kushtia, its overall density is higher in Kushtia. This is probably because there are more medium highland areas in Kushtia than in Ishurdi. Further, the higher demand for fuel to cure tobacco in Kushtia might inspire farmers to grow *A. nilotica*, primarily a fuelwood tree. *P. sylvestris* and *A. heterophyllus* are mainly grown for juice and fruit, respectively. Water stagnation caused by an irrigation project might contribute to the lower jackfruit tree density at Kushtia. At Ishurdi, *Acacia catechu* density increases in the highlands. In the northernmost region of Charghat, an average of 39

Table 3 .   Number of planted (P) and voluntary (V) tree species/ha on highland sites in the Ganges floodplain.

Location	P. syl-vestris		B. fla-bellifer		A. nilo-tica		A. catechu		M. indica		A. hetero-phyllus	
	P	V	P	V	P	V	P	V	P	V	P	V
Bagherpara	60	42	37	25	20	40	10	-	10	-	30	-
Narail	9	42	60	7	-	-	-	-	-	-	-	-
Kushtia	-	-	-	-	2	29	-	-	15	-	8	-
Ishurdi	10	40	-	11	20	18	-	20	8	-	50	-
Charghat	51	30	8	-	-	9	16	23	17	2	35	-

Table 4 .   Number of planted (P) and voluntary (V) tree species/ha on mid-elevation sites in the Ganges floodplain.

Location	P. syl-vestris		B. fla-bellifer		A. nilo-tica		A. catechu		M. indica		A. hetero-phyllus	
	P	V	P	V	P	V	P	V	P	V	P	V
Bagherpara	45	39	-	20	8	7	-	-	-	-	-	-
Narail	4	8	38	16	-	-	-	-	-	-	-	-
Kushtia	12	2	12	4	5	22	-	-	8	1	25	-
Ishurdi	-	8	13	-	-	8	-	-	7	-	-	-
Charghat	-	12	-	6	-	7	-	9	-	-	12	-

Table 5. Average number of planted (P) and voluntary (V) trees owned per farm by farm size.

Farm Size	P. sylvestris[1]		B. flabellifer[1]		A. nilotica[2]		A. catechu[3]		M. indica[2]		A. heterophyllus[2]	
	P	V	P	V	P	V	P	V	P	V	P	V
Marginal	3	2	2	1	.3	3	0	1	.1	0	2	0
Small	3	2	1	1	1	2	.1	.2	1	.2	5	0
Medium	16	13	2	.3	5	4	.3	12	2	.2	6	0
Large	14	11	4	2	.3	5	2	4	3	0	4	0

[1] available at all locations.
[2] unavailable at Narail.
[3] unavailable at Narail, Bagherpara, and Kushtia.

Table 6. Farmers' uses of *P. sylvestris* and *B. flabellifer* (by percent).

Location	P. sylvestris						B. flabellifer					
	Fruit	Fuel	Juice/ Molasses	Mat	Bldg. material	Fencing	Fruit	Fuel	Juice/ Molasses	Bldg. material	Canoe	Fan
Bagherpara	100	100	100	100	100	4	100	100	100	100	56	85
Narail	100	78	100	68	33	38	100	83	93	63	86	83
Kushtia	93	85	100	83	33	13	100	73	95	25	55	75
Ishurdi	100	87	100	80	54	33	100	74	94	50	50	64
Charghat	100	100	100	96	52	68	100	88	80	64	50	80

Table 7. Percentage of farmers using *Acacia catechu* and *Acacia nilotica* for various purposes.

Location	*Acacia catechu*					*Acacia nilotica*							
	Fruit	Fuel	Wood extract	Bldg. material	Furniture	Fruit	Fuel	Fencing	Furniture	Bldg. material	Cart wheel	Plow	Boat
Bagherpara	-	-	-	-	-	11	96	4	81	22	100	96	-
Narail	-	-	-	-	-	-	-	-	-	-	-	-	-
Kushtia	-	-	-	-	-	63	83	50	46	67	83	79	-
Ishurdi	13*	54	100	8	17	83	79	39	54	23	78	88	29
Charghat	44*	88	100	40	60	72	92	80	80	80	88	92	44

* used only as cattle feed.

36

Table 8. Percentage of farmers using *Mangifera indica* and *Artocarpus heterophyllus* for various purposes.

Location	*Mangifera indica*						*Artocarpus heterophyllus*					
	Fruit	Fuel	Fenc-ing	Furni-ture	Bldg. mate-rial	Boat	Fruit	Fuel	Fenc-ing	Furni-ture	Boat	Bldg. mate-rial
Bagherpara	100	100	74	100	7	--	100	100	52	96	19	14
Narail	--	--	--	--	--	--	--	--	--	--	--	--
Kushtia	100	94	--	85	17	--	100	93	--	93	4	25
Ishurdi	100	86	18	75	--	29	100	96	--	86	--	--
Charghat	100	100	16	100	64	68	100	94	15	92	40	78

37

trees/ha are found. The density of *P. sylvestris* in the Charghat highlands is similar (81 trees/ha) to that of Bagherpara. *Albizia procera, Bombax malabaricum, Azadirachta indica, Tamarindus indica, Eugenia javanica, Cocos nucifera* and *Litchi chinensis* also grow in the crop fields of the study area.

Although larger farmers own more trees grown on crop fields than do smaller farmers, marginal and small farmers own a comparatively higher proportion (Table 5).

Multiple Uses

The trees are not monopurpose, although different ones are planted or grown for various primary uses. Mango, for example, is grown mainly for fruit and timber, but is also used for fuel and boatmaking. The six tree species discussed here have at least two common uses, including fuel and building materials (Tables 6-8), but the primary uses are as follows:

o *Phoenix sylvestris* - Molasses, fruits, mat making

o *Borassus flabellifer* - Fruit, juice, molasses, fans

o *Acacia arabica* - Cart wheels, plows, fuel

o *Acacia catechu* - Tannin (Khair), fuel

o *Mangifera indica* - Fruit, fuel, furniture

o *Artocarpus heterophyllus* - Fruit, furniture, fuel

The uses of and preferences for these trees vary among locations. At Narail, where boats are important during the wet season, making canoes is as important a use of *Borassus flabellifer* as is making juice.

These trees provide considerable amounts of fuel (Tables 9 and 10). Use of cow dung as fuel is comparatively lower at all locations (7-22%) and farm categories (14-21%) than the national average.

The trees also support various crafts and cottage industries. Thus, employment opportunities for women, artisans, and small businessmen have been generated. *Phoenix sylvestris* leaves, for

example, support mat making for household use and for sale by women, particularly those from poorer households (Aktar and Haque 1986). Juice extraction and molasses making also support a special group of technicians, who are usually poor farmers. Making boats from *Borassus flabellifer* requires skilled artisans to work about 8-10 days per boat. Collection and marketing of its fruit also generate employment.

Making cart wheels and plows from *Acacia nilotica* wood also requires skilled artisans. Small traders market cart wheels in different parts of the country. The Charghat area is famous for making cart wheels. *A. nilotica* is probably the most desired wood for making both cart wheels and plows, as farmers believe the wood is not easily damaged by prolonged use in wet conditions.

The juice of *Acacia catechu* is extracted by boiling the wood chips of the tree to make Khair (tannin), used as an additive for chewing betel leaves. Small cottage industries have been developed to prepare this product.

The array of tree uses is expanding. For example, making boats using *Borassus flabellifer* trunks was uncommon 100 years ago. The high cost of more traditional wooden boats has necessitated widespread use of this tree for this purpose. Use of *Acacia nilotica* wood for making cart wheels was uncommon in the Bagherpara-Jessore area 20 years ago. But the current higher price at Charghat and the cheap and relatively under-exploited Acacia trees in the Bagherpara-Jessore area has influenced artisans of Charghat to move to Jessore to make the wheels there.

Because of economic pressures and increased demand for fruit, fuel, fodder, and construction materials, farmers plant more trees today than previously, even non-traditional species such as *Acacia catechu* or *A. nilotica* (Table 11). At the same time, many trees are cut before they attain the farmer-defined optimum age (Tables 12-13). Although farmers define the optimum age of *A. nilotica* for wood at 30-40 years, about 38% of the trees are felled at less than 10 years and 53% felled between 10-20 years. Though large farmers maintain the trees for longer periods, the difference is not striking among farm categories.

Farmers fell trees for various reasons (Table 14). Marginal and small farmers use trees to meet

Table 9. Sources of household fuel.

| | Wood and lopping[1] | | | | | Leaves/ twigs | Jute stick | Cow dung | Husk/ bran | Other[2] |
Location	Ps	Bf	Mi	Ah	An					
Bagherpara	12	4	4	4	9	18	16	19	4	10
Narail	10	11	-	-	-	28	33	7	4	7
Kushtia	5	4	13	12	10	6	18	22	6	4
Ishurdi	10	2	4	2	10	21	19	15	10	7
Charghat	15	4	14	7	12	10	15	10	7	6

[1] Ps = *Phoenix sylvestris*, Bf = *Borassus flabellifer*, Mi = *Mangifera indica*, Ah = *Artocarpus heterophyllus*, and An = *Acacia nilotica*

[2] includes rice and wheat straw, collected root stocks of jute and sugarcane, and other tree species.

Table 10. Percentage of household fuel sources by farm size.

| | Wood and lopping* | | | | | Leaves/ twigs | Jute stick | Cow dung | Husk/ bran | Other |
Farm Size	Ps	Bf	Mi	Ah	An					
Marginal	9	3	2	1	6	20	18	21	5	15
Small	7	3	3	1	7	19	21	19	7	13
Medium	9	2	1	1	6	20	25	19	8	9
Large	4	2	16	2	7	16	23	14	10	6

* Ps = *Phoenix sylvestris*, Bf = *Borassus flabellifer*, Mi = *Mangifera indica*, Ah = *Artocarpus heterophyllus*, and An = *Acacia nilotica*.

Table 11 .   Percentage of trees planted more or less than 10 years ago.

Site	P. sylvetris		B. flabellifer		A. nilotica	
	< 10 yr	> 10 yr	< 10 yr	> 10 yr	< 10 yr	> 10 yr
Bagherpara	45	55	98	2	100	0
Narail	36	64	28	72	-	-
Kushtia	57	43	100	0	85	15
Ishurdi	100	0	50	50	67	33
Charghat	97	3	54	46	50	50
Mean	67	33	66	34	75	25

Site	A. catechu		M. indica		A. heterophyllus	
	< 10 yr	> 10 yr	< 10 yr	> 10 yr	< 10 yr	> 10 yr
Bagherpara	-	-	33	67	81	19
Narail	-	-	-	-	-	-
Kushtia	-	-	11	89	64	36
Ishurdi	100	0	43	57	NR[*]	NR[*]
Charghat	69	31	8	92	73	27
Mean	85	15	24	76	80	20

Note: Acacia spp. usually occurred voluntarily and rarely were planted before the last 10 years.
[*] NR = not reported.

Table 12. Age at which species are felled (by farm category).

Farm Size	P. sylvestris			B. flabellifer			A. nilotica		
	< 10 yrs	10-20 yrs	> 20 yrs	< 10 yrs	10-20 yrs	> 20 yrs	< 10 yrs	10-20 yrs	> 20 yrs
Marginal	0	0	100	0	0	100	30	60	10
Small	0	29	71	0	0	100	50	50	0
Medium	0	39	61	NR[*]	NR[*]	NR[*]	55	42	3
Large	0	34	66	0	0	100	17	62	21

Farm Size	A. catechu			M. indica			A. heterophyllus		
	< 10 yrs	10-20 yrs	> 20 yrs	< 10 yrs	10-20 yrs	> 20 yrs	< 10 yrs	10-20 yrs	> 20 yrs
Marginal	100	0	0	50	50	0	0	0	100
Small	NR[*]	NR[*]	NR[*]	0	40	60	0	0	100
Medium	21	71	8	0	63	37	14	57	29
Large	NR[*]	NR[*]	NR[*]	0	0	100	NR[*]	NR[*]	NR[*]

[*] NR = not reported.

41

Table 13 . Farmers' perception of minimum and optimum age and production duration for various species (in years).

Use	P. syl- vestris	B. fla- bellifer	A. nilo- tica	A. catechu	A. indica	A. hetero- phyllus
Juice						
Minimum age	6-7	15-17	-	-	-	-
Optimum age	15-25	25-33	-	-	-	-
Duration	6-45	15-45	-	-	-	-
Fruit						
Minimum age	8-10	16-19	8-10 *	-	8-10	7-10
Optimum age	15-25	30-40	20-30	-	20-35	20-35
Duration	8-45	8-60	8-50	-	8-60	7-60
Wood						
Minimum age	15-20	20-25	8-10	10-15	15-20	15-20
Optimum age	25-35	35-45	30-40	35-45	30-45	30-45
Duration	15-50	20-60	8-60	10-60	15-65	15-65

* used only for cattle feed.

Table 14. Percentage of farmers who fell trees for various purposes by farm size.

Farm Size	Repay loan	Buy bullock	Culti- vation cost	Cere- mony	Mate- rial	Bldg. expense	Buy land	Farm tool
Marginal	19	19	10	4	24	19	0	0
Small	28	5	10	5	28	14	5	0
Medium	31	16	11	5	5	11	5	9
Large	11	21	22	14	5	7	5	5

household expenses (including purchase of food), repay loans, and purchase bullocks. Medium and large farmers use trees to purchase bullocks, meet cultivation costs, and as part of social ceremonies. All groups, especially marginal and small farmers, use trees to meet contingencies. At Bagherpara, income from date palm sustained poor families 5-6 months in a year (Aktar and Haque 1986). Small, medium, and large farmers also use trees to purchase land.

Trees and Crops

Trees are found growing in association with almost all predominant crops. Variations in different cropping patterns are associated mostly with land, soil, and socioeconomic factors. At any particular location, cropping patterns do not differ because of the tree species grown. The crops and cropping patterns found associated with different trees, locations, and land types are presented in Tables 15-20. Differences are also associated with age and canopy structure. Turmeric cultivation is linked to relatively older trees, as turmeric grows well under shade.

Planted trees are usually well-spaced in lines. Wild trees are allowed to grow scattered in the field. Crop production and productivity need to be studied with respect to varying tree density and plot location and arrangement. Though direct benefits of trees in crop fields have not been measured by farmers, they understand that litter fall adds organic matter to soil. *Acacia nilotica* and *A. catechu* add nitrogen to the soil, which probably offsets some adverse effects of competition for light, water, and nutrients. Furthermore, birds living in trees may help reduce the insect population.

Farmers select trees to fit crops and vice versa (Table 21). Existing crops may dictate what tree species will be planted. However, in an existing orchard, the crop introduced must be able to produce satisfactory yield in association with the tree species present. For example, turmeric may be preferred in a shaded jackfruit orchard.

Conclusions and Recommendations

To the best of the authors' knowledge, this is the first organized study of trees grown on crop fields in Bangladesh. But several limitations prevented these scientists from thoroughly studying intricate issues related to the agronomic and socioeconomic aspects of the agroforestry practices. Because of deteriorating food, fuel, fodder, and timber production and environmental degradation, growing improved trees with crops must receive adequate attention. Agroforestry practices are crucial in areas where establishment of block forests is almost impossible. Future research on growing multipurpose trees on crop fields should consider the following issues:

o Detailed analysis of crop-tree interaction in relation to soil fertility, tree density, canopy structure, and root growth pattern.

o Improved management packages for both trees and crops.

o Determination of rotations of various species for different farm categories (small farmers may desire and need shorter rotations than larger farmers)

o Selection of appropriate species (development of improved varieties of traditional species versus introduction of new species).

o Identification of effects of tree-crop competition for light, water, and nutrients.

o Research on how trees provide small farmers food security.

o Research on how trees affect crop pests and diseases.

o Development of appropriate tree-crop farming systems and the minimum area needed for small farmers to survive.

o Understanding social implications/conflicts of growing trees.

o Research on present tenurial arrangements.

o Documentation and research on the historical changes in species distribution and density in particular areas.

o Economics of growing trees in combination with various crops.

Table 15. Tree-crop associations with *Phoenix sylvestris* Roxb.[*]

Location	Highland	Medium Highland
Bagherpara	Aus rice/fallow/wheat Aus rice/fallow/chickpea + mustard jute/fallow/lentil + mustard	Aus rice/Aman rice/ chickpea or wheat jute/Aman rice/mustard or potato
Narail	jute/fallow/mustard or lentil Aus rice/fallow/mustard Aus rice/fallow/chickpea or linseed	Aus rice/Aman rice/ chickpea
Kushtia	jute/fallow/winter vegetables Aus rice/fallow/wheat	Aus rice/Aman rice/ winter vegetables Aus rice/Aman rice/wheat
Ishurdi	Aus rice/fallow/mustard or lentil jute/fallow/mustard or lentil	fallow/Aman rice/wheat Aus rice/Aman rice/ mustard or lentil Aus rice/Aman rice/wheat jute/Aman rice/wheat
Charghat	sugarcane sugarcane + lentil turmeric Aus rice/fallow/wheat or lentil	Aus rice/Aman rice/wheat

[*] Note: Aus rice is direct seeded, Aman rice is transplanted, and Boro rice is winter transplanted.

Table 16 .   Tree-crop associations with *Borassus flabellifer* L.[*]

Location	Highland	Medium Highland
Bagherpara	Aus rice/fallow/mustard + lentil Aus rice/fallow/chickpea + linseed jute/fallow/chickpea + linseed	Aus rice/Aman rice/chickpea Aus rice/fallow/Boro rice jute/fallow/Boro rice
Narail	jute/fallow/mustard jute/fallow/wheat	jute/Aman rice/lentil or chickpea fallow/Aman rice/grass pea
Kushtia	Aus rice/fallow/winter vegetables jute/fallow/winter vegetables	Aus rice/Aman rice/wheat
Ishurdi	Aus rice/fallow/mustard jute/fallow/mustard	Aus rice/Aman rice/mustard fallow/Aman rice/wheat
Charghat	sugarcane sugarcane + lentil Aus rice/fallow/wheat	Aus rice/Aman rice/wheat

[*] Note: Aus rice is direct seeded, Aman rice is transplanted, and Boro rice is winter transplanted.

Table 17. Tree-crop associations with *Acacia nilotica* Willd.[*]

Location	Highland	Medium Highland
Bagherpara	Aus rice/fallow/mustard + lentil jute/fallow/radish or country bean eggplant/fallow/country bean taro/fallow/lentil	jute/Aman rice/chickpea Aus rice/Aman rice/chickpea
Kushtia	Aus rice/fallow/wheat jute/fallow/winter vegetables	Aus rice/Aman rice/wheat jute/Aman rice/wheat
Ishurdi	Aus rice/fallow/lentil or chickpea jute/fallow/mustard turmeric	Aus rice/Aman rice/wheat Aus rice/Aman rice/ Boro rice
Charghat	Aus rice/fallow/wheat or mustard jute/fallow/wheat or mustard sugarcane	Aus rice/Aman rice/fallow

[*] Note: Aus rice is direct seeded, Aman rice is transplanted, and Boro rice is winter transplanted.

Table 18. Tree-crop associations with *Acacia catechu* Willd.[*]

Location	Highland	Medium Highland
Ishurdi	Aus rice/fallow/lentil Aus rice/fallow/wheat	-
Rajshahi	Aus rice/fallow/mustard Aus rice/fallow/lentil Sugarcane	-

[*] Note: Aus rice is direct seeded.

Table 19. Tree-crop associations with *Mangifera indica* Linn.[*]

Location	Highland	Medium Highland
Kushtia	jute/fallow/winter vegetables Aus rice/fallow/winter vegetables	-
Ishurdi	Aus rice/fallow/mustard Aus rice/fallow/lentil	Aus rice/Aman rice/wheat fallow/Aman rice/Boro rice fallow/Aman rice/wheat
Charghat	turmeric sugarcane	-

[*] Note: Aus rice is direct seeded, Aman rice is transplanted, and Boro rice is winter transplanted.

Table 20. Tree-crop associations with *Artocarpus heterophyllus* L.[*]

Location	Highland	Medium Highland
Bagherpara	Aus rice/fallow/linseed eggplant country bean	Aus rice/Aman rice/ chickpea
Kushtia	Aus rice/fallow/wheat jute/fallow/winter vegetables	-
Ishurdi	Aus rice/fallow/wheat Aus rice/fallow/mustard or turmeric	Aus rice/Aman rice/wheat Aus rice/Aman rice/ mustard or chickpea
Charghat	Aus rice/fallow/wheat sugarcane turmeric	Aus rice/Aman rice/wheat Aus rice/Aman rice/ mustard

[*] Note: Aus rice is direct seeded and Aman rice is transplanted.

Table 21. Basis for selecting species in tree-crop mixtures.

Location	Crops based on trees		Trees based on crops	
	Yes	No	Yes	No
Bagherpara	85	7	78	4
Narail	79	8	13	4
Kushtia	33	38	67	8
Ishurdi	8	50	17	67
Charghat	48	48	48	16

Acknowledgments

The authors gratefully acknowledge the assistance of A.M. Musa, A. Harun, Mafizul Islam, and A. Quddus of the BARI On-Farm Research Division for their assistance in conducting this survey and compiling survey data.

REFERENCES

Aktar, M.S. and F. Haque. 1986. Analytical Framework for Understanding Rainfed Agroforestry System. Paper presented at the Workshop on Social Forestry in Bangladesh, September 12-14, 1986, Dhaka, Bangladesh. Organized by Bangladesh Centre for Advanced Studies and University of California, Berkeley.

Gupta, R.K. 1978. Ecological distribution and amplitude of some indigenous species vis-a-vis their role in afforestation of the arid regions of N.W. India. In *Environmental physiology and ecology of plants*, ed. David N. Sen, pp. 423-436. Dehra Dun, India: Bishen Singh and Mahendra Pal Singh.

Joshua, W.D and M. Rahman. 1983. *Physical properties of soils in the Ganges river floodplain of Bangladesh*. Strengthening of Soil Resources and Development Institute, FAO/UNDP project BGD/81/023. Dhaka, Bangladesh: Department of Soil Survey.

Manalo, E.B. 1976. *Agroclimatic survey of Bangladesh*. Joydebpur, Bangladesh: Bangladesh Agricultural Research Institute.

Multipurpose Tree Species for Small-Farm Use in Nepal

Maheshwar Sapkota

Institute of Agriculture and Animal Science
Tribhuvan University
Rampur, Chitwan, Nepal

The current demand for food, fodder, fuelwood, shade, and timber increases the urgency to improve multipurpose tree species. Although the concept of planting woody species on farms and marginal lands is not entirely new to Nepal, the planting of multipurpose trees is still limited. Since farmers believe that trees compete with agricultural crops for water, light, nutrients, and space, they are not fully motivated to plant trees. The research carried out in this regard indicates that improperly managed trees depress crop yields, primarily by competing for light. However, with proper species selection and crop tree management, such as hedgerow planting of Leucaena leucocephala *and intercropping of Dhaincha (*Sesbania aculeata*) with maize, farmers can obtain fodder and fuelwood without damaging cereal crop production. The soil is also enriched with biofertilizers. Scientists still need to introduce and establish more promising multipurpose tree species and generate technologies appropriate for small farmers.*

Nepal is predominantly an agrarian country that occupies a large part of the central Himalaya and its foothills. Of the nation's total land area (141,180 km^2), the Terai plains occupy 17%, the hill and mountain region 68%, and the Himalayan region 15%. Cultivable land occupies 18% of the total land area. From 1975-1980, the forested area decreased and cultivable land area increased. From 1980-1985, however, the forested area increased from 29.06 to 37.60%, largely because of major afforestation projects. However, these projects have not greatly affected the food, fodder, firewood, and timber situation because of an increasing population, especially in the Terai.

Therefore, various agencies have begun planting multipurpose, fast-growing tree species, such as *Leucaena leucocephala, Acacia auriculiformis, Cassia siamea, Eucalyptus camaldulensis, Dalbergia sissoo, Artocarpus lakoocha,* and *Pinus roxburghii.* About 92,833 ha of land have been afforested, and the current five-year plan calls for

planting an additional 175,000 ha, and 1.1 million ha more by the year 2000.

Farm Forestry Project

Supported by the International Development Research Centre of Canada, the Institute of Agriculture and Animal Science and the Institute of Forestry have collaborated for the past five years to determine which multipurpose trees are suitable for various agroclimatic regions of Nepal. Agroforestry studies, species elimination trials, and spacing experiments have been carried out. In addition, feeding studies on goats and buffalo calves have been conducted. Multipurpose tree seedlings of promising species are cultivated and distributed to farmers. Three districts (Chitwan, Parsa, and Sarlahi) have been selected as research and extension sites, and another three districts (Gorkha, Kaski and Bara) as research sites.

When *Leucaena leucocephala, Dalbergia sissoo* and *Eucalyptus camaldulensis* were planted at three different spacings (1 x 1 m, 1 x 2 m and 1 x 4 m) at Rampur, the mean height and mean dbh of *L. leucocephala* and *E. camaldulensis* were more than double the height and dbh of *D. sissoo* (Table 1). However, *D. sissoo* did equally well in sandy soil with a high water table at Parsa, Chitwan. In another trial, *L. leucocephala* performed better than *D. sissoo* and *Cassia siamea* after six months, but the three species were still too young to have established a proper growth pattern (Table 2). At Tamagarhi, all seedlings of *Heynia trijuga* and *Acacia decurrens* died within six months after planting. *L. leucocephala, Ceiba pentandra, Acacia catechu* and *Morus integrifolia* survived and achieved the greatest heights (Table 3). *Ceiba pentandra* also performed well in terms of height and survival achieved in six months at a trial in the hills of Gorkha. (Table 4).

Table 1. Mean height and dbh of species in two-year-old block plantation at Rampur, Nepal.

Species	Spacing (m)	Ht (m)	dbh (cm)
L. leucocephala	(1 x 1)	6.32	4.20
D. sissoo	(1 x 1)	3.48	1.94
E. camaldulensis	(1 x 1)	7.60	5.10
L. leucocephala	(1 x 2)	6.64	5.42
D. sissoo	(1 x 2)	3.56	2.12
E. camaldulensis	(1 x 2)	6.84	5.14
L. leucocephala	(1 x 4)	6.26	4.70
D. sissoo	(1 x 4)	3.16	2.88
E. camaldulensis	(1 x 4)	8.02	6.46

Source: Third Annual Report, Farm Forestry Project (1985-86).

Table 2. Height (cm) of various six-month-old tree species at Rampur, Nepal.

Plant-plant distance (cm)	L. leuco-cephala	D. sissoo	C. siamea
30	72.5	45.0	44.5
60	95.0	15.0	25.0
90	92.5	30.0	52.5
120	85.5	30.0	37.5
150	35.0	57.5	48.5
180	95.5	77.5	52.5
210	97.5	52.5	35.5
240	75.0	37.5	45.5
270	60.0	37.2	35.0

Source: Third Annual Report, Farm Forestry Project (1985-86).

Table 3. Height and survival of six-month-old multipurpose trees planted at Tamagarhi, Nepal in 1985.

Species	Height (m)	Survival (%)
Dalbergia sissoo	.75	63
Bauhinia variegata	.80	56
Leucaena leucocephala	1.31	96
Albizia lebbek	.52	75
Samanea saman	.89	65
Ceiba pentandra	1.43	100
Casuarina cunninghamiana	.63	71
Albizia amara	1.05	65
Cedrella odorata	.27	37
Prosopis juliflora	.87	40
Acacia catechu	1.37	96
Bauhinia purpurea	1.61	93
Albizia procera	.86	96
Acacia decurrens	0	0
Bauhinia malabarica	.94	75
Cassia sophera	.65	56
Heynia trijuga	0	0
Morus integrifolia	1.06	100

CV - 23 %

LSD - 0.05   0.29

Source: Third Annual Report, Farm Forestry Project (1985-86).

Table 4. Height and survival of six-month-old multipurpose trees planted at Gorkha, Nepal in July 1985.

Species	Height (cm)	Survival (%)
Pinus roxburghii	37	47.4
Ceiba pentandra	34	92.8
*Acacia auriculiformis**	27	83.3
Dalbergia sissoo	26	65.2
*Eucalyptus camaldulensis**	26	28.6
*Ficus roxburghii**	25	71.5
Leucaena leucocephala	23	50.8
*Albizia procera**	23	39.6
Morus integrifolia	18	63.5
Melia azedarach	18	63.5
*Cassia siamea**	18	85.7
Albizia lebbek	16	74.5
*Bassia butyracea**	13	73.8

CV - 20 %

LSD - 0.05   8.26

* augmented species

Source: Third Annual Report, Farm Forestry Project (1985-86).

Table 5. Uses, propagation techniques, and germination of some multipurpose trees. [*]

Species	Use	Propagation	Germination
Leucaena leucocephala	Good fodder, green manure, firewood, thatching material	Pods should be collected Nov.-Jan. Seeds should be scarified 22,500/seeds/kg	80-95%
Dalbergia sissoo	Furniture & veneer, fuelwood, good charcoal, fodder	By cuttings and seeds. Seeds ripen in Dec.-Jan.	95% when fresh, decreases later on
Acacia catechu	Timber, firewood, charcoal, medicine, coloring material goat fodder	Seeds. Seeds ripen in Nov.-Dec.	80-90% start in 1-2 weeks
Artocarpus lakoocha	Good fodder, edible fruit, firewood, fruit wine (local)	Seeds, fruits ripe in July; stump cuttings. Seeds viable only 1 week	80%, 3-5 fruits, 4-5 seeds per fruit
Bauhinia purpurea	Fodder, edible seed, fuelwood, casual timber	Seeds, seedlings	85-90% 4,250 seeds/kg
Bauhinia variegata	Fodder, edible flower, pickles, edible seed, firewood, casual timber	Seeds, pods ripe in Feb.-May. Seedlings are transplanted July-Aug.	90% fresh, decreases to 40%
Cassia siamea	Fodder for goats and sheep (but contains alkaloid), firewood, timber, green manure mulch, bedding	Direct sowing of seeds or transplanting of seedlings	80%

[*] More species listed in Singh 1982.

Use of Multipurpose Tree Species

Of the numerous tree species used to meet the needs of people and livestock, a few are mentioned here. *Leucaena leucocephala* and *Dalbergia sissoo* are fast-growing species that have many uses for small farmers. The leaves and tender twigs are fed to animals, branches are used as thatching materials, and stems are used for timber. *D. sissoo*'s valuable timber is used for furniture and construction material. *L. leucocephala* stems are used for making the handles of sickles and for constructing temporary houses. Because of its nitrogen-fixing characteristics and the high percentage of protein in its leaves, it is becoming popular in the Farm Forestry Project areas and nationwide. *Eucalyptus camaldulensis* is a fast-growing species used for timber and firewood in Nepal and in match industries in India. In Nepal, crops such as finger millets and pulse seeds can be grown under the trees without reducing crop yield, e.g., the community-managed trial site in Gunja Nagar, Chitwan. *Cassia siamea* and *Acacia auriculiformis* are drought-resistant, fast-growing species. Their thick leaves can be used for green manure and mulch. The leaves of *A. auriculiformis* are eaten by goats. Twigs and branches are used for firewood, and the stems are used for thatch and temporary house construction.

Local species like *Artocarpus lakoocha, Bauhinia purpurea, B. variegata* and various *Bambusa* spp. are liked by small farmers because of their many uses. Some common multipurpose species and their uses are given in Table 5. Although *Sesbania aculeata* is not a perennial tree, it is one of the most popular species of small farmers in the Terai. It is used as a green manuring crop, adding 50 kg/N kg to the soil for the use of subsequent paddy crops (Pandey 1983). Twigs and leaves are used as fodder, bark as fiber for making roaps, and dried stems as firewood and thatching material. *Acacia catechu* is a valuable multipurpose tree species used for timber, fodder, medicine, and for producing Kattha, a traditional betel nut chewing mixture.

Agroforestry at Rampur

When maize was intercropped with *Leucaena leucocephala, Dalbergia sissoo,* and *Eucalyptus camuldulensis* in 1 x 1 m, 1 x 2 m, and 1 x 4 m spacings, the crop yield was substantially reduced in all but the 1 x 4 m spacing. When sesame and wheat were intercropped in the hedgerow (alley) cropping pattern of 2 x 5 m, there were no differences between yields and controls without trees. In addition, the growth of *L. leucocephala* increased (Farm Forestry Project 1987).

REFERENCES

Farm Forestry Project. 1987. *Third annual report.* IOF/IAAS. Kathmandu: Tribhuvan University.

Pandey, S.P. 1983. Green manuring paddy with *Sesbania aculeata* (Dhaincha) at various levels of fertilizer nitrogen. *J. Inst. Agri. Anim. Sci.* 2(1 & 2):35-39.

Singh, R.V. 1982. *Fodder trees of India.* New Delhi: Oxford and IBH Publishing Co.

Multipurpose Trees for Small-Farm Use in The Central Plain of Thailand

Suree Bhumibhamon

Kasetsart University
Bangkok, Thailand

The alluvial plain of Central Thailand is a highly fertile area for growing agricultural crops. The forests in this region have decreased continuously at an alarming annual rate of about 2%. Only 9 of the 18 provinces in the region have varying degrees of existing natural forest. The total forested area in the region is about 26%, which is less than the national average (29%). This paper discusses land use and tree growing in the Central Plain, and includes a list of trees grown around farm houses. Future tree-growing trends are presented based on studies of the attitudes of administrators and policymakers. The studies clearly demonstrate the great opportunity that exists to grow multipurpose trees to serve the growing regional demand.

The Central Plain of Thailand was created after other regions of the country during the Jurassic period (about 181 million years ago). Its floristic structure varies according to topographic structure. The lowland rain forest occurs in the large area of alluvial soil deposition. The western mountainous zone contains most of the remaining forests, including dry evergreen, dipterocarp, and mixed deciduous types. Mangrove and beach forests are located near the sea. The Royal Forest Department (1984) estimates areas covered by tropical evergreen, mixed deciduous, dry dipterocarp, and mangrove forests at about 12,449 km^2, 5,192 km^2, 540 km^2, and 335 km^2, respectively.

The total area of the Central Plain is 67,398 km^2 and covers 18 provinces. As in other regions of the country, its forests have been continuously depleted--from 35,660 km^2 (1961) to 23,970 km^2 (1973), 21,830 km^2 (1976), 20,420 km^2 (1979), 18,520 km^2 (1982), to 17,228 km^2 (1985). The present annual depletion rate of forest resources is 1.9%, exceeded only by the northern region. Nine of its 18 provinces have no forested areas. Only five provinces in the western mountains have large areas of remaining forest. The other four provinces have only small forested areas.

The forests of the Central Plain can be classified as protected or productive. Protected forests include two wildlife sanctuaries (6,447 km^2) and four national parks (668 km^2). The remaining protected forests consist of botanical gardens, arboreta, nature parks, and nonhunting areas. This means that only 10,000 km^2 of forested area can be exploited.

Exploited forest includes concessions, as well as reserved and protected areas. About 30% of the reserved forest has been cleared to grow cash crops and develop new settlements. The volume of growing stock in the exploited forest is estimated at only 20.25 million m^3 out of the total national growing stock of 184.88 million m^3, or about 10%.

While sawmills are scattered throughout the country, other factories that produce plywood, veneer, fiber board, particle board, and paper are located mainly in the Central Plain. Urapeepatanapong and Mungkorndin (1983) surveyed the supply and demand of charcoal in the Central Plain. Results showed that most charcoal is made using small kilns (41.95%), graveyard type (32.25%), big oven (16.13%), sawdust mound type (6.45%), and Mark V (3.23%). Each type has a different capacity and uses mainly miscellaneous hardwood species. Bambusa, Dendrocalamus, Gigantochloa, and Thyrsostachys are also used as sources of charcoal.

Of the total charcoal consumption in the Central Plain, about 30% was for household use (3.07 million ton/yr.), about 80% for industrial use (556,932 kg/yr.), and about 50% for the service sector (0.3 million ton/yr.). The total charcoal consumption in the Central Plain was about 30% of the nation's total. This indicates that a high percentage of illegal charcoal from the northeast and the lower northern region is transported to the Central Plain. In the remote areas of the Central Plain, Combretum species on marginal lands are also used to produce charcoal.

The present paper addresses the possibility of growing trees for small-farm use to help create regional self-sufficiency and reduce the amount of illegal charcoal taken from other regions.

Climate

Climate in the tropical savannah of the Central Plain is influenced by northeast and southwest monsoons. Based on precipitation, three zones can be identified. The first is a large, moist area covering 14 provinces with a high precipitation rate of about 1,300 mm and about 121 rainy days per year. The second zone is in the western mountainous area (Suphan Buri, Kanchanaburi, and Ratchaburi) with an average annual precipitation of 1,174 mm and about 110 rainy days. The third zone is the dry area of Prachuab Khiri Khan, having an average precipitation of 1,031 mm and 119 rainy days. The average temperature over the area ranges from 26-30°C. The relative humidity ranges from 56.7-89.7%.

Geological formations consist mainly of sedimentary and metamorphic rock. Alluvial soils and river gravels were formed two-three million years ago. Soils of the Krabi, Ratchaburi, and Tanausi groups are scattered in the western parts of the region. There are basalts, granite, diorite, quartzodiorite, andesite, rhyorite, tuff granodiorite, porphyry, gneiss, and schist. The Chao Phya River Delta can be divided into the old and new deltas. Fan complexes are also found in the alluvial terraces (Department of Land Development 1985).

Soil resources in the Central Plain include sandy, loamy, clayey, and skeleton types. There are about 30 great soil groups. The large groups are slope complex (27.6%), clayey sulfic tropaquepts (10.89%), and clayey tropaquepts (10.12%). Each soil group has varying potential to grow crops.

Problem soils are found in at least 37.3% (2.6 million ha) of the region's total area. There are shallow soils with gravel (6%), saline soil near the beach (1.8%), sandy soil near the beach (1.5%) and uplands of more than 35% slope (27.6%).

About 54% of the total land area is used for agricultural farming, including paddy fields (2.1 million ha), cash crops (1.36 million ha), intensive horticulture (8,780 ha), and fruit trees (0.24 million ha, of which one third is coconut plantation). The available area for tree growing in the Central Plain is about 0.82 million ha.

While growing more trees on private land is a possibility, only 47.18% of the agricultural land is farmed by its owners. This, of course, affects the possibility of growing trees in farm areas.

Man and Trees

For more than 600 years, the Central Plain of Thailand has been considered the center of Thai civilization. The relationship between man and trees was mentioned in a stone pillar inscription in which farmers were encouraged to grow *Areca catechu, Mangifera indica, Cocos nucifera, Tamarindus indicus*, and *Corypha umbraculifera*.

Existing trees on farm lands indicate the preferred choices by farmers. However, introduction of new species into the region has slightly altered the evolution of plant species. Any plant that provides maximum satisfaction to the land owner will remain in the farm area.

In the early days, bamboo houses were built near waterways. Plants grown in farm areas were selected by tradition, e.g., *Cassia fistula, Mangifera indica, Tamarindus indicus*, and *Diospyros siamensis*. Various bamboos were commonly grown as shelterbelts surrounding home areas. Bhirom (1972) wrote that three cooking stoves were built in the kitchen and only fuelwood sticks collected from nearby forests were used.

Behind the houses were farming areas, followed by bush forest and, beyond that, the natural forest. When a canal was constructed, a strip village was formed. In the Central Plain, scattered villages were also formed by farmers building houses near their paddy fields. Trees were planted in most temples, which were community centers.

Bhirom (1972) discussed some restrictions in Thai culture about growing trees on farms and around houses. For instance, it was believed that growing *Caesalpinia pulcherrima* or *Plumeria acuminata* in farm areas would bring misfortune. Even today, Thais in the Central Plain still maintain the traditional practice of planting *Phyllanthus emblica* in front of the house and *Artocarpus heterophyllus* in the backyard.

Future Tree Growing

To determine the tree-growing trend in the Central Plain, 60 district chiefs in the area were interviewed. Most indicated they had served in their posts for less than 10 years. Their degree of participation in tree growing ranged from slight to moderate. Only six said they were fully involved in tree growing in their districts. The size of each district varies greatly, but most district chiefs are responsible for about 150 villages.

About 37 districts claimed to have fuelwood scarcities. District chiefs concluded that greater opportunities for growing trees existed in farm areas rather than in industrial plantations or large farm woodlots. They recommended growing the following trees. In order of preference, they are: *Eucalyptus camaldulensis, Casuarina junghuhniana, Cassia siamea, Pterocarpus macrocarpus, Azadirachta indica, Acacia auriculiformis, Leucaena leucocephala, Mangifera indica, Tamarindus indica, Cocos nucifera, Lagerstroemia calyculata, Lagerstroemia speciosa,* and *Cassia fistula*. District chiefs recommended that the government should supply seedlings to farmers. They said that farmers still lack basic knowledge about agroforestry and that extension was considered a major activity for the responsible government agencies.

The majority of parliamentary members in the Central Plain indicated the need for tree planting, particularly in abandoned areas, infertile zones, and on farms. They believed farmers should set up self-help programs to supply wood from forest resources, as they are renewable resources. Farmers should decide by themselves which multipurpose species should be grown in their farm areas. Members of Parliament also thought that farmers would grow multipurpose trees if they provided more cash earnings. They recommended that village chiefs take the lead in developing more green areas.

Farmers Make Their Own Choice

Two groups of farmers were interviewed, one in the old farming area of Bangkok Noi on the outskirts of Bangkok and the other in a remote village area in Sara Buri.

The first group earn their living by growing mixed fruit tree species in the farm area, e.g., *Durio zibethinus, Baccaurea sapida, Eugenia javanica, Artocarpus heterophyllus, Cocos nucifera, Garcinia mangostana, Garcinia schomburgkiana, Azadirachta indica, Bouea burmanica,* and *Citrus hystrix*. Growing cuttings of these species is economically feasible. *Leucaena leucocephala, Bambusa,* and *Eucalyptus camaldulensis* are scattered throughout the area. Expansion of the settlement area has decreased the number of trees.

Farmers in two villages of Sara Buri comprised the second group. The first village is located in a poorly productive forest, while the second is on the edge of a natural forest. Farmers in both villages grow various fruit crops, including *Mangifera indica, Cocos nucifera,* and *Artocarpus heterophyllus*. Fuelwood and charcoal are collected mainly from farm area cuttings of *Leucaena leucocephala, Mangifera indica, Tamarindus indicus,* and *Annona squamosa*. Villagers said the state should plant more trees and provide seedlings of trees they are interested in, preferably multipurpose trees species.

REFERENCES

Bhirom, S. 1972. *Thai houses in the Central Plain*. Bangkok: Silpakorn Univ., Dept. of Fine Arts.

Department of Land Development. 1985. *Land use in the Central Plain*. Bangkok: Dept. of Land Development.

Royal Forest Department. 1984. *Yearly report*. Bangkok: RFD.

Royal Forest Department. 1986. *Yearly report*. Bangkok: RFD.

Urapeepatanapong, C. and S. Mungkorndin. 1983. Demand and supply of charcoal in urban and rural areas in Thailand. In *Proc. the third seminar on silviculture-forestry for rural community, no. 33, pp. 1-33*. Bangkok: Kasetsart University Faculty of Forestry.

Multipurpose Trees for Small Farmers in India

Narayan G. Hegde

The Bharatiya Agro Industries Foundation
Pune, India

Small farmers dependent on fragments of degraded, dry land for their livelihood represent 74.5% of all agricultural holdings in India. In the absence of assured irrigation, the present agricultural system can barely provide employment for 100 days, often with the risk of crop failure. In addition to year-round employment and food grains, these families also need fodder and fuel to meet their household needs. Observations in rural areas indicate that planting of multipurpose tree species is a profitable proposition under farm forestry and agroforestry systems. However, in the absence of water resources, protection from stray animals, and investment capacity, small farmers hesitate to initiate new ventures. They need to select suitable, fast-growing tree species that can generate more employment and profits without affecting crops. They also need to develop a basic infrastructure for extension, input supply, and marketing surplus produced under this program. Some of the tree species ideally suited for small farmers are Acacia nilotica var. cupressiformis, Albizia lebbek, Azadirachta indica, Cassia siamea, Casuarina equisetifolia, Dendrocalamus strictus, Derris indica, Eucalyptus hybrid, Melia azedarach, Moringa oleifera, and Leucaena leucocephala. However, these species must be screened by farmers, depending on their resources, ability to protect seedlings from livestock, and local produce demand. Species such as Agave sisalana, Jatropha carcus and Prosopis juliflora make good living hedges while generating additional income.

Fodder and fuelwood shortages are steadily increasing as production programs become less able to cope with increasing demand. An important reason for the slow rate of reforestation is lack of cooperation among rural people in planting and protecting trees. Until recently, fuelwood supply was the responsibility of the state forest departments. Farmers never bothered to plant trees, as fuelwood was easily available from the forests. Moreover, the tree species promoted by the forest department had slow growth rates and long gestation periods of 80-100 years. Farmers did not perceive any direct, short-term benefits in planting these tree species. Furthermore, under government regulations, farmers were not permitted to fell the trees grown on their own land.

The forest departments established during British rule to carry out wood extraction and conservation needed further guidance to cope with the changing situation. Some serious problems faced by the forest officers were poor growth of traditional species on heavily eroded forest lands, biotic interference beyond the tolerance limit, and lack of concern and cooperation by local communities in protecting the plantations. There was a need for an immediate change in the forest development policies to meet the growing needs of the people, while protecting remaining forests. This serious concern led the government to establish and implement a separate social forestry program in the field.

Social Forestry Programs

For the seventh National Five-Year Plan (1986-90), the Government of India has earmarked a budget of Rs 45,450 million (US $ 3,636 million) for forestry in which Rs 37,950 million (US $3,326.9 million) is for social forestry and Rs 7,500 million (US $ 600 million) is for production forestry, as compared to Rs 4,521.9 million (US $ 361.75 million) spent during the sixth Five-Year Plan (1980-85), through both social forestry (US $ 281.5 million) and production forestry (US $ 80.2 million) programs. While aiming to promote tree planting through social forestry programs, it is necessary to make sure that tree planting will not in any way suppress food production. This is not difficult with the availability of 93.69 million ha of wastelands throughout the country. In addition to these wastelands, about 45 million ha of forest area are barren.

Although the government is aware of its inability to develop and protect the plantations, many policymakers are opposed to parting with the ownership of these wastelands. Recently, however, several new schemes have been developed recently to lease these wastelands to poorer farmers on a usufruct basis. With that, many small farmers can soon obtain part of the wastelands to plant trees. In addition to afforestation of wastelands, farmers can be motivated to initiate farm forestry on their marginally productive fields, bunds, borders and backyards.

The government's major objective in promoting this program was to ensure that these farmers could meet their fodder and fuelwood requirements from these plantations. However, the schemes did not attract the target group. Most of the small farmers were unaware of these schemes, although they suffered the most from fodder and fuelwood scarcities. They worried more about their daily wage earning, of which more than 80% was spent on food, than about future supplies of fodder and fuel. Even today, they procure their fuelwood free of cost either from community wood lots or illicit cutting of tree branches and bushes along roads.

Fodder production is considered unnecessary as most cattle will not produce more even if they are fed good-quality fodder. These livestock are able to survive on whatever grass is available.

Farm Forestry Schemes

Many farmers benefitted by participating in farm forestry programs. Most of these were large-scale holders who could realize the advantages of farm forestry over agricultural crops alone. This program was further boosted when wood-based industries entered the field to ensure a support price and contracts for the purchase of wood as raw material for paper, pulp, and plywood production. The farmers could correctly assess the marketability of the produce and found that selling wood as timber, poles, and industrial raw material would ensure higher prices in a ready market. In contrast, people needing fodder and fuelwood over a widely scattered area did not have this purchasing power. In the absence of a market infrastructure for fodder and fuelwood, farmers with surplus produce had to waste these valuable commodities. Recognizing this problem, most farmers shifted to marketable species, such as Eucalyptus and Casuarina. Another important

reason for selecting such species was the problem of providing protection from livestock. It is almost impossible to prevent browsing when fodder species are cultivated in isolated plots. Control of stray animals in the field often creates tensions among various village groups. In such a situation, Eucalyptus and Casuarina can be managed easily, although these are browsed by sheep and goats in the absence of other forage.

Feeling the pressure from farmers to supply seedlings of commercial species, the social forestry departments of the state governments started raising these seedlings for free distribution. This helped the departments reach their seedling distribution goal, but it did not directly promote the production of fodder and fuelwood. The profitability of this farm forestry was high, and most of the early adopters earned their fortunes through Eucalyptus cultivation. These farmers felt it more profitable to shift their farm forestry operations from degraded wastelands to fertile agricultural land where biomass yield could be increased severalfold in shorter time under good management with optimal use of inputs, such as fertilizer, irrigation, and protection. Farm forestry became a boon for the absentee landlords to retain their ownership, as they did not have to depend on local managers who might claim ownership under the prevailing tenancy act.

This style of farm forestry development has become somewhat controversial among policymakers. The public and social forestry departments were seriously criticized for deviating from their objectives. The major objections were that the program was decreasing food grain production and not helping the local poor meet their fodder and fuel needs. The allegations are not totally true, and social forestry officials are not entirely to blame. However, these developments provide an excellent insight into the entire fodder and fuelwood crisis and the modifications required to motivate farmers to cultivate fodder and fuelwood species.

The conclusions are as follows:

o Although the severe scarcity of fodder and fuelwood affects small farmers in rural India, they do not seriously consider producing these commodities on their farms until the alternative free supply is exhausted.

o Farmers like to grow trees mainly to earn more money rather than to meet their household needs.

o Farmers like to grow the produce that can be sold easily at a higher price.

o Farmers need marketing back-up in terms of minimum support prices and assurance of a market for the entire quantity produced.

o Organized community programs to control stray grazing are needed.

If the above points are given due consideration, large-scale social forestry can be promoted to benefit small-scale farmers.

Increasing the Profitability of MPTS Cultivation

Tree planting is designed not only to supply fodder and fuelwood, but also to generate additional employment throughout the year. For a sound, long-term program, activities should be profitable to sustain the interest of the intended beneficiaries. This justifies the need for developing strategies to make tree planting profitable. While there is no additional expenditure involved in adopting agroforestry, farm forestry requires a sizable investment not only to carry out field operations but also to sustain families until their plantations start generating income. The government has recognized this by arranging bank loans at lower interest rates for small farmers. This enables them to use all the necessary inputs to increase the biomass production.

It is possible to increase profitability by improving the growth rate of trees and yield through improved agronomic practices. It is generally believed that tree crops do not require interculturing, irrigation and manuring, as they are capable of obtaining nutrients and water at greater soil depths. However, recent field trials worldwide have proven the beneficial effects of these inputs.

Water is a critical input that affects the establishment and growth of trees. Biomass yield can increase by 300-400% with regular irrigation in dry areas. It may not be possible to irrigate all plantations because of water scarcity or undulating topography. In such areas, innovative approaches such as contour bunding, soil mulching, use of water-retaining chemicals (polymers), and drip irrigation can help conserve available moisture. Efforts can also be made to tap available water resources through open wells, bore wells, lift irrigation schemes, and by plugging gullies. In most areas, it should be possible to carry out watershed surveys and store the rainwater in ponds. Water collected in such ponds may not remain until the end of summer, but can provide at least 3 or 4 irrigations and significantly reduce the effects of dry spells. However, this technique can be practiced only when all farmers owning land in a watershed unite to develop their area.

Most wastelands have poor soil fertility, and plants respond well to the application of fertilizers. Analysis of the soil for pH and macro- and micro-nutrients would help in scheduling fertilizer application.

Selection of Suitable Tree Species

The key factor contributing to the success of afforestation is the selection of suitable tree species. The establishment, survival, growth, ability to withstand adverse conditions and browsing, yield and market value for the produce are some important factors related to species that affect profitability.

Small farmers in India do not have the ability to invest in long-term projects. In the absence of wage-earning opportunities, poor farmers harvest immature grain to avoid starvation. Such families prefer to plant short-duration trees.

The selection of tree species also depends on the demand for tree products. Naturally, farmers like to select species that bring them more revenue. The wide choice of available species produce fodder, fuelwood, poles, edible and non-edible oils, raw material for pesticides and medicines, fruits, nuts and vegetables, honey, lac, gum, silk, timber for general construction, handicraft wood, and raw material for a variety of wood-based industries. In a given situation, farmers should receive help in identifying all species that can adapt to the local agroclimatic conditions. The next step would be to prepare a short list of these species based on the demand and market value for the produce.

Farmers often make mistakes in analyzing the long-term demand and supply for wood products. In Gujarat State, farmers started cultivating Eucalyptus in anticipation of selling poles 5-6 years ago, when Eucalyptus poles of 8-10 m length and 15 cm diameter sold in the local market for Rs 90 (US $7.20) each. This raised high hopes and many farmers planted Eucalyptus. But now the prices have fallen to Rs 20 (US $1.60) per pole, well before these trees are ready for harvest. Planting for pulpwood would have been more profitable.

Further selection of species can be based on capital requirements and a farmer's ability to invest that amount. Other aspects, such as biotic interference and ability of farmers to provide protection, should also be considered before making the final selection.

It has been observed that small farmers like to maintain a few plants of different species that can provide income at different times, whereas large farmers prefer large-scale plantations of few species. Small farmers, who generally have surplus manpower in their households, can intensively cultivate a variety of species, which demand labor on different occasions. This would help generate additional employment and increased income. Small families can also sell produce directly in local markets at higher prices. Such operations would be expensive for the large-holders, as they have to employ outside labor at higher wages.

Before establishing a plantation, it is also necessary to ensure that the species selected do not harm crop production on neighboring farms. Farmers often complain that trees affect crops adversely through moisture stress, shading, and obstruction in agricultural operations due to the spreading of roots. When rich farmers take up farm forestry, neighboring poor farmers are often too afraid to complain. But when poor farmers plant trees, rich neighbors start complaining even before they observe any stress on their crops. Poor farmers need to select the species carefully to avoid any later resistance.

Another way to ensure safety of the trees is to promote block plantations, where all the farmers owning land in a block of 50-100 ha of degraded or poor soils are persuaded to plant trees, irrespective of their income status. Under such situations, the poor farmers will have access to all the improved techniques available to the elite farmers. These farmers can jointly organize and protect the plantation for the community at a lower cost. In such situations, it should be possible to select more profitable tree species, even those prone to browsing.

Marketing Infrastructure

As profitability is directly linked to marketing, a program cannot succeed without setting up a suitable infrastructure for marketing the produce. Fuelwood disposal is a big problem in rural areas, even though the national surveys report a widening gap between the demand and supply. In the absence of organized selling arrangements, farmers must depend on middlemen known for exploitation. Marketing networks would help reduce the cost of handling, transportation, and storage as the produce can be dispatched directly to the consumers from production site.

Introduction of post-harvest processing techniques in villages can contribute to the value of products before selling. This arrangement provides additional employment opportunities for people, low overhead cost of establishment, and reduction in the handling loss of the raw material, which helps ensure higher profit margins.

Above all, a marketing infrastructure can increase the confidence of farmers about the salability of their produce at a remunerative price. This setup can also take up input distribution services and motivate the farmers to plant trees on a larger scale. In addition to the government departments, several voluntary agencies are presently engaged in the promotion of tree planting. Some of these agencies can be entrusted with the responsibilities of training, motivation, supplying input, developing nurseries, and organizing marketing to strengthen the program.

With an initial increase in profitability, small farmers are motivated to work hard and make efficient use of the time of all family members. At this stage, they may find it better to use the fuelwood and fodder grown on their own farms than to waste their time and energy in collecting these materials elsewhere. Moreover, it is the women of only the poorer families who gather fuelwood free of cost. Poor farmers would prefer to end this hardship faced by their women as soon as possible.

A similar trend was observed in the cattle development program undertaken by The Bharatiya Agro-Industries Foundation in India. Farmers wanted to sell most of the milk they produced, while policymakers wanted this milk to be consumed by the farm families. The farmers initially sold most of the milk, but they started retaining enough for home consumption as soon as the anxiety of collecting more money subsided. The same logic applies to farmers taking up tree planting in rural areas.

Suitable Species for Small Farmers

Tree planting on private lands by farmers to generate additional employment and income is a fairly new concept in India. In earlier years, farmers established fruit orchards under assured irrigation. Deliberate planting and cultivation of multipurpose trees as an economic activity was restricted to very few species.

Cultivation of *Sesbania sesban* on the sugarcane field bunds is an old tradition in the south and central parts of India. Farmers sow the seeds while planting sugarcane and maintain them for three years. During this period, farmers regularly lop the side branches for fodder and finally harvest the main trunk for fuel. Its foliage is a popular livestock fodder, and there is good demand for this fodder in the towns, particularly by goat and buffalo owners. The only disadvantage of this species is that the plants lose vigor if the foliage is harvested continuously for more than 1 or 2 years. The burning quality of the wood is satisfactory. However, this species is confined to irrigated areas.

Thespesia populnea (Portia) is another tree grown in humid regions, particularly on paddy field bunds in central India. The growth rate of this species can be enhanced by regular pollarding. Portia foliage makes a good fodder, but it is more profitable to allow these shoots to grow as straight poles that can be harvested in about 3 years. Other species grown by the farmers in humid areas are *Dendrocalamus strictus* (male bamboo), *Albizia procera* (white siris), *Erythrina indica* (Coral tree), *Derris indica* (Pongamia), and *Melia azedarach* for various purposes, especially household needs.

In dry areas, species like *Acacia leucophloea, Acacia nilotica* var. *indica, Acacia nilotica* var. *cupressiformis, Acacia senegal, Ailanthus excelsa,* *Azadirachta indica, Pithecellobium dulce,* and *Prosopis cineraria* grow naturally on the field bunds and uncultivated areas.

In addition to the above species, farmers in certain areas plant small numbers of trees like *Albizia lebbek, Dalbergia sissoo, Tectona grandis,* and *Terminalia arjuna* in their backyards to generate additional income. Food species like *Artocarpus heterophyllus, Emblica officinalis, Mangifera indica, Moringa oleifera, Syzygium cumini,* and *Tamarindus indica* are also planted in their backyards for home consumption and sale in local markets. These trees also serve as cash reserves for poor farmers, who sell the wood as timber for meeting family obligations. Such tree planting programs by farmers had been low key until the government introduced the concept of social forestry in 1951.

At that stage, the forest departments promoted the large-scale cultivation of several other species, such as Eucalyptus hybrids and *Casuarina equisetifolia*. Farm forestry was undertaken on degraded lands to meet the growing demands for pole timber and pulpwood. Because these species need a regular source of moisture for straight growth and higher biomass production, cultivation could not be extended to large areas. For dry areas, species like *Acacia tortilis, Cassia siamea, Parkinsonia aculeata,* and *Prosopis juliflora* have been recommended, as these species are drought resistant and not browsed by livestock.

In the late 1950s, the Department of Agriculture started promoting *Gliricidia sepium* as a green manure crop in heavy rainfall areas. This species has been included in the social forestry program as a fodder and fuelwood species. Gliricidia looks tender, but trials in Karnataka and Maharashtra states have proved its ability to withstand severe drought conditions. When this species is introduced into a new area, livestock initially do not browse its unfamiliar foliage and trees grow without any interference. Some of the wood based industries in northern India have promoted the cultivation of Poplars (*Populus euphratica*) in an agroforestry system, with an arrangement for buying wood at a reasonable price. This scheme is becoming popular among farmers, as the trees do not interfere with crop production and a fair market price is assured.

not interfere with crop production and a fair market price is assured.

Leucaena leucocephala is a recent introduction in India, although the Hawaiian type has long existed in some parts of the country. During the past 10 years, Leucaena has been planted extensively under social forestry programs in most states with an encouraging response. Currently, many exotic species are being screened for suitability under various agroclimatic conditions in India. Some worth mentioning are Acacia mangium, Albizia falcataria, Calliandra calothyrsus, and Leucaena diversifolia. Further trials are needed before these species can be promoted under social forestry programs.

All the species listed above are not necessarily ideal for small farmers. It also is not necessarily true that farmers select these species to maximize their earnings. Except for large-scale farm forestry operations, most farmers grow these trees for specific benefits instead of selecting species with higher returns. The time has now come to determine the direct and indirect benefits of different species so farmers can select the best ones for their socioeconomic needs.

Evaluation of Important MPTS Suitable for Small Farmers

Keeping in view the present marketing infrastructure, demand for produce, farmers' household requirements, and agroclimatic conditions, the cultivation of some of the species presented in Table 1 could profit small farmers.

Commonly known as Ramkanti babul, Acacia nilotica var. cupressiformis grows erect and reaches 25 m in height with a slightly crooked trunk of 30 cm in diameter. This species is ideally suited for dry areas with annual rainfalls of 200-400 mm and a long dry season. The species, being less thorny, can be managed easily in agroforestry and farm forestry systems. Except for goats, livestock generally do not browse the plants. Saplings start branching soon after planting. Branches grow straight, almost parallel to the main trunk, and close to each other. As goats cannot approach the branches in the central portion, growth is generally not affected even after browsing. Its wood is heavy, strong, and commands a good price as timber for general construction and farm implements. The annual biomass increment can reach 10-12 m^3 per ha when planted at closer spacings.

Table 1. Selection of tree species for small farmers.

Rainfall range	Species	
	for which browsing can be prevented	for which control of browsing is difficult
200-400 mm	Acacia nilotica var. cupressiformis Azadirachta indica	Acacia nilotica var. cupressiformis Cassia siamea
400-800 mm	Azadirachta indica Acacia nilotica var. cupressiformis	Cassia siamea Derris indica
800-1200 mm	Leucaena leucocephala Albizia lebbek Moringa oleifera	Derris indica
1200 mm and above	Leucaena leucocephala Melia azedarach Dalbergia sissoo Dendrocalamus strictus	Casuarina equisetifolia Eucalyptus hybrid

Commonly known as neem, *Azadirachta indica* is one of the most popular plants in India. Neem tolerates severe drought and a variety of soils, including alkaline, shallow, and gravelly land. The annual biomass increment can reach 10 m^3 with adequate moisture. The wood is hard, durable, and makes good timber. Farmers usually lop the side branches of trees grown on field bunds every year to reduce shading on field crops, using the loppings for fodder and fuel.

The bitter compounds present in neem seeds have insecticidal and bactericidal properties. Recent research recommends the use of neem oil and neem cake for controlling certain pests and diseases of several important crops including rice, cotton, fruits, and vegetables. Use of neem products as pesticides can reduce the cost of insect control. In India, neem oil is presently used for making soaps and brings Rs 11-12 (US $ 1.0) per liter.

Neem trees can be established easily on wastelands at a spacing of 4 x 4 m to maintain 625 trees per ha. The trees mature in 6 to 8 years and yield an average 30-40 kg seeds every year. At the current price, the gross annual income from seeds alone is Rs 12,500 (US $ 1,000) per ha, almost a fivefold increase over the income from food crops grown in drought prone areas. Thus, neem can generate more employment, income and protect the environment by reducing the use of pesticides.

It has also been reported that mixing urea with neem cake (10%) can reduce the activities of nitrifying bacteria in the soil and help reduce fertilizer use and costs by about 25-30% This species can be used to establish both farm forestry and agroforestry systems by small farmers. People are slowly beginning to understand and appreciate the direct and indirect benefits of neem. It will be a popular tree in social forestry programs.

Cassia siamea (Kassod) is a beautiful evergreen flowering tree that is drought tolerant and adapted to a wide range of soils. Being a non-browsed species, it can be planted in scattered fields in drought prone areas under agroforestry as well as farm forestry systems. The annual biomass increment may reach 15 m^3. Its wood is hard and useful as timber and fuelwood.

Derris indica (Pongamia) is a native species, deciduous or evergreen. The tree grows fast and reaches 12 m in height with a crooked, branchy trunk. It prefers humid conditions but withstands long drought periods. The annual biomass increment may reach 15 m^3 per ha. Its wood can be used for furniture, carving, timber, and fuel.

Derris seeds contain an oil usable for leather dressing, soap making, lubrication, and varnish production. Its oil and cake have pesticidal properties. Livestock generally do not browse the foliage. The species also can be established in agroforestry systems in agricultural fields.

Leucaena leucocephala is a fast-growing tree species useful for fodder, fuel, pole timber, and pulpwood production. These trees prefer humid conditions, but can tolerate semi-arid conditions. Protection from livestock damage is the major problem, particularly during the first year. Fortunately, Leucaena has not yet been attacked by psyllids in India. The annual biomass increment can reach 50 m^3 per ha with adequate moisture supply. Leucaena is ideally suited for small farmers in agroforestry. Paper mills have offered to pay Rs 650 (US$50)/t wet wood (minimum diameter 6 cm) as compared to Rs 450/t for Eucalyptus wood.

Commonly known as Siris in India, *Albizia lebbek* is a popular timber tree that can be grown in semi-arid areas with an annual rainfall of 500-2,000 mm in farm forestry and agroforestry systems. This species prefers well drained and deep soils. Annual biomass increment ranges from 3 to 8 m^3, depending on planting density. Side branches can be lopped for feeding livestock, allowing the main trunk to grow straight. Saplings require protection from livestock during the first two years.

Moringa oleifera (drumstick) is a popular vegetable throughout India. It is an ideal species for agroforestry in areas where moisture supply is adequate. Trees can be established easily by planting poles. Fruiting starts within 2 or 3 years. Pods command a good price. One tree can generate Rs 60-80 (US $5) each year. Tender foliage and flowers are also used as vegetables. Its wood is light and burns fast.

Casuarina equisetifolia is a fuelwood and pole timber species that grows mostly in the southern part of the country and in coastal areas. It requires deep sandy soil with an adequate moisture supply. Except for goats, its foliage is not browsed by

Table 2. Methods of propagation and products of species used as living fences.

Species	Common Names	Method of Propagation	Products
Acacia rugata	Shikekai	Seed	Detergent
Agave sisalana	Sisal agave	Bulbet	Fiber
Agave vera-cruz	Agave	Bulbet	Fiber
Caesalpinia crista	Molucca bean, Sagargotha	Seed	Medicine
Caesalpinia decapetala	Chillar	Seed	Biomass
Erythrina suberosa	Coral tree	Seed, cuttings	Fodder, Biomass
Euphorbia spp.	Kalli, Thor	Stem cuttings	Latex
Jatropha carcus	Jatropha/ Mogali-erand	Stem cuttings	Oil
Lantana camara	Lantana	Seed, root and stem cuttings	Biomass
Pandanus odoratissimus	Kewda	Stem cuttings	Flowers
Parkinsonia aculeata	Horse bean	Seed	Biomass
Prosopis juliflora	Vilayati, Babul	Seed	Biomass

Table 3. Adaptability of living fence species to various soils

Species	Soil Type					
	Humid	Saline	Alkaline	Dry	Gravelly	Sandy
Acacia rugata	Vg	Av	Pr	Av	Av	G
Agave sisalana	G	Av	G	G	G	G
Agave vera-cruz	G	Av	G	G	G	G
Caesalpinia crista	Vg	Av	Av	Pr	Av	G
Caesalpinia decaptela	Vg	Av	Vg	Av	Av	G
Erythrina suberosa	Vg	Pr	Av	Pr	Av	Vg
Euphorbia spp.	Vg	Av	G	Av	G	G
Jatropha carcus	G	Av	G	Av	G	G
Lantana camara	Vg	G	Vg	G	G	G
Pandanus odoratissimus	Vg	Av	Av	Pr	Pr	G
Parkinsonia aculeata	G	G	Vg	G	G	Av
Prosopis juliflora	G	Vg	Vg	G	G	G

Vg = very good, G = good, Av = average, Pr = poor.
Source: Hegde, 1987.

livestock. It is good for planting in farm forestry and agroforestry systems.

The Eucalyptus hybrid is the most widely grown farm forestry species in India. However, most plantations are owned by rich farmers. It is not advisable to grow Eucalyptus on farm bunds or in fertile fields. However, on saline and alkaline wastelands where moisture is adequate, Eucalyptus can be cultivated to reclaim soils and generate additional income. Eucalyptus wood has an assured market in the timber and pulpwood industries. Livestock generally do not browse the foliage.

Melia azedarach (chinaberry) is a popular tree in the central and northern parts of India. It grows well in deep soils where moisture is adequate. Melia is an ideal agroforestry species. Its straight trunk commands a high price as timber, its foliage makes good fodder for sheep and goats, and its seeds contain an oil with pesticidal properties.

Commonly known as shishum, *Dalbergia sissoo* grows fast. It makes an excellent timber, and side branches can be lopped for feeding livestock. It requires deep loamy soil and adequate moisture to grow well.

Dendrocalamus strictus (male bamboo) is a popular species in humid regions, but farmers rarely plant large areas with it. It grows well on canal bunds. Trees begin producing income in the fifth year.

Other species

Acacia senegal is another drought resistant species grown in dry areas of northern India. Biomass yield is low, but gum can be tapped from this tree. With the development of the necessary infrastructure for gum extraction and processing, plantations of this species can be a regular source of high income to small farmers.

Terminalia arjuna is another important species, as it can support a silk industry. In the absence of necessary infrastructure, however, farmers are unable to take advantage of this species.

Plantation Protection

Providing protection from livestock is a major problem. Small farmers cannot afford to spend money on fencing their fields. To overcome this problem, particularly under farm forestry systems, establishment of live hedges is being promoted in rural areas. Before planting such hedges, a trench 1 m wide and 75 cm deep is dug along the boundary, and the soil is used to build a mound along the inner side of the trench to prevent livestock from crossing the boundary. The hedge plants are grown on the mound where loose soil facilitates better establishment.

While a wide range of species is available for establishing hedges, income generation should be a consideration in the selection. A list of species, their adaptability, and uses are given in Tables 2 and 3. Species such as *Agave sisalana, Jatropha carcus,* and *Prosopis juliflora* are superior because of their respective contributions of fiber, oil, and biomass.

REFERENCES

Hegde, N.G. 1987. *Handbook of wastelands development.* Pune, India: BAIF.

A New Farming System--Crop/Paulownia Intercropping

Zhu Zhaohua

Chinese Academy of Forestry
Wanshoushan, Beijing, China

Intercropping of Paulownia elongata *trees with agricultural crops is being adopted by an increasing number of farmers in the plains of North China. There are now over 1.5 million ha of farmland in a Paulownia intercropping system. About 500 million Paulownia trees in this region provide local farmers with an annual timber growth of 5.7 million m^3 with a value of 1.7 billion RMB yuan (US$ 460 million) and 75,000 tons of tree leaves with a potential 2,868 tons of nitrogen. This paper analyzes the ecological and economic effects of Paulownia planted at different spacings with various crops. It also introduces the extension system that made possible the rapid adoption of this technology by farmers in the North China Plain.*

The North China Plain is the alluvial plain of the Huanghe (Yellow) and Huaihe River valleys. The climate is dry between October and early June. The rainy season is from mid-June to mid-September due to the influence of the monsoon winds from East Asia. It is dry and extremely cold in winter because of the winds from Siberia and the Mongolian Plateau. Mean annual rainfall is 500-900 mm, increasing gradually from the northwest to the southeast. Almost 80% of the rain falls during the wet season. The soil in this area is alluvial and much of it is quite poor. It is sandy, saline, and alkaline. Before liberation, the people led a very poor life due to serious natural calamities, mainly sandstorms, droughts and floods. From May to June, a hot and dry wind often blows with speeds over 3 m/sec., temperatures over 300 C, and a relative humidity of less than 30%. This is quite harmful to agricultural crops, especially wheat (the main crop of the area), and 20-40% of the production is often lost. Both timber and firewood for farmers are quite scarce in this area of more than 300 million people.

In the early 1960s, an interesting phenomenon was observed by farmers in Lankao County, Henan Province. *Paulownia elongata* trees scattered in the fields did not reduce crop yields markedly and appeared to minimize damage from sandstorms and soil erosion. The phenomenon then caught the attention of the local government and foresters, and some small-scale plantations were established. On the basis of farmers' experiences, systematic research was conducted on experimental fields, and research support was received from the International Development Research Centre (IDRC) of Canada. At present, there are 1.5 million ha of farmland intercropped with Paulownia. What follows briefly introduces the main research results and the extension methods used to make possible the rapid development of the farming system in this region.

Major Research Contents and Objectives

Research has concentrated on determining optimum spacing and management methods. In spacing trials (5 x 6 m, 5 x 10 m, 5 x 20 m, 5 x 30 m, 6 x 30 m, 5 x 40 m, and 5 x 50 m), the biomass and yield of the main crops (wheat, cotton, maize, sweet potato, and tea) and microclimate changes in the fields have been measured, and economic analyses have been completed. Most experiments have been conducted on plains along the old course of the Huanghe River in Minquan County, Henan Province and in Dangshan County, Anhui Province. Paulownia/tea intercropping research took place in Tongling City, a hilly area in the lower reaches of the Yangtse River.

Paulownia as an Intercropping Species

Paulownia elongata is a deep-rooting species. In sandy and other soils, an average of 76% of the absorbing root system is at a depth of 40-100 cm. Only 12% is in the cultivated layer of 0-40 cm where the root systems of major food crops are primarily distributed. Therefore, competition for water and fertilizer between the trees and food crops is minor. Water in the deeper layer, as well as fertilizers moved into the deeper layers, may be absorbed by the roots of Paulownia trees. In the dry season, Paulownia can absorb underground water from the deeper layers and humidify the air by transpiration, beneficial for growing food crops.

The crowns of Paulownia, especially of Paulownia elongata, are sparse and much light can pass through. Light penetration through the crowns of 7- and 8-year-old trees in early summer (June) is 40-50%, and around 20-40% in middle and later growth periods. Since the branching angle of Paulownia is large, leaves spread systematically and seldom overlap, so food crops may obtain much light at any time. Light penetration through P. elongata crowns is 20% higher than that of poplars (Populus tomentosa) and 38% higher than black locust (Robinia pseudoacacia).

The leaf renewal and defoliation periods of Paulownia are later than for other trees (about 20 days later than for poplar), and the defoliation period is later than for most other timber species. Late leaf renewal favors the growth of summer crops, and late defoliation protects wheat seedlings from early frost damage.

Microclimate Changes under Intercropping

When fields are intercropped with Paulownia, changes brought about in temperature, humidity, wind velocity, and evapotranspiration are generally favorable to food crops. For example, comparisons made between an intercropped field and a control plot show that intercropping can reduce wind velocity by 21-52%, and reduce ground evaporation (surface water) by 9.7% during the day and 4.3% at night. The moisture content of the soil from 0-50 cm is 19.4% higher than that of the control land. Intercropping also influences temperature. In late autumn, winter, and early spring, the wind velocity is reduced by the resistance of tree branches. As a result, the temperature is 0.2-1°C higher than the control land. In summer, the temperature in the intercropped land is reduced by 0.2-1.2°C during the day. The trees can thus help protect crops against natural disasters such as drought, wind, sandstorms, dry and hot winds, and early and late frosts. Changes in the hydrological balance also are favorable to crops. When evapotranspiration was measured by 5 methods, total evapotranspiration of the control land was evidently higher than that of the intercropped land (Table 1). Intercropping thus reduced the consumption of water in the field.

Table 1. Evapotranspiration in mm/day in intercropped and control plots.

	Age	Spacing	EBBR	P-M	B-R	D-P	Pn
May 23	3	5 x 6m					
Control			5.19	7.47	7.96	7.89	10.05
Intercrop			3.19	5.19	5.18	5.76	7.47
May 21	3	5 x 10m					
Control			5.51	6.05	7.59	7.26	9.38
Intercrop			4.42	6.06	6.73	6.04	8.35
March 31	11	5 x 30m					
Control			5.11	7.07	6.02	7.48	7.79
Intercrop			4.85	6.89	5.72	6.79	7.19

EBBR = energy balance, PM = Penman-Monteith, BR = Brown-Rosenberg, Pn = Penman, DP = Doorenbos-Prait

As trees age and planting density increases, however, radiation declines in intercropped fields. The average solar radiation in a field with 6-year-old trees planted at a density of 5 x 10 m was only 45.7% of the control. For fields with 11-year-old trees planted at a spacing of 5 x 30 m, average radiation was 80.6% of the control. The horizontal distribution of solar radiation differs greatly, with crops near the tree rows receiving only about half the radiation of crops in the middle of tree rows planted at a 6 x 30 m spacing. This horizontal distribution of solar radiation is closely associated with the horizontal distribution of crop yield. Thus, a priority is obtaining the greatest economic benefit through planting and silvicultural measures that affect solar radiation intensity.

Economic Effects of Intercropping

The main summer harvest crop in this region is wheat. Numerous studies indicate that at tree densities of 5 x 20 m to 5 x 40 m with a rotation of 10 years, the wheat yield could be 6-23% higher than the control. But at greater tree densities, solar radiation is not sufficient, especially for the late growth of wheat, and yields decrease correspondingly. At Dangshan Experimental Station, wheat yields dropped in fields with 4-year-old Paulownia trees spaced at 5 x 6 m and 5 x 10 m (Table 2).

The major summer-sown crops in this region are maize, cotton, soybean and sweet potato. Since yearly precipitation is concentrated during the growing season for these crops, those in intercropped fields have a stronger need for solar radiation than does wheat. According to the reports from Heze Prefecture of Shandong Province, 8-year-old trees (average dbh 29.8 cm and 11.5 m height) spaced at 6 x 25, 5 x 30, 5 x 40, 5 x 50, and 5 x 60 m, reduced yields of sweet potato, peanut, maize, soybean and cotton by 37.6, 32.2, 25.9, 27.7, and 14.5%, respectively, below the controls.

Cotton yields showed no significant differences when the intercropped trees were less than 5 years old. But when the density increased, apparent differences occurred (Table 3). For 3-year-old Paulownia trees planted at 5 x 6 m, the yield of cotton was about normal, with little decrease. When the trees are 4 years old, the yield of cotton was merely 13.8% of the normal yield. At a density of 5 x 10 m, cotton yields decreased by about 20%.

At a density of 5 x 20 m, studies indicate that yields should exceed more than 90% of the control.

The effect of intercropping on yield is intimately related to the levels of farmland management and soil improvement. In the 1950s and 1960s, when sandstorms were not effectively controlled, crop damage in the region was rather serious. The local farmers know that on such land crops would yield nothing without being protected by trees.

If farmers do not want crop yields on intercropped fields to be lower than yields on non-intercropped fields, trees should be spaced no closer than 5 x 20 m for the first 6-7 years. After that time, trees should be thinned to a spacing of 5 x 40 m. New rows of trees should be planted to replace those in thinned rows. When the newly planted trees reach 4 or 5 years of age, the remaining 10 and 11-year-old trees could be harvested. This is considered the optimum intercropping pattern if agricultural production is the main objective.

Economic Effects of Paulownia

Under intercropping, intensive management of the agricultural crops inevitably creates better growing conditions for Paulownia than is found in a pure Paulownia forest. Under intercropped conditions, 10-year-old trees reach a mean diameter of about 35-40 cm and a volume of 0.4-0.5 m^3, with the better ones reaching up to 1.5 m^3. If 60 trees are planted per ha, 25-30 m^3 of timber will be produced in 10 years. This volume will be worth 8,750-10,500 Yuan RMB (US$1 = 3.7 RMB yuan), and the income from timber is an average of 900 Yuan/ha/year. This represents a significant increase in income and an improved living standard for people. Paulownia timber alone generates 15-25% of the economic income by unit area. At present, about 500 million Paulownia trees are growing on the North China Plain (most are young). It is estimated that they can provide local farmers over 5 million m^3 timber, which will alleviate the critical timber shortage in the region.

The whole Paulownia tree is valuable. Besides timber, branches, leaves and flowers can be used. One 10-year-old tree can produce 350-400 kg of branches for fuel, allowing crop stalks to be returned to the field as compost. Leaves and flowers are rich in nutrients, and are suitable for feeding pigs, sheep, and rabbits. The leaves are

Table 2. Wheat yield in intercropped and control plots.

Tree age yrs	Spacing m	Yield	
		Intercropped kg/ha	Control kg/ha
3	5 x 6	3,630	3,630
	5 x 10	4,058	
4	5 x 6	3,666	4,730
	5 x 10	4,293	
5	5 x 6	2,930	4,208
	5 x 10	3,555	

Table 3 . Effect of tree age and spacings on cotton yields.

Age yrs	Spacing m	Yield	
		Control kg/ha	Intercropped kg/ha
3		--	
	5 x 6		3,425
	5 x 10		4,256
	5 x 20		--
	5 x 40		4,010
	5 x 50		5,256
4		3,672	
	5 x 6		506
	5 x 10		3,008
	5 x 20		3,600
	5 x 40		3,736
	5 x 50		3,683

rich in nitrogen (3.09% dry weight). The content of coarse protein in Paulownia seedlings is 26%. A single 8 to 10-year-old tree normally produces 28 kg of dry leaves. If it is assumed that each of 500 million planted Paulownia trees produces 5 kg of dry leaves annually, the yield would be 2,500 million kg of dry leaves, 650 million kg of coarse protein, and 75,000 tons of nitrogen. The trees also offer much fodder and organic fertilizer for animal husbandry and agriculture. Paulownia intercropping can thus combine agriculture, animal husbandry, and forestry into a new farming system in which the three components stimulate one another.

Extension of Paulownia Intercropping to Farmers

Paulownia intercropping has been developed on the basis of farmers' experiences. But such quick and large-scale extension with consistent improvements has been possible because of the close cooperation between scientists, government officials, and farmers. What follows is a brief account of this author's personal experience and understanding.

Paulownia research began in 1973 when there were about 20,000 ha of farmland under intercropping. Paulownia intercropping impressed this author immediately. He noticed that people were rich and crop yields were high and stable wherever Paulownia intercropping was practiced well. He believed that more scientists should study this technology and extend it to more farmers. In 1974, Wu Zhonglun and he initiated and established the National Paulownia Research Coordination Group. In 1976 the first meeting for Paulownia research coordination was attended by 230 scientists, marking the beginning of a relatively systematic study of Paulownia on a national scale. In the beginning, this coordination was organized spontaneously. Since 1979, this research has been listed as a key project in the Ministry of Forestry. In 1983 it was listed as a key research project by the State Science and Technology Commission and supported by IDRC. To promote the development of Paulownia, research personnel have employed various means, including writing articles and reports and popularizing the idea to political leaders and farmers through television, newspapers, and journals. Several large training courses have been held for forestry technicians, who then train local government officials and

farmers. Three scientific films have been made and shown throughout China. Demonstration areas have been set up to show silvicultural and utilization techniques, as well as provide government officials, technicians, and farmers with good seeds. In 1986-1987, nearly 10,000 government officials, technicians, and farmers from all over the country visited these demonstration areas.

The Chinese Government has played a decisive role in promoting large-scale crop/Paulownia intercropping. Based on farmers' experiences, this author prepared a proposal in 1974 and submitted it to the forestry minister and the leaders of the State Council. In 1977, the forestry minister conducted an on-the-spot meeting, attended by 100 county directors from the North China Plain. Many prefectures and counties have made Paulownia intercropping an integral part of the farmland infrastructure construction and include it in their overall programs. For the sake of forest development in the plain region, similar on-the-spot meetings have been held since 1977.

Technical guidance is offered to farmers by local technicians. There is a forest bureau in every county and a forestry station in every district that provide guidance to farmers. Some local technicians take part in the research project directly, and some farmers are trained as technicians.

REFERENCES

Zhu Zhaohua, Lu Xinyu, and Xiong Yaoguo. 1982. *Research on Paulownia*. (In Chinese.) China Forestry Press.

Zhu Zhaohua and Zhao Qingru. 1985. *Paulownia in China: Cultivation utilization*. Singapore: Asian Network For Biological Sciences and IDRC.

The Research Institute of Forestry, Heze Prefecture of Shandong Province. 1985. *Study on economic effects of crop/Paulownia intercropping*.

Session II: The Role of Eucalyptus on Small Farms-- Boon or Bust?

Chairman: Suree Bhumibhamon
Discussant: Patrick Robinson

Session II Summary

Suree Bhumibhamon

Department of Silviculture, Faculty of Forestry, Kasetsart University
Bangkok, Thailand

Growing Eucalyptus remains a crucial issue in tropical countries. In Thailand, a national seminar on Eucalyptus was held in June 1987, and recommendations on growing the species were proposed. It was concluded that Eucalyptus has had no significant negative impact on the environment but that large-scale monocultures require proper assessments. Growing Eucalyptus as a community plantation or in agroforestry systems requires open discussions with farmers. Growing Eucalyptus in upland watershed areas and in conserved forests is forbidden in Thailand.

The first speaker, Amar Nath Chaturvedi of the Tata Energy Research Institute, New Delhi, discusses the silvicultural requirements of Eucalyptus for small farms. Since 1790, various Eucalyptus species have been introduced in India, including *E. globulus*, *E. citriodora*, *E. grandis*, *E. tereticornis*, *E. camaldulensis*, and a promising hybrid (*E. tereticornis* x *E. camaldulensis*). The hybrid is grown as a cash crop to serve industrial needs. Chaturvedi indicates that, contrary to common belief, Eucalyptus uses water efficiently. The wood produced per unit of water transpired is higher than for most other tree species. It also improves alkaline sites by lowering the pH and the percentage of sodium and potassium, as well as increasing the organic carbon. Chaturvedi also provides planting, management, and rotation guidelines to maximize yields and income, as well as decrease competition with intercrops.

Zheng Haishui, research engineer from the People's Republic of China, discusses the role of 0.7 million hectares of Eucalyptus plantations in southern China. Among the many species planted, *E. citriodora* and *E. leizhou* No. 1 (hybrid) are planted intensively. They commonly are planted on the sides of homesteads and wastelands. They are used in shelterbelts, and also help conserve soil and water. Eucalyptus provides construction wood, poles, fuelwood, oil, and honey.

The last paper is presented by Kevin J. White, plantation management consultant of Sagarnath and Nepalganj Forest Projects in Nepal. He describes a bright future for Eucalyptus. There is a large and increasing demand for wood and other tree products in Asia, he says, and Eucalyptus species and provenances are available to fit individual sites. Furthermore, Eucalyptus can serve as a monocrop and in agroforestry systems. Advances in vegetative reproduction will make this future even brighter. He recommends establishing a regional policy and a coordinating unit operating with major soil types and environmental zones.

The consensus following these presentations indicates it is too early to determine whether growing Eucalyptus will prove a boon or bust. Participants recommend that environmental impact studies be carried out on farmlands.

Silvicultural Requirements of Eucalyptus for Small Farms

A.N. Chaturvedi

TATA Energy Research Institute
New Delhi, India

Eucalyptus camaldulensis, E. citriodora, E. grandis, E. globulus and E. hybrid (Mysore gum) are important plantation species in India. Large-scale plantings of the hybrid have occurred on large and small farms. The species has been both praised and criticized in India. For good growth, the hybrid needs well-drained, deep, fertile soil with a pH of 6.5-8.5. The maximum solar energy efficiency of these trees is between 5 to 9 years, when rapid growth occurs. The wood produced per unit of water transpired is higher than for most other tree species. Much of the water consumed by Eucalyptus is recycled as dew or raindrops. Eucalyptus wood is heavy and has a ready market for local and industrial use. Hybrid plantations show poor response to thinnings, and diameter growth slows at close spacings. The hybrid is associated with mycorrhizae, which affect tree growth. Iron and zinc deficiencies have induced chlorosis and retarded growth. The quality of seeds affects wood production. Methods of clonal propagation from coppiced shoots have given encouraging results.

Eucalyptus trees have been planted in various countries to increase wood production. Between 1961 and 1975, the annual production of Eucalyptus wood on plantations outside Australia increased to about nine times the volume harvested annually in Australian natural forests (FAO 1979). Eucalyptus was first introduced to India around 1790 when various species obtained from Australia were planted by Tipu Sultan in the state palace gardens at Nandi Hills near Mysore (Karnataka). In 1843, Captain Cotton of the Madras Engineers successfully introduced *Eucalyptus globulus* in the Nilgiri Hills of Tamil Nadu State. This species is the main source of firewood in the township of Ooty, and the leaves support many small oil distilling units. *Eucalyptus citriodora* has been planted along avenues in several regions of India. *Eucalyptus grandis* has been cultivated on several thousand ha in the hilly parts of Kerala State for pulpwood and fuel for the local tea industry. The most widely planted species is Mysore gum (popularly called the Eucalyptus hybrid), which developed as a natural cross of the species planted in Nandi Hills. Boland (1981) believes that the Eucalyptus hybrid is principally a typical Australian *Eucalyptus tereticornis*, probably from the southern provenances, and that many Eucalyptus hybrid plantations in India contain some mixture of *Eucalyptus camaldulensis*. Eucalyptus hybrids were first grown on a plantation scale in 1952 in Karnataka. Since then, Eucalyptus has been planted extensively on plantations to meet the demands for fuelwood, small timber, poles, and pulpwood. This species has also been grown along highways, canals, and railroad tracks in single and multiple rows. Large-scale plantations (both manual and mechanized) in several states are estimated at 210,000 ha.

Eucalyptus on Farmlands

Since 1972, Eucalyptus has been planted increasingly by farmers in the Punjab, Haryana, Gujrat, Maharashtra, Andra Pradesh, Karnataka, and Uttar Pradesh states. In Maharashtra alone, 22 district Eucalyptus-growers societies have been registered. Trees are planted alone or in combination with agricultural crops. Farmers plant trees to increase the profitability of their farmlands. Eucalyptus species have an established market, and wood is readily accepted for pulpwood, poles, sporting goods, activated charcoal, packing cases, small timber, firewood, and charcoal. Large-scale plantings of Eucalyptus have led some people to believe the tree is an environmental hazard. Unfortunately, the technical knowledge that should have accompanied the planting of this species is lacking. Much of the criticism has been caused by disappointed expectations rather than ecological effects. People have blamed Eucalyptus for the failure of some plantations, rather than blaming the real culprit, poor forestry practices.

Water Uptake

Many people argue that Eucalyptus trees absorb too much soil moisture and thus ruin the soil. Soil water is of three types:

o Gravitational - not usually available for plant growth,

o Capillary - primarily available for plant growth, and

o Hygroscopic - not available for plant growth.

When plants permanently wilt under drought conditions and do not recover even when watered, the Permanent Wilting Percentage (PWP) for the soil has been reached. It varies between 3-15% water (dry weight basis), and is much higher for clayey than for sandy soils. This author examined many sites where Eucalyptus died due to drought. In none of these sites were the soils totally deprived of moisture. Mortality was higher at sites with clayey soils. No tree can completely dry out any soil, and Eucalyptus is no exception (Davidson 1985).

Eucalyptus species grow fast and need more water than some other tree species. But they also produce more wood, and the wood/water ratio is higher for Eucalyptus than for many other tree species (Ghosh 1978). These studies can be carried out only under controlled conditions and for short periods of about one year. To determine the moisture uptake of a particular tree species alone under field conditions is not possible because many factors are involved. Similarly, the movement of water to and from a site cannot be controlled. The results of studies carried out on six tree species clearly showed that the average consumption of water/g of biomass was 0.51 for the Eucalyptus hybrid compared to 1.30 for *Pongamia pinnata* (see Table 1) (Chaturvedi 1984a).

Another important aspect of water consumption is recycling of water. The water transpired by trees as water vapor is returned to the ground as dew or rain, which is again available to the plant. Therefore, water is not actually consumed by plants but is recycled (Tiwari and Mather 1983).

Growth and Water Consumption

Study of the yield tables of the Eucalyptus hybrid shows that its current annual increment peaks in the fifth year and then drops (Chaturvedi 1983). The mean annual increment peaks in the sixth or seventh year. During this period, growth rates are high, and trees demand more moisture and nutrients. Growth rates begin to decline after the sixth year, and the demand for moisture and nutrients also decreases. During periods of fast growth, sites do not increase appreciably in nutrient and moisture status. Once the growth rate slows, however, nutrient uptake decreases and their recycling increases. If trees are harvested before maximum growth is reached, the sites will be poorer in moisture and nutrient status. This author carried out studies on chemcial changes brought about by some tree species in saline alkaline soils. Four years after being planted, the Eucalyptus hybrid improved the sites by reducing the pH, increasing the organic carbon, and reducing the percentage of sodium and potassium (Table 2). However, *Prosopis juliflora* had a greater beneficial effect.

Under Planting

Some people often complain it is difficult to plant any tree under the shade of Eucalyptus. This factor, however, is actually related to rates of growth. When Eucalyptus trees are growing fast (3-7 years of age), other species planted underneath Eucalyptus have little chance of surviving. However, certain agricultural crops can be successfully grown in the first three years, and many tree species can be successfully planted under Eucalyptus if they are introduced after the seventh year. This author has planted *Shorea robusta* under the shade of the Eucalyptus hybrid. The species has been growing for 7 years and has reached a stage where the Eucalyptus canopy can be removed. When mixing tree species, it is essential to know their rates of growth and to specify when the species can be introduced.

Spacing

Much confusion exists about spacing and growth. Some suggest that closer spacing of tree species (sometimes referred to as high density planting) can increase biomass production. Such studies do not consider changing growth rates. Also, the biomass reported at these sites is not the

Table 1. Average consumption of water/g of biomass for six tree species.

Species	Water consumed l	Biomass produced g	Water consumption/ g biomass l
Pongamia pinnata	679	520	1.30
Albizia lebbek	1,371	2,355	0.58
Syzygium cumini	1,460	2,386	0.61
Acacia auriculiformis	1,475	1,713	0.86
Dalbergia sissoo	1,794	2,005	0.89
Eucalyptus hybrid	2,662	5,209	0.51
Treeless area	502	--	--

Table 2 .   Initial chemical composition of soil and changes after four years of three tree species at various depths.

Species	Depth cm	pH	Ec mmoh	Org. Carbon %	Na	K	Ca	Mg
					me/100 g			
Eucalyptus hybrid	0-15	9.51	0.407	0.33	5.85	0.59	12.85	4.40
		8.94	0.340	0.39	3.66	0.45	19.45	16.55
	15-30	9.65	0.516	0.12	8.69	0.50	10.23	4.92
		8.93	0.500	0.14	6.53	0.40	21.56	7.37
	30-45	10.12	0.980	0.09	11.76	0.62	11.25	3.00
		9.12	0.583	0.10	7.19	0.42	17.30	4.03
Terminalia arjuna	0-15	9.37	0.399	0.30	5.92	0.55	13.00	4.86
		9.00	0.275	0.40	3.86	1.18	17.85	13.45
	15-30	9.84	0.046	0.10	8.78	0.48	11.35	4.10
		9.10	0.410	0.16	7.00	1.00	19.55	13.28
	30-45	10.00	0.956	0.08	10.85	0.59	10.50	3.50
		9.19	0.470	0.12	8.14	1.15	16.45	4.04
Prosopis juliflora	0-15	8.75	0.430	0.30	5.67	0.60	11.85	4.90
		8.00	0.485	0.42	1.98	0.59	24.60	12.20
	15-30	9.20	0.495	0.18	8.64	0.52	10.05	4.44
		9.93	0.538	0.27	1.45	0.53	22.35	5.82
	30-45	9.95	0.500	0.10	12.28	0.76	10.35	4.02
		8.98	0.441	0.15	5.25	0.50	23.00	6.34

* Changes occur after four years.

Table 3. Strength properties of Eucalyptus hybrid at different ages.

Age	Specific gravity	Modulus of rupture in static bending test kg/cm^2	Maximum crushing stress in compression parallel to grain test kg/cm^2	Izod value cm/kg
4	0.529	343	179	136
5	0.539	515	253	147
6	0.541	562	282	163
7	0.584	644	291	177
8	0.623	768	340	218

type sold in markets. Such studies take biomass samples during active growth periods and often exaggerate the amount produced. These biomass production figures are irrelevant to small farmers, whose major consideration is production of salable wood.

When planted at close spacings, Eucalyptus grows fairly rapidly. The wood density is generally low, and the moisture content of wood and the percentage of bark are high. The younger wood has low strength properties and does not burn well. As a genus, Eucalyptus responds poorly to thinnings (Chaturvedi 1986). Once suppression sets in, subsequent opening of the crowns does not increase the growth rate, especially with regard to diameter. It has been clearly shown that trees planted at wider spacings grow better in average diameter and height than do densely planted trees that are later thinned to the same stocking level. These thinnings also increase growth stress and reduce the quality of wood (Kubber 1987).

Price/Size Gradient

A log of large dimensions demands a higher price per unit volume than does a smaller log of the same species. This is true of most Eucalyptus species in India. Thus, farmers who produce trees of smaller dimensions earn less per unit area than they would if they produced wood of larger dimensions. Consequently, it is unwise to plant trees at close spacings, even though the total

quantity of wood produced at the age of harvesting may be the same as for widely spaced plants. Trees should not be harvested as long as the annual increment is more than that of the preceding year and more than the banking rate of interest. Farmers can easily calculate this using the following formula:

$$\text{Accretion \%} = \frac{D2^2 - D1^2}{D1^2} \times 100$$

D2 is the present diameter of a tree and D1 is the diameter of the tree in the previous year. The increment of wood should be weighed by the increase of price due to size differences. Several farmers in India who have made such calculations before harvest have consistently decided to harvest trees after 10 years or more.

Other Characteristics

Wood Density

Fast growth does not cause a reduction in wood density. The wood density of the Eucalyptus hybrid increases with age. The tree reaches maximum density at age 13-14 (Table 3). At this age, the wood acquires its maximum strength properties and, consequently, its marketable quality improves. The market rate changes accordingly. Farmers clearly get a higher economic return if they harvest trees at this age.

Pulpwood

In many countries, Eucalyptus is used as pulpwood for the rayon and paper industries. For the paper industry, a high lignin content is not desirable. Paper industries in India prefer Eucalyptus wood with a specific gravity of less than 0.6. The Eucalyptus hybrid crosses the specific gravity limit of 0.6 in the ninth year. Most government plantations are dedicated to pulp production, and the felling rotations of such plantations are 7-9 years. Many farmers also believe that this is the best age for harvest. However, Eucalyptus wood commands a higher per unit volume price when used as small-sized timber. Consequently, for farmers, it is economically more viable to fell trees at about 14 years. At this age, the recycling of nutrients by Eucalyptus is significant and the price obtained is comparatively higher.

Coppicing

Eucalyptus responds well to coppicing. In several plantation projects, 4-5 coppice yields are projected. Studies carried out in government plantations in Uttar Pradesh, Karnataka, and several other states have clearly shown that yield declines in the second coppice rotation and drops drastically in the subsequent coppice rotations. In certain plantations from which data are available, the drop in yield for the second and third coppice was 30% and 70%, respectively. This is probably due to the short harvest rotations, which deplete soils of certain nutrients. Even though coppicing vigor decreases after the tenth year, it is advisable to increase the rotation age and to restrict the coppice cycle to only one coppice rotation.

Clonal Propagation

A trial of rooting young coppiced shoots in mist chambers had about a 70% success rate. Young, five-year-old trees gave better results. Harvesting trees during winter months and planting the coppiced cuttings during February and March showed fair success even in open field conditions (Zobel et al. 1984).

Chlorosis

In several plantations, Eucalyptus has shown signs of chlorosis. In one case, this was caused by deficiency in iron due to excess calcium in the soil.

In some cases, foliar spray of plants with 1% ZnSo4 + 1% urea removed the symptoms of chlorosis.

Mycorrhizae

Preliminary studies have shown that associations of mycorrhizae and Eucalyptus benefit growth rates.

Crop Rotation

It has also been observed that the health of Eucalyptus suffers from repeated replanting. It is necessary to rotate species to maintain the health of soils. In ongoing trials, nitrogen-fixing trees like *Acrocarpus fraxinifolius, Albizia lebbek, Cassia siamea, Pongamia pinnata, Butea frondosa, Dalbergia sissoo,* and *Acacia nilotica* show good growth and ability to improve the soil. A rotation with *Popular deltoides* has also given good results.

Planting Guidelines

Eucalyptus species are excellent for planting on farmlands alone or in combination with agricultural crops. However, it is necessary to match the edaphic and climatic conditions of a site with the silvicultural requirements of the species. Eucalyptus hybrid planting will be a boon if the precautions listed below are followed (Chaturvedi 1984b):

o Trees should be planted on well-drained, deep, and porous soils.

o Trees should not be planted on soils below pH 6.5 or above pH 8.5.

o Under rainfed conditions, Eucalyptus hybrid should be planted in warm climatic zones where annual rainfall is 1500- 2400 mm; under irrigated conditions, rainfall should be above 1000 mm.

o Trees should not be planted on eroded lands with low organic carbon in the soils.

o Trees should not be planted on water-logged sites.

o Trees should not be planted closer than 3 x 2 m, i.e., a stocking of about 1,500 stems per ha.

o Trees should not be harvested earlier than 8
  years except where growth stops, which
  happens on unsuitable sites.

o After the second coppice rotation, all
  Eucalyptus should be uprooted and the area
  planted with non-Eucalyptus species.

o Only good-quality seed or planting material of
  known origin should be used.

o Eucalyptus trees should not be pruned during
  growth.

REFERENCES

Boland, D.J. 1981. *Indian Forester*, vol. 107, no. 3.

Chaturvedi, A.N. 1983. *Eucalyptus for farming.* U.P. Forest
Bulletin, no. 48.

_____. 1984a. Water consumption and biomass
production of some forest trees. *Commonwealth For. Rev.*
63:3.

_____. 1984b. Do's and dont's in eucalyptus
farming. *My Forest*, vol. 20.

_____. 1986. Growth in eucalyptus hybrid
plantations and stocking. *Van Vigyan*, vol. 24, no. 182.

Davidson, J. 1985. *Setting aside the idea that eucalyptus are
always bad.* UNDP/FAO Project BGD/79/017.

FAO. 1979. *Eucalyptus for planting.* FAO Forestry Services
Bulletin, no. 11.

Ghosh, R.C. 1978. Some aspects of water relations and
restrictions in Eucalyptus plantations. *Indian Forester*, July
1978.

Kubber, Hans. 1987. Growth stresses in trees and related wood
properties. *Forestry Abstracts*, vol 8., no. 3.

Poore, M.E.D. and C. Fries. 1985. *The ecological effects of
Eucalyptus.* FAO Forestry Paper, no. 59.

Tiwari, K.M. and R.S. Mather. 1983. Water consumption and
nutrient uptake by Eucalyptus. *Indian Forester*, December
1983.

Zobel, J. Bruce, L.B. Garcia, K.I. Yara, and E. Campinhos, Jr.
1984. *The new Eucalypt forest.* Symposia proceedings, p.
13. Falun, Sweden: Marcus Wallenberg Foundation.

The Role of Eucalyptus Plantations in Southern China

Zheng Haishui

Research Institute of Tropical Forestry
Chinese Academy of Forestry
Long Dong, Guangzhou, People's Republic of China

Fast-growing, multipurpose Eucalyptus species have been cultivated as ornamental trees since they were introduced to China in 1890. In the 1960s, large-scale Eucalyptus forests were planted for timber and other products. Now more than 300 species have been planted in 600 counties of 15 provinces and autonomous regions. Of the 0.7 million ha planted, more than 0.2 million are located beside farm houses and along roads and waterways. The most common plantation trees are E. exserta, E. citriodora and E. leizhou No. 1. In southern China, Eucalyptus wood is used for pillars, furniture, farm tools, and pulpwood. Leaves are used to produce tannin extract and oils. Flowers help support apiculture. However, some problems with declining soil fertility and yields have occurred. Some work has been done to mix forests of Eucalyptus with other species like Acacia auriculiformis.

Eucalyptus species are fast-growing, multipurpose, adaptable trees. Nearly 100 years of cultivation in southern China have proven they can grow in vast areas of poor, eroded soils where other species, such as *Pinus massoniana*, *Cassia siamea*, *Gmelina arborea*, *Zinnea* sp. and *Albizia lebbek*, perform badly. Since Eucalyptus thrive on such soils, they are widely planted. Their plantation area is increasing, and they play an important role in crop production and the lives of farmers.

Development of Plantations

Eucalyptus were first introduced to China in 1890 as ornamental and garden trees. They have gradually become plantation trees as timber demand has increased. In the early 1950s, Yuexi Eucalyptus Forest Farm was established to study techniques of raising and planting seedlings. More than 10 state forest farms were set up to plant Eucalyptus alongside railroad tracks in southern China. In the 1960s, when Eucalyptus were planted throughout southern China, selecting and breeding a natural hybrid began. The hybrid *Eucalyptus*

leizhou No 1. was selected. In the 1970s, the Eucalyptus Cooperation Committee of the five southern provinces--Guangdong, Guanxi Zhuang Nationality Autonomous Region (GZNAR), Yunnan, Fujian, and Jiangxi-- was formed to promote cooperation and information exchange. The Committee now includes nine provinces and regions. Increasingly, Eucalyptus is being introduced and cultivated in southern China. According to a 1982 census, over 300 species are distributed in more than 600 counties in 15 provinces and regions. More than 200 species are planted for timber. The most widely planted species are *Eucalyptus exserta*, *Eucalyptus citriodora* and *E. leizhou* No.1. Tables 1 and 2 show the growth of some Eucalyptus species in southern China. Eucalyptus plantations now cover about 700,000 ha in southern China. About 200,000 are planted near villages, in fields, and along roads and waterways. The state owns about 200,000 ha (30%), and 500,000 ha (70%) belong to individuals and collectives. Most plantings are in the provinces of Guangdong and the GZNAR.

Small Farmer Plantations

Because of their rapid growth, multiple uses, and adaptability, Eucalyptus are widely planted in the countryside. The majority of farmers in Sichuan, Zhejiang, Jiangxi, and Yunnan provinces like to plant 300-500 Eucalyptus trees in wastelands, beside houses, and along roads. About 10 years after planting, they can harvest 8-10 m^3/ha of timber and 3-4 tons/ha (dry weight) of fuelwood, half of which is consumed and the other half sold. The average yearly income is about 100 RMB yuan ($1 U.S. = 3.7 yuan). Families that collect leaves for distilling oil or beekeeping can gain another 200 RMB yuan annually.

Farmers in Guangdong, GZNAR, and Fujian provinces like to plant 0.3-0.4 ha of Eucalyptus in wastelands near villages or surrounding crop land. In coastal areas where typhoons occur, these

Table 1. Growth of some Eucalyptus species in southern China.

Species	Age yrs	Height m	Diameter cm	Volume m^3	Location
E. amplifolia	52	20.0	45.0	1.27	Shipai, Guangzhou
E. bicostata	40	20.0	100.0	5.81	Dengxian, Sichuan
E. botryoides	52	25.0	70.0	3.84	Shipai, Guangzhou
E. camaldulensis	48	38.6	96.5	8.52	Leizhou, Fujian
E. camaldulensis	40	40.0	150.0	26.15	Fuzhou, Fujian
E. citriodora	60	30.0	85.0	6.81	Kangle, Guangzhou
E. citriodora	40	38.0	96.0	10.95	Liaozhou, Guangxi
E. dealbata	58	26.5	70.0	4.08	Shamian, Guangzhou
E. exserta	40	30.0	80.0	6.03	Shipai, Guangzhou
E. globulus	70	39.7	120.0	16.61	Baoshan, Yunnan
E. globulus	40	47.0	110.0	17.23	Kunming, Yunnan
E. maculata	60	30.0	65.0	3.98	Kangle, Guangzhou
E. microcorys	52	20.0	34.0	0.72	Shipai, Guangzhou
E. paniculata	52	30.0	76.0	5.08	Shipai, Guangzhou
E. robusta	52	24.5	98.4	6.32	Nanan, Fujian
E. rudis Endl.	70	36.6	153.4	18.30	Kuiqi, Fuzhou
E. saligna	55	30.0	68.0	4.36	Shipai, Guangzhou
E. seeana	60	25.0	50.0	1.96	Kangle, Guangzhou
E. tereticornis	60	35.0	82.0	7.39	Shipai, Guangzhou
E. tereticornis	50	20.0	91.0	5.59	Ganzhou, Jiangxi

Source: Qi and Zhang, 1983.

Table 2. Eucalyptus growth in southern China.

Species	Mean height/Mean dbh		
	yr 1	yr 2	yr 6
E. var. ABL no. 12	3.1/	4.6/4.2	11.3/9.9
E. alba	--	--	--
E. botryoides var. lyne	2.9/	4.0/4.4	9.0/11.4
E. camaldulensis	3.5/	5.1/4.3	11.8/7.1
E. citriodora	2.8/	3.9/2.9	9.3/6.7
E. cloeziana	2.3/	3.8/3.8	13.5/13.3
E. exserta	4.6/	6.1/4.7	11.5/8.0
E. grandis	4.0/	5.4/5.3	11.7/8.3
E. houseana	2.5/2.7	3.2/4.2	--
E. leizhou no. 1	4.7/	6.7/5.3	12.7/9.3
E. maculata	6.0/	4.4/3.3	9.9/7.6
E. melanoxylon	2.7/2.4	4.4/4.0	--
E. pilularis	3.2/3.1	4.0/3.8	--
E. reeana var. constricta	3.9/	6.5/8.4	11.4/16.0
E. robusta	3.9/	4.1/5.1	11.1/10.6
E. saligna	4.3/4.2	6.5/5.1	--
E. staigeriana	3.8/	4.0/3.1	13.1/12.5
E. tereticornis	3.8/	4.7/5.1	11.0/8.3

Source: Qi and Zhang, 1983.

forests serve as shelterbelts and provide timber and fuelwood in a 8-10 year rotation. Prior to the 1970s, poor planting methods (digging 20 x 20 x 20 cm holes, not applying fertilizer, and poor maintenance) and wind damage (a loss of 3-5% yearly) significantly reduced survival and production (Xi 1980). Generally, 30 m^3 of timber and 10 tons/ha (dry weight) of fuelwood could be harvested in a 10-year rotation.

In the 1980s, planting and management methods have greatly improved and intensified. Tilling by tractor, applying basal and additional fertilizers, rational spacing (1.0 x 1.5 m or 1.0 x 2.0 m), and short rotations have been adopted by some farmers. The growth, yield, and income of fuelwood for short-rotation plantations are shown in Tables 3 and 4. Differences in growth and economic efficiency of non-intensive and intensive management are shown in Tables 5 and 6. In a seven-year-old stand, intensive management produces over 29.28 m^3/ha timber, 21.19 m^3/ha fuelwood, and 996.94 RMB yuan/ha more than non-intensively managed stands.

Continued planting of Eucalyptus over time can deplete soil fertility. To improve the soil, mixed plantings of Eucalyptus with *Acacia auriculiformis* has been studied. Some preliminary results have been obtained, and some farmers have begun planting mixed forests. According to one study, the amount of litter under mixed forests is 50-85% higher than in a pure Eucalyptus forest. The top soil in mixed forests begins changing color and its humus increases. The growth and economic efficiency of mixed forests are shown in Tables 7 and 8. The biomass and economic efficiency of mixed forests are 16 and 20% higher, respectively, than those of pure forests. *Acacia auriculiformis* can usually be cut for fuelwood and green manure 2-3 years after planting, and 20-30 tons/ha (fresh weight) of fuelwood and 8-10 tons/ha of leaves (used for manure) can be gathered. A coppice harvest of *A. auriculiformis* can be obtained after another two or three years, and the yield is slightly higher than that of the first harvest. After three or four harvests of *A. auriculiformis*, Eucalyptus timber can be harvested.

Uses in the Southern Countryside

Wood

Eucalyptus wood is hard, strong, and durable, and has a wide range of construction uses (girders, poles, doors, windows, furniture, farming tools, posts, and fence stakes). But it lasts only three to five years as a pillar species. The total wood usage by farmers in southern China exceeds 300,000 m^3 each year. In the coastal provinces, *E. citriodora* and *E. globulus* make excellent fishing boats because of their resistance to decay in sea water. Fishermen in Fujian Province are willing to exchange two-three m^3 *Cunninghamia lanceolata* wood for one m^3 *E. citriodora* wood.

Fuelwood

Because Eucalyptus wood has a high caloric value (4700-4850 kcal/kg) and burns readily, it makes a good fuelwood. To help meet the serious need for timber and fuelwood in the countryside, 400-500 kg of fuelwood (dry weight) can be obtained from one m^3 timber. A 10-year-old plantation usually produces two or three tons/ha of fuelwood.

By-Products

Since 1958, Eucalyptus leaves have been refined into oils. More than 3,000 tons are produced annually for spices, perfumes, and medicines. Recently, China has begun to export 1,000 tons of Eucalyptus oils annually. Phytohormones refined from Eucalyptus leaves can also improve the growth of vegetables and fruits. Eucalyptus flowers are a good resource for apiculture. In southern China, over 3,000 tons of honey are obtained annually from bees that feed on Eucalyptus flowers.

Shelterbelts

Typhoons often strike the coastal areas of southern China and do great damage to rubber trees, rice, peppers, and other crops. A typical typhoon damages 40-50% of the crops, and yields decline by 20-30%. After establishment of windbreak forests of Eucalyptus, however, only 5-10% of the crops are damaged when a heavy typhoon occurs. At present, the shelterbelt forest area covers 800,000 ha on Hainan Island and in Guangdong Province. A shelterbelt can reduce wind speed by 40-60% at a distance twice the

Table 3. Growth, yield, and income of *Eucalyptus leizhou* No. 1 planted at 1 x 2 m spacing.

Age	Mean Diameter cm	Mean Height m	Volume m^3	Timber Yield m^3/ha	Wood Yield t/ha	Income Yuan
1	2.9	4.4	12.96	--	--	
2	4.8	6.9	31.11	--	--	
3[a]	5.7	8.6	56.60	7.8	2-3	1,200
4	6.4	10.5	67.90	--	--	
5[a]	7.0	12.5	85.90	10-12	3-4	1,755
6	7.5	12.8	91.50	--	--	
7[b]	8.0	13.2	97.95	60-68	20-22	13,100

[a] intermediate cutting of 20% of the trees
[b] harvest cutting
Source: He and Zheng, 1986.

Table 4. Growth and yield of *Eucalyptus leizhou* No. 1 coppices planted at a density of 3,400 trees/ha.

Age	Mean Diameter cm	Mean Height m	Timber Yield m^3	Fuelwood Yield t/ha
0.5	3.2	1.8	--	--
1	4.5	3.1	--	--
1.5	6.1	4.2	--	--
3	8.5	5.2	31.34	13.0

Table 5. Growth and yield of *Eucalyptus leizhou* No. 1 in non-intensive and intensive management systems.

Management	Survival trees/ha	Mean height m	Mean diameter cm	Volume m^3/ha	Timber Yield m^3/ha	Fuelwood Yield m^3/ha
Intensive	4,125	11.4	9.0	130.93	80.23	50.70
Non-intensive	2,265	10.8	9.7	80.45	50.95	29.50

Table 6. Income and profit (in yuan/ha) from *Eucalyptus leizhou* No. 1 plantations managed in intensive and non-intensive systems.

	Income		Profit		Total
Management	Timber	Fuelwood	Timber	Fuelwood	Yuan
Intensive	22,691	1,521	1,861	761	2,621
Non-intensive	14,408	885	1,188	443	1,624

Table 7. Growth comparison of *Eucalyptus exserta* monoculture with mixed plantation of *E. exserta* and *Acacia auriculiformis.*

Type	Age	Dbh	Height	Volume	Biomass (dry wt.)
Pure forest					
E. exserta	3	3.25	5.60	34.23	53.26
Mixed forest					
E. exserta	3	3.71	6.14	18.56	30.78
A. auriculi-formis	3	3.95	5.70	23.74	30.93
E. exserta + *A. auriculi-formis**	3	3.82	5.80	42.39	61.71

* weighted average.

Table 8. Economic returns (in yuan/ha) from a three-year-old pure plantation of *Eucalyptpus exserta* and a mixed plantation of *Eucalyptus exserta* and *Acacia auriculiformis.*

Type	Income	Profit
Pure forest		
E. exserta	1,720	620
Mixed forest		
E. exserta	992	108
A. auriculiformis	992	108
E. exserta + *A. auriculiformis**	1,984	884

* weighted average.

height of the windbreak. With the help of shelterbelts, farmers on Hainan Island are now obtaining stable harvests. In 1967, farmers in Tunpan community, Hepu County, GZNAR planted 328 ha of timber and windbreak forests with Eucalyptus. Rice output increased from 1,500 kg/ha to 6,000 kg/ha, sweet potatoes from 3,000 kg/ha to 7,500 kg/ha, and the total output of crops doubled. The plantations also provided fuelwood and timber for farmers.

Soil and Water Conservation

There are vast barren and eroded lands in southern China where Leucaena, Gmelina, Calliandra, Cassia, and Zinnia species perform badly but where Eucalyptus species grow normally. Eucalyptus are often planted as pioneer trees for soil and water conservation. More than 80 km of a shelterbelt composed of 6-8 rows of Eucalyptus were established along the banks of the Nanliu River, Hepu county, GZNAR. Banks with shelterbelts lost less soil in storms than bare banks, whose lack of protection led to destruction of homes and fields.

Negative Impacts

Long-term planting of Eucalyptus decreases soil fertility and damages the soil's physical properties, which reduce crop production. A second generation *E. exserta* plantation produced only 52.3% output of the first generation in Shilin Forest Farm, Leizhou Forest Bureau, Guangdong Province (Ling 1982). A first-generation, non-fertilized *E. exserta* forest yielded 8-10 m^3/ha annually in the Baishiling Forest Farm, Hainan Island, but the second generation, which was fertilized, yielded only 5-8 m^3/ha per year. Without intensive management, pineapples, peppers, and other economic crops perform badly on land formerly planted with Eucalyptus.

In addition, Eucalyptus land becomes increasingly drier because of the tree's tendency to consume large quantities of soil moisture. According to one study, a *Eucalyptus torelliana* forest transpired 206.5 mm soil moisture per year while only 138.5 mm moisture permeated the soil (Tan 1982). To maintain water balance, *E. torelliana* had to obtain water from deeper soil layers, which caused ground water levels to drop. In Haikang County, Guangdong Province, a spring dried up in the 1960s when a secondary forest was

felled and replaced with a Eucalyptus forest (although the Eucalyptus forest grew well). After suffering years of drought, farmers in the Shangyong Community of Hainan Island complained that Eucalyptus used too much soil moisture. In view of this problem, we are trying to establish forests of mixed species.

REFERENCES

He, Kejun and Zheng Haishui. 1986. Cultivation and management of *Eucalyptus leizhou* no.1. (Unpublished, in Chinese).

Ling, Changfa. 1982. A survey and research on degradation of soil fertility on eucalyptus land. *Science and Technology of Eucalyptus*, 3:15-19. (In Chinese).

Qi, Shuxiong and Zhang Hanhua. 1983. Eucalyptus in China. (Mimeograph, in Chinese).

Tan, Shaoman. 1982. A study on transpiration of *Eucalyptus torelliana*. *Science and Technology of Eucalyptus*, 3: pp. 19-25. (In Chinese).

Xie, Cailing. 1980. Growth of introduced eucalyptus in Leizhou Forest Bureau. *Science and Technology of Eucalyptus*, vol. 2. (In Chinese).

Eucalyptus on Small Farms

Kevin J. White

Sagarnath and Nepalganj Forest Projects
Forest Products Development Board, Ministry of Forests
Kathmandu, Nepal

The use of Eucalyptus tree crops in the farmland sector on small or large holdings is in its infancy. With increasing privatization and diminishing regional forest land resources, it is logical to expect and to promote the transfer of responsibility for wood production to farmers. Eucalyptus, with its wide range of end uses and present and future yield potential, will play a major role in this process. This paper discusses species and provenance selection, seedling production, field establishment, spacing, rotation length, use of intercrops, growth and yield, and utilization. Certain problem areas, including the impact of Eucalyptus on the environment, water supply, and fertility, are cited. To increase wood yield, investment of capital for clonal stock production is urgently needed. Also needed is support for extension services in both training and field operations.

Both this workshop's agenda and the commonly reported attitudes of people who plant Eucalyptus emphasize the tree's role as a cash crop, particularly for wood. There is nothing wrong with this attitude. Farmers can decide to grow whatever crops they like to sustain and improve their quality of life. Wood should be and is becoming regarded as any other cash crop in the region. The goal, however, should quite definitely be multiple end uses. This paper does not focus on an exclusive tree-farming objective, but, in general, favors a combination of trees and other crops for optimum diversified and sustained farm income.

Multiple end uses may not be apparent for Eucalyptus, an exotic species familiar to few and grown exclusively for wood throughout the region. The answer to whether it can be grown as a multipurpose tree is strongly affirmative, although some end uses need to be developed and enhanced through research. Most additional end uses require promotion and physical demonstration projects to ensure wide-scale adoption.

For example, *Eucalyptus camaldulensis* has the potential to supply a high-quality, commercial grade pharmaceutical oil. However, there is great variability in total leaf oil yield among provenances and in yield of the valuable 1.8-cineole content. These can be greatly influenced and adjusted through relevant screening of seed sources, selecting quality progeny, and multiplication of high-yielding strains in clonal vegetative production.

Clearly, tree-farming research must be closely related to end products, which are extremely diverse. In the case of wood pulp industries, individual pulping procedures require wood with quite specific qualities. The possible uses of Eucalyptus are as follows:

o wood products - round timber, fuelwood, charcoal, pulpwood, fiber and flake board, and converted timber.

o leaves - oils (residues as fuel or mulch), cattle bedding (compost).

o flowers - nectar production for honey.

o seed - seed sale

o bark - tannin production (residue to fuel).

Because of its rapid growth rate, Eucalyptus wood can be regarded as a farm crop. There is little reason to confuse tree farming with what is generally accepted as forestry. Indeed, with rotations ranging from 9 months to 8 years, trees must be treated like any other cash crop. With such short rotations, it becomes apparent that a market economy approach should be taken, under which wood production would be transferred from the public sector to private farms. Privatization of wood production would lead to cost effective investments, large-scale supplies, and a self-regulatory price structure.

The title of this session, "The Role of Eucalyptus on Small Farms - Boon or Bust?" is extremely premature. Generally, the use of Eucalyptus on farms is a recent concept. Additionally, end uses have been restricted to relatively few wood products (fuel, posts, and pulp). Little basic research has focused on identifying species for specific environments to obtain maximum yield. Even less research has worked on matching provenances and sites, which is essential for maximizing yield. Even more limited is the clonal selection of trees of identified desirable characteristics, their vegetative reproduction, and actual distribution and use. Further advances in sexual recombination of desirable characteristics are only a possibility at the moment because of the general lack of forest tree research.

The genus's potential has been only partially realized. To answer the session's question, Eucalyptus will likely become a boon once its potential is developed. It would be appropriate to plan and implement responsible programs using this quite remarkable genus and sensibly developing its capacity to benefit humanity.

Eucalyptus Utilization

Brief reference is made to some multiple uses of Eucalyptus biomass. The actual use of any product depends on market potentials (local or export), though many Asian nations have large built-in advantages of high populations and potentially high demand.

Uses of round wood are legion in a rural setting. Marketing opportunities also exist for charcoal, replacement of traditional bamboo in scaffolding, telephone or electric energy poles, feed stock, pulp, and reconstituted board industries.

Solid wood products are available from young trees; some 40% of the tree volume could be converted to sawn timber using simple farm techniques (pit sawing or its equivalent) at rotations of 8-10 years. This can be carried out by underemployed rural labor in seasonal down times, such as winter. Products can be marketed from the farm or in urban centers, either individually or through cooperative ventures. The opportunity exists to retain wood of considerable upgraded value on farms.

Rural industrialization could follow large-scale plantings on large or small farms. Local processing to sawn wood can provide more benefits. Some feasible industries include production of electrical components (blocks, switch frames, and battens) and furniture, which would provide the rural economy employment and higher quality products. Regional industries, such as pulp and fiber board, could improve the economy on an even larger scale.

Obtaining oil from Eucalyptus leaves offers an interesting possibility. Leaf collection could be readily organized and use of a simple steam distillation processing technique has potential for village industry. Oil yields vary among species, provenances, and progeny of half-sib populations. Oil characteristics are considered strongly genetically controlled, and selection and vegetative reproduction of clones with desirable yields and qualities are both practical and ideally suited to farm regimes.

Yields of up to 3.5% of fresh leaf weight with 70-75% cineole are reported with *Eucalyptus dives* (not suited to the subtropics of this region), while yields of *Eucalyptus citriodora* in India are reportedly 4.5% oil with 82% citronellel (Willis 1986). Experience with *Eucalyptus camaldulensis* of Petford origin from northern Queensland at the Sagarnath Project, Nepal, shows a commercial yield of 0.7% with 60% cineole content. Individual trees have shown up to 2.7% yield and cineole content of up to 72%. The need for and the productive capacity of selective cloning programs are evident.

Tannin is a potential end use of Eucalyptus bark. While Eucalyptus tannin is of a lesser quality than that of Acacia, it does have potential regional demand (Hills 1986).

Honey is an unusual potential by-product. In lowland Nepal, the honey industry is limited due to the lack of a nectar flow in winter after the oilseed crop (rape) flowers. There are 4-6 months of nectar "drought" until the spring flowering of mango and litchi trees. During this period, bees must be fed or subsist on the stored honey, negating commercial collections. *Eucalyptus camaldulensis* flowers at this time, providing an acceptable honey. Observations indicate a wide range of tree flowering time and selection among many flowering types is feasible for a long winter nectar yield.

Apiculture at the small-farm level would provide a useful economic input.

Eucalyptus seed is a potential cash crop. Currently, seed from Australia costs $200-$400 per kg depending on quality. The northern Queensland natural forest yielded about 7 tons in 1985. It is uncertain how long this can continue, though the demand for export and local use will remain high and assuredly increase. High-viability seed can be harvested from 2.5-year-old trees, and if reasonable attention is paid to the parent tree origins and diversity, seed cropping could be a commercial farm venture. Seed harvests of quality randomized seed stands of selected clone plants would attract high prices and are feasible on small farms.

These products are not necessarily derived from a single crop of trees, but can be obtained along with agricultural crops as discussed further on.

Which Species To Plant?

The genus provides an overwhelming range of species, and the options are vastly extended by the large number of provenances. It is indeed fortunate that plant researchers can rely on the stored wisdom of the CSIRO Division of Forest Resources, Seed Section at Canberra, which can quickly list promising species and provenances for testing if station, soil, climate, altitude, and data are supplied. Another advantage is that the Seed Section provides free research quantities of seed.

Personal observations indicate that most large-scale plantings in the region use either species or provenances of lower productivity than the better seed stock now available. One can only wonder at the management inertia that allows this to happen. In both Bangladesh and Nepal, my experience in the tropical and subtropical lowlands leads me to recognize the outstanding contribution and preeminence of *Eucalyptus camaldulensis* over other species. Particularly good are some of its provenances from the Walsh River watershed on the western fall of the Dividing Range, about 17 degrees South. Some of these provenances, such as Emu Creek Petford and Wrotham Park, are world famous.

Interestingly, the next best species, *Eucalyptus tereticornis*, comes from several hundred miles northeast, on the eastern fall 15 degrees South. Again, some provenances of this species, such as

Kennedy River and North of Lakeland Downs, are widely known and respected.

Table 1 presents a short list of highly recommended species and provenances for the regional moist, non-waterlogged, non-saline soils of the tropical and subtropical lowlands 10-30 degrees North.

These provenances (except Katherine and Gibb River) are northern Queensland selections of the Walsh River Watershed (or nearby) and of the area 200 miles north of Cairns. The localities are well known to commercial seedsmen. Species determined as unsuitable in the trials included *Eucalyptus brassiana*, *E. exserta*, *E. pellita*, and *E. robusta*. Another species of interest is *E. urophylla*, for which continuing research is planned.

Table 1 and the discussion above are only a guide but have been proven effective in the described area. Anyone contemplating planting in difficult areas, such as waterlogged, saline, or excessively dry soils, should consult the CSIRO Seed Section. More detailed observations on species, provenance, and site interactions may be found in White (1979, 1986a).

Seedling Production

Eucalyptus seedlings take only 5-6 weeks to raise. There are constraints, of course. Nursery fungal diseases may require a fully sterile environment until the plants are 3 weeks old, and the soft, fragile nature of the germinants requires tender care. Heavy rains can destroy stock less than 1 month old, and a misting spray system and sufficient water are essential.

Nurserymen should be careful not to extract the last germinating seed from the tray. They should take the early germinants and the better second growth ones and discard the rest (seed germination counts are unreliable in producing quality results). Fast-growing tree selection starts in the seed tray. Slow-growing plants should be culled. Ideally, plants should be transferred to the field before any roots penetrate beyond the plant container. They should be in a fully unhardened condition for planting in a weed-free, freshly prepared site with ample soil moisture (White 1979, 1986b).

Table 1. Recommended species and provenances for the regional moist tropics and subtropics.

Species	Provenance
E. camaldulensis	Emu Creek Petford; Lappa Junction; CSIRO 13849 8.1 km from Petford; 16 km E of Petford; 7 km NW of Irvinebank; Wrotham Park; Katherine N.T., Gibb River; Gilbert River
E. tereticornis	North of Lakeland Downs; Kennedy River; South of Laura; Morehead River

Table 2. *Eucalyptus camaldulensis* growth in various spacings at Sagarnath, Nepal after 33 months.

Spacing m	Trees/ ha	Average Diameter cm	Mean Annual Increment m^3
1 x 1	9,500	4.2	24.8
3 x 2	1,534	7.7	15.8
4 x 2	1,138	8.5	15.0
5 x 2	930	9.5	15.7

Field Establishment

A basic concept that must be considered with fast-growth, high productivity, short-rotation tree crops is that such trees should be treated like other agricultural crops. This means trees must be free of weed competition, should be protected from grazing and fire damage, and that standard agricultural practices apply.

A few points on appropriate techniques are needed. It is usual to plant during the early rains (pre-monsoon if sufficiently reliable) to maximize the growing season before drought or lower winter temperatures slow growth. Weeds can best be minimized by using compatible intercrops. Some tree growth will be lost, but reduced weeding costs make an acceptable trade-off. In any case, the combined tree/crop yield will be greater than either singly. Grazing protection will be required for 6 months.

Tree spacing markedly affects both yield and product dimensions. Close spacing (1 X 1 m) provides high yields of small products suitable for homes or nearby markets, including fuel, household timbers, and fencing. Wider spacings, e.g., 2 x 5 m, reduce total yield, which can be compensated by increased individual tree unit volume (with higher monetary impact) and possibly shorter tree crop rotations (see Table 2). Spacing selection is a management decision that must be closely linked with marketing intentions.

In paddy areas, planting is restricted to the bunds with one or two alternating lines of trees planted 1 x 1 m on each side of the bund. In dry land, routine line plantings are possible. Initially, the establishment can include many trees (1 x 1 m or 2 x 2 m) with the intention of thinning stems early. Spacing also will be affected by intercrop objectives. Crops that demand more sunlight will be spaced wider if second- and third-year crops are planned.

Eucalyptus species readily fit into existing village plantings and would be suitable for inclusion in home/village forests in Bangladesh. The "mixed forest" direction in such plantings would eliminate the alleged disadvantages of planting only Eucalyptus.

Intercropping

Ordinary farming in many soil and climate conditions offers an unusual rural income option in which both trees and other complementary crops can be grown simultaneously on the same land. While intercropping will probably reduce the yields of both trees and agricultural crops, the total return is considerably more than that of either system alone. Under extreme environments (high temperatures, strong winds) the presence of trees in crop lands can increase crop yields. A preferred intercrop system in a Nepal subtropical lowland forestry project is designed to provide optimum benefits to tree growth and fire protection of the tree crop (Table 3). This could be applied in a farm system. Although it does not take monetary yield into account, it is considered near optimum. The cropping system may be varied in the first plantation year to take advantage of soil types.

Sesame is used to control weeds, particularly in years 2 and 3. Cash returns should be approximately Rs. 11,000 ($500) per ha for 3 years of crops. In the fourth year, agricultural crops might still be planted, with lower yields expected, or farmers could turn to fodder crops. In later years, pineapples or fodder crops could be planted.

In this case, the depression of tree growth by the traditional maize crop is a tolerable loss because tree volume growth in an intercropped system can be up to 16 times as much as in monoculture plantations, with the complete transfer of weeding costs from the forestry to the agricultural sector. An example from two similar adjacent areas of contrasting treatments in Nepal illustrates this point (Table 4).

The effect of trees on agricultural crops was not quantified, but 1,250 trees per ha with a 50 cm radius clear space around the tree took up 450 m^2 per ha. The crop yield in the first year was estimated at 1.5 tons per ha for ground nuts and maize, and 0.3 tons per ha for oilseed. The maize yield was reduced by 30% and 60% in the second and third years, although in the third year it was used as a forage crop only. Since crops vary widely throughout the region, this regime is only a recommended schedule for a forestry project in subtropical, lowland Nepal.

Table 3. A preferred intercrop system in Sagarnath, Nepal, with Eucalyptus planted at a 4 x 2 m spacing.

Season	First Crop	Second Crop	Third Crop
Summer	Ground nuts* (sandy soil) or Maize (silt soil)	Early maize for both soil types	a. Maize as a fodder crop b. Stylo Verano, fodder
Winter	Rape (tori) and Mussar dhal after maize	a. Sesame or niger or b. Rape (tori) and Mussar dhal	a. Sesame, to provide a small oilseed yield and to control weeds

* Ground nut, *Arachis hypogaea*; Maize, *Zea mays*; Mussar dhal, *Lens esculenta*; Niger, *Guizotia abyssinica*; Rape, *Brassica campestris* var. *toria*; Sesame, *Sesamum indicum*.

Table 4. Effect of intercropping on tree growth.

Block Treatment	Age yrs	Survival %	Annual Volume Increment m^3
No intercrop	4.7	67	0.91
Intercropped	2.7	91	14.60

A cautionary note essential for successful intercropped tree farming is that only compatible trees and crops should be used. These must be determined locally. In Sagarnath, incompatible crops that strongly depress tree growth have been identified as sesame in the first year and sorghum in the first and possibly second years.

Where possible, it is sensible to retain traditional agricultural local crops as their culture and yield are understood. The opportunities for higher value and longer-term horticultural crops like pineapple or medicinal crops like reserpine (*Rauwolfia serpantina*) and plau-noi (*Croton sublyratus*) should be considered in planning, as should production of quality cattle fodder in older stands (White 1986c).

Environment

Eucalyptus plantations in Asia have often attracted hostile media attention. Comments range from the uninformed to serious scientific uncertainty about research results and their implications for land-use options and policies. This seems to be a problem in Asia alone. The general fear of ecological changes in converting forest to monoculture plantations may not be an issue for farmlands, but use of Eucalypts on farms can be influenced by the debate over their uses in plantations.

Eucalyptus species are considered to consume large amounts of ground water. They produce a high volume of wood, which requires a large water supply. However, the cell metabolism of Eucalyptus is highly efficient, and they need less water than many species to produce a gram of biomass (Chaturvedi, Sharma, and Srivastava 1984). Whether this affects ground water supply depends on the local environment. At Sagarnath, Nepal, this point is often raised. Here the plantation tree crops are grown under rainfed conditions. The water tables in over 99% of the area are inaccessible at depths of 80 m or more. When the available soil moisture content is exhausted, trees, like other plants, stop growing and die.

Those concerned with Eucalyptus and soil water problems are advised to consider the local environment carefully and to modify tree plantings to optimize local benefits.

In areas of high or rising water tables (and consequent salinization), Eucalyptus species have been used to reverse the situation. The drainage of the Pontine marshes near Rome has been attributed to the use of Eucalyptus. The extraction of soil water and its release into the atmosphere through tree systems in West Australia is used to control rising water tables and salinization problems. In Pakistan, India, and Thailand, there are vast areas of degraded soil of alkaline/saline/raised water table origin (Thomson 1987). Experience indicates that Eucalyptus species tolerate these conditions. They will quite probably be among the leaders in wood and other asset productivity, and will be responsive to simple breeding applications to greatly increase yields. Thomson (1987) considers that the current growth of 5-83 of *Eucalyptus camaldulensis* plantations in saline land plantations in the Sind Province of Pakistan could be doubled by using more salt-tolerant provenances. Thus, Eucalyptus water relations are more likely to be beneficial than harmful.

There is also a fear that intensive tree crops will cause excessive depletion of soil fertility. This should not be a problem. Evidence in forestry systems from older plantations in India give no proof that this will occur and suggest the reverse. Tree roots deeply explore soil profiles and trap and return leached minerals to the surface litter. The minerals lost in wood removal are limited, and most leaf litter is expected to be recycled. Furthermore, in the tree farming concept, losses of essential minerals can be economically replaced in fertilizers. The removal of nutrients in wood harvests of 8 to 10-year-old tree crops is minimal when compared to the total annual removal of agricultural crops and residues from adjacent farmlands.

In farming systems that may involve the total removal of the tree crop, it is possible that a fertility drain would occur. However, if the first principle is followed (to treat wood production in farm systems as any other crop), this loss would be covered by fertilizer application if wood crops are economically viable.

Soil toxification by Eucalyptus root secretion or litter is not considered relevant, as is witnessed by healthy, vigorous intercrops and the strong growth of understory species. The long dominance of Eucalyptus over a major part of the Australian scene and their non-toxic impact must also be considered a relevant testimony.

Concern is expressed that Eucalyptus plantings on small and large farms cause social problems, such as lowering production of food and fruit crops, decreasing rural employment opportunities and, and contributing to adverse environmental conditions. Some of these fears (decreased employment and crop yields) can be overcome by the intercrop option, which can lead to greater employment opportunities, a wider crop range, and considerably increased farm incomes.

Large-scale programs should note these concerns and take steps to ameliorate them. Additionally, attention should be paid to the use of marginal lands (saline/alkaline) to minimize the amount of high-quality agricultural land diverted to tree growth. With marginal lands, the need for investment in species and provenance trials is urgent, as is the need to select superior genetic stock.

Growth, Yield, and Rotations

Table 5 shows an expected yield value from agricultural soils in the subtropical and tropical lowland farms of the region. The higher yields are from the reasonably stress-free soils and environments. The lower yields indicate soil and environment limitations. In every case, these are expectations of unselected seedling stock, and the yields relate to fully stocked stands (e.g., 1,000 stems/ha, 3-5 years old). Naturally, lower stocking rates reduce yields.

The biomass yield of reasonably stocked stands (1,000 or more trees/ha), including the wood yielding stem, is of great interest when considering other end uses. Table 6, which is useful in determining such yields, is extracted from data by Hawkins (1987). A practical local estimate of the yield of existing plantings can be rapidly and simply determined based on average tree diameter and stocking rate.

At Sagarnath, Nepal, oil production from a 6-year-old, fully stocked stand is about 200 liters/ha. A similar yield could be expected from older crops. The over-the-counter price of this oil in Nepal is currently $8 per liter.

The above yields relate to unselected seedling origin tree crops. Selected clonal plantings could double these yields.

The question of rotation or harvesting age is often raised, provoking a range of replies. End use of the wood produced and the time to reach desired dimensions govern spacing. Where small-sized wood is intended for local fuel or farm use, short rotations of 1-2 years are practical. These also fit into farm management practices where there are fears of trees interfering with other crops. Under these circumstances, the wood crop is coppiced before it affects the other crops. Larger end-use crops, such as poles and scaffolding, may need 3-4 year rotations, and bigger poles and sawlogs will need 8-10 years. The larger trees will still be multipurpose trees, with some 40% of the butt volume usable as saw logs or large poles, 20% as posts, poles or pulpwood, and perhaps the remaining 40% as commercial fuelwood. The branches may be a commercial or social-use fuel, the bark could yield tannin, and the leaves could produce oil.

Clonal Planting

Although still in its infancy, clonal production of Eucalyptus is of great global interest. This resulted from the development of simple vegetative reproduction pioneered in large-scale forestry projects in Brazil and the Congo, and in the recognition of an immense diversity within the seedling populations of many Eucalyptus species. Combining the reproduction technique with selections from the broad spectrum of seedlings and their genetic characteristics is now being increasingly used in mass selection programs in fields as diverse as wood production and amenity floral displays. This is only the beginning.

There is a worldwide trend to clonal plantations in wood production. The objective is to convert to vegetative reproduction instead of using seedling stock. This aspect is not new. Cyptomeria, Populus, and Salix have been vegetatively reproduced, perhaps for centuries. What is new is the emphasis on parent stock selection and later breeding of superior stock so that genetically superior plants grown by vegetative reproduction will be established in plantations. As relatively few parents will enter this stream, their progeny (clones), carrying the full genetic complement of the parent stock plant, will replicate this parent.

Table 5. Annual Eucalyptus production per ha for regional subtropical lowlands.

Site Quality[a]	Weight[b] t	Volume m^3	Gross Value[c] US$
1	15-23	20-30	274-420
2	8-15	10-20	146-274
3	1-8	1-10	18-146

[a] 1 = moderate to high fertility, adequate soil moisture

2 = moderate fertility, soil moisture and/or alkalinity a limiting factor

3 = low to moderate fertility, adverse soil qualities, i.e., alkaline/saline, stony/sandy, and soil water a limiting factor.

[b] 30% moisture content, air dried during dry season; all yields are overbark.

[c] Value of standing tree of which weight is equated to a local market price of fuelwood at R 400 per ton ($18.26). No costs of growing, harvest, transport, or sale have been deducted.

Table 6. Green weight (kg) biomass table for lowland, subtropical Nepal.

Dbh	Stem	Branch	Leaf	Total
2	1.73	.23	.30	2.26
4	7.85	.95	1.11	9.91
6	19.11	2.35	2.51	23.97
8	37.50	4.53	4.56	46.59
10	63.78	7.59	7.28	78.65
12	98.85	11.63	10.71	121.19

Source: Hawkins, 1987.

Eucalyptus has become a priority species for this because of its worldwide distribution as a fast-growing wood crop. Because of its seedling genetic diversity, it has extremely high potential for rapid selection of desired characteristics and increasing this production. Yields 3-5 times higher than unselected seedling stock are accepted as common. The objective of an annual increment yield of 100 m^3/ha/yr is quoted in the literature.

The impact of this revolution will be extraordinary. There could be a global shift of wood production to countries endowed with the land and environment for rapid tree growth. A more localized effect would be raising local wood production levels (important in fuel-deficient Asia) or reducing the land area and production costs needed to produce wood for a specific purpose, such as wood input to a pulp mill. The effect of this will be similar to the effects of rice and wheat breeding programs. Indeed, it must be seen that no other wood production management option possesses a similar potential for rapidly increasing the wood supply and dramatically lowering production costs.

Although the wood production aspect has been stressed, many of the other end uses are amenable to equally rapid improvement. The percentage of leaf oil could be doubled and the 1.8 -cineole content lifted to commercially acceptable levels without further refinement. The quality and quantity of bark tannins could be directed to acceptable commercial grades. Flowering times can be spread to provide a winter-long nectar flow. Tree form can be modified to provide narrowly crowned trees more suitable to intercropping practices. Wood quality can be selected for individual industries, e.g., to supply wood of less than 550 kg/m^3 for pulp production.

The potential of multiple-use Eucalyptus clones offers exciting challenges as well as promising rewards for rural development.

Research Directions

Due to the generally unrealized potential of Eucalyptus on farmlands and to the large impact that would follow its expanded use on farms, a coordinated regional research body should be established to promote field research and administration, farming of Eucalyptus in major soil and environment land areas, and using existing facilities or establishing new ones in collaboration with national authorities. The charter would include promotion of multiple uses and establishment of demonstration industries and marketing. Its parallel could be the program at the International Rice Research Institute.

A primary general direction of Eucalyptus research would identify (or confirm) the most likely species usable for rapid wood production. Combined with this and complementary to wood production would be increased yields of commercial by-products. More specific topics for research are mentioned in the following subsections of this paper.

Species/Provenances

o Identification from the wild of appropriate species for the region's major site environments and the selection of matching provenances in this range.

o Mass selection of quality genetic clone banks.

o Selection of genetically improved progeny (half sibling) from families raised from selected mother trees from the wild, favoring rapid wood or other asset production.

Vegetative Reproduction

o Establishment of macro- and micro-vegetative reproduction techniques. National staff should be trained in these methods. Early high-yield selection attention would be directed to:

-- increasing wood growth of identified provenances to twice the volumes of Table 5 with a long-term objective of attaining 100 m^3/ha on Site Quality 1 areas.

-- manipulation of wood density to produce pulp grades of under 550 kgs/m^3 and of higher density wood (more than 600) for fuel and construction timber.

-- doubling the total leaf oil of *Eucalyptus camaldulensis* to 2% and increasing the 1.8-cineole content to more than 75%.

-- development (in conjunction with increased wood yields) of improved nectar yields and spread of flowering time for rural,honey-based industries.

-- research on bark tannin yield, uses, and economic feasibility in selected programs.

Other important features affecting tree culture on farms and requiring early investigation and selection relate to tree form. A long-term intercrop combination can be successful only if tree roots have a strong vertical penetration rather than a spreading net that causes problems of plowing and excessive competition with other crops. Crown form also is important. A narrow tree crown of either short or pendulous branches minimizes sunlight interception and promotes vigor in the intercrop.

Field experience indicates that *Eucalyptus camaldulensis* (Emu Creek Petford provenance) at Sagarnath, Nepal has an acceptable root system. Furthermore, some vigorous narrow-crown selections are identified and are being reproduced as cuttings for field trials.

Conclusion

There is an immense and increasing demand for wood and other tree products in the region. In general, this demand must be met from tree-planting schemes in the public and the private sectors. Many of these needs can be provided from short-term tree crops in time periods that are agricultural or horticultural in scope. There are many opportunities for using intercropping techniques so that trees and compatible crops can be grown simultaneously on the same land, which diversifies and increases farm yield and income.

Eucalyptus occupies a key position in this proposal as it offers a range of species for the tropics and subtropics adaptable to many environmental conditions. These species are supported by a range of provenances that can match individual sites, including difficult soils and environments. Eucalyptus species have the desired characteristic of fast growth for wood products, and combine this attribute with many other economically important products. The different species and provenances exhibit a diverse genetic make-up at quite early ages that greatly facilitate selection of superior clones to reproduce desirable traits in remarkably shorter rotations.

These assessments apply to seedling crops from unselected seed sources. Advances in vegetative reproduction and the combination of this with selected superior genetic stock can quickly provide astonishing gains in yields, production cost reduction, and higher incomes. Early attention to capitalize on these gains is highly recommended. This could best be obtained through a regional policy and coordinating unit operating with major soil types and environment zones. In the overall assessment, there can be no thought of "boon or bust." Regarding use of Eucalyptus, the star is only rising.

REFERENCES

Chaturvedi, A.N., C.S. Sharma, and R. Srivastava. 1984. Water consumption and biomass production of some forest trees. *Comm. For. Rev.* 63:3.

Hawkins, T. 1987. *Biomass and volume tables for* E. camaldulensis, Dalbergia sissoo, Acacia auriculiformis *and* Cassia siamea *in the central bhabar of Nepal.* OFI Occasional Paper No. 33. Oxford, UK: Oxford Forestry Institute, Department of Plant Sciences.

Hillis, W.E. 1986. *Technical communication No. 27.* Guangxi Zhuang Autonomous Region, China: China Australia Afforestation Project, Dongmen State Forest Farm.

Thomson, L.A.J. 1987. *Report to ADAB on seeds of Australian trees for developing countries.* Canberra: CSIRO Division of Forest Protects.

White K.J. 1979. *Fast growing tree species for industrial plantations: Eucalyptus, Pinus and others.* Field Document No. 18. Bangladesh: Forest Research Institute.

_____. 1986a. *Tree farming practices in the Bhabar Terai of central Nepal.* Manual No. 2. Kathmandu: Sagarnath Forest Development Project, Ministry of Forestry.

_____. 1986b. *Forest nursery strategies and practices in the Bhabar Terai of central Nepal.* Manual No. 3. Kathmandu: Sagarnath Forest Development Project, Ministry of Forestry.

_____. 1986c. *Intercrop strategies and practices in the Bhabar Terai of central Nepal.* Manual No. 1. Kathmandu: Sagarnath Forest Development Project, Ministry of Forestry.

Session III: Nitrogen-Fixing Trees as MPTS for Small-Farm Use

Chairman: Kenneth G. MacDicken

Discussants: Narayan Hegde
Vajira K. Liyanage

Session III Summary

K.G. MacDicken, N.G. Hegde, and V. Liyanage

Forestry/Fuelwood Research and Development (F/FRED)Project, Thailand
The Bharatiya Agro-Industries Foundation, India
Coconut Research Institute, Sri Lanka, respectively

The question of why special emphasis is placed on nitrogen-fixing trees is raised by several participants. Several genera are mentioned that do not fix nitrogen but have higher fodder value (i.e., higher nutrient content and digestibility) as well as multiple uses. Nitrogen-fixing trees may well be suited to degraded sites low in nitrogen. Many of these trees are pioneer species that are aggressive colonizers of degraded sites in their centers of origin.

The dangers of using a narrow genetic base for introducing nitrogen-fixing trees are described. One example is the reproduction of large quantities of Leucaena leucocephala seed from a limited population of mother trees. Advantages of using a broad genetic base, particularly that of increased population resistance to pest and disease problems, are recognized as is uncertainty about how this might best be done with a self-crossing species like Leucaena leucocephala.

The assumption that systems that work with Leucaena can work with other nitrogen-fixing trees is not sound. Many systems based on nitrogen-fixing trees and the economic analysis of these practices are derived from experience with Leucaena leucocephala. It has generally been assumed that practices based on Leucaena are both biologically and economically valid with other nitrogen-fixing trees. This is not a valid assumption.

Current problems with the Leucaena psyllid have led some to consider simply switching over to a new species, such as Gliricidia sepium. The following lesson should be underscored: reliance on a single species for major MPTS programs may expose small farmers to unacceptable risks. Using alternative species, with reliance on indigenous species when possible, should help reduce (but not eliminate) the risk of failure of planting programs.

Limitations in the methodologies used to assess fodder value of trees are recognized. The in vitro digestibility approach used alone is questioned. Other fodder characteristics, including palatability and tannin content, are noted as important factors to be evaluated.

Long-term research is critical. Lack of data for many MPTS is indicative of a hurried, short-term approach to produce quick results. Long-term, methodical research is seldom planned and implemented. Interests are more often directed to short studies that provide "quick-fix" solutions to complex problems.

Nutritional constraints for the growth of MPTS are highlighted by Palmer. As most small farmers own degraded, low-productive land deficient in several nutrients, fertilizer application might help increase tree growth and yields. Although resource constraints prevent small farmers from using all the inputs at optimum levels, it is advisable to recommend fertilizer applications, including the macro- and micro-nutrients, in view of the low cost involved.

It can also be suggested that regional and blockwise soil classifications should be made on the basis of mineral deficiencies to enable extension workers to focus on the application of specific nutrients in the area.

In the paper presented by Relwani, several promising nitrogen-fixing tree species are identified for wastelands with shallow, gravelly soils found in semi-arid and arid regions. Leucaena leucocephala has been reported as superior to all other nitrogen-fixing species tested in three trials. Some of the naturalized species, including Acacia nilotica var. cupressiformis, Acacia tortelis, Acacia senegal, Dalbergia sissoo, Albizia lebbek, Gliricidia sepium, Parkinsonia aculeata, Sesbania grandiflora, Sesbania sesban and newly introduced species like Acacia deamii and Sesbania formosa are also promising.

Shakya discusses 12 nitrogen-fixing trees indigenous to Nepal, of which six are found on farmlands of the Terai and middle hills. Few nitrogen-fixing trees are found on farms in the high hills. The only indigenous species to be studied in the areas of propagation, growth, and yield is *Alnus nepalensis*, which has attracted significant international interest. However, nitrogen-fixing trees are widely distributed in terraces and on the edges of non-irrigated farmlands in the hills and the Terai. Priority species suggested include *Dalbergia sissoo*, Albizia species and *Acacia catechu*.

Laquihon highlights the importance of nitrogen-fixing trees as MPTS for soil conservation. Serious ecological problems, such as soil erosion caused by deforestation and heavy rains, are being solved by developing Sloping Agricultural Land Technology (SALT), an integrated, diversified hillside farming system that uses thick hedgerows of nitrogen-fixing trees as soil binders, fertilizer generators, and moisture reservoirs. This system allows small farmers to grow annual and perennial food crops between the hedgerows. While this system can reduce soil erosion and may restore degraded hilly land to a profitable farming system, initial soil erosion rates should be quantified.

The significance of nitrogen-fixing tree species for small farmers is based largely on their use in management systems (e.g., lopping for fodder, fuelwood, and soil improvement).

Some Nutritional Constraints for the Growth of Shrub Legumes on Acid Infertile Soils in the Tropics

Robert Bray, Brian Palmer, and Tatang Ibrahim

CSIRO Division of Tropical Crops and Pastures, St. Lucia, Australia
CSIRO Division of Tropical Crops and Pastures, Townsville, Australia
Balai Penelitian Ternak, Sei Putih, Sumut, Indonesia, respectively

As part of a program to determine the suitability of shrub legumes for infertile acid soils, experiments have been established at two sites in Indonesia and two in Australia. All four soils are low in exchangeable calcium, and in each country one soil has high aluminum saturation, the other low. The responses of Leucaena leucocephala, Calliandra calothyrsus, *and* Gliricidia sepium *to the application of phosphate, calcium as a nutrient, lime, and trace elements are being assessed. Preliminary data show a response at all sites to phosphate, with an additional response to calcium as a nutrient.* Leucaena responds *positively to lime, whereas* Calliandra's *and* Gliricidia's *responses to added lime are masked by an induced trace element deficiency. This reduction in yield due to lime addition was alleviated by application of trace elements. The data could indicate a tolerance in* Leucaena *to low levels of trace elements.*

The use of shrub legumes for a variety of purposes is widespread. They can provide valuable sources of feed for animal grazing, cut-and-carry feed, fencing, fuel, shade, and soil improvement and stabilization. Clearly, it is unreasonable to expect any one species to fulfill every role in all environments. Thus, while *Leucaena leucocephala* produces high-quality feed and has an excellent coppicing ability, it does not flourish on acid soils (Hill 1971). *Calliandra calothyrsus*, a valuable source of firewood in Indonesia, is thought to grow better than Leucaena in cooler climates (Panjaitan et al. 1986), and *Gliricidia sepium* is considered more adaptable to acid soils (Chadhokar 1982).

There is, however, a paucity of information on the comparable performance of these legumes over a range of sites and environments, particularly regarding their response to soil nutrients. Since there are vast areas of acid soils in the tropics (to which Leucaena is not presently well adapted), it is of some importance to determine which aspects of acid soil nutrition effect differential species response, both to enable the selection of suitable species for particular sites and to derive criteria for future selection programs. This paper presents preliminary data from a series of field trials designed to examine the performance of Leucaena, Calliandra, and Gliricidia over a range of acid, infertile soils in the tropics and their response to a range of nutrients.

Materials and Methods

Three species of shrub legumes were used in this study--*Leucaena leucocephala* cv Cunningham, *Calliandra calothyrsus* (CPI 110395) and *Gliricidia sepium* (CPI 110398). Seedlings were prepared at each site by sowing pregerminated scarified seed into plastic bags of unamended soil from that site. Seedlings were inoculated with an appropriate rhizobium strain at transplanting.

Site Location

Four sites were chosen for this work, two in Indonesia (Sei Putih and Sembawa) and two in Australia (Utchee Creek and Silkwood). Relevant soil characteristics are given in Table 1. The Putih site is 50 km ESE of Medan in North Sumatra, where the annual rainfall is 1,900 mm. The parent material of this soil is probably an old alluvium that was derived from redistributed acid volcanic tuff. The soil appears to be well-drained and is more than 150 cm deep. The soil is classified as a Tropudult. The site at Sembawa is located approximately 40 km NW of Palembang in South Sumatra and has an annual rainfall of about 2,000 mm. The soil is classified as a Paleudult. Both of these Indonesian soils are referred to locally as Red Yellow Podzolics and represent a unit used to describe all unproductive acid red and yellow soils in Indonesia. They cover a large portion of the Indonesian land surface, approximately 30%, or 51 million ha (Driessen and Soepraptohardjo 1974).

Table 1. Soil properties of experimental sites.

Site	Depth	pH	Exchangeable Cations (meg/100 g)				%Al sat
			Ca	Mg	Na	K	
Sei Putih	0-10 cm	5.3	1.76	0.50	<0.02	0.41	8
	30-40 cm	5.1	0.72	0.17	<0.02	0.12	51
Sembawa	0-10 cm	4.9	0.42	0.32	<0.02	0.11	85
	30-40 cm	5.0	0.12	0.04	<0.02	0.03	97
Utchee Creek	0-10 cm	5.3	1.44	0.60	0.12	0.23	11
	20-30 cm	5.1	0.53	0.19	0.10	0.09	14
Silkwood	0-10 cm	5.3	0.34	0.23	0.11	0.15	77
	20-30 cm	5.0	0.03	0.05	0.06	0.03	91

Table 2. Rates and types of fertilizer applied to field experiments.

Fertilizer	Product	kg element/ha	Lime (t/ha)
P	Diammonium phosphate[a]	33	
	Monoammonium phosphate[b]	33	
Ca (nutr)	Gypsum	40	
Ca (lime)	Calcium hydroxide		
Sembawa			2.0
Silkwood			2.0
Sei Putih			0.4
Utchee Creek			0.4
Trace Elements (TE)			
Cu		3.0	
Zn		3.0	
Mo		0.2	
B		1.5	
Mg		20.0	
Basal Treatment K;S		36/18; 16/8[c]	

[a] Indonesia
[b] Australia
[c] split dressing

The two sites in Australia are located approximately 200 km north of Townsville and have an annual rainfall of 3,500 mm. The Utchee Creek site is classified as an Oxisol and is a highly weathered uniform textured soil of basaltic origin. The Silkwood soil is classified as an Inceptisol and is described as a mottled gradational soil with a dark A horizon formed from a well-drained alluvium. These soils are described in detail by Murtha (1986).

Sites in the two countries were chosen so that all were acidic and infertile. There were similarities between the soils at Sei Putih and Utchee Creek in that both have low exchangeable calcium and low aluminum saturation. The soils at Sembawa and Silkwood both have low exchangeable calcium and high aluminum saturation.

Fertilizer Regimes

Rates of fertilizer application are given in Table 2. All fertilizers were applied in a band 50 cm wide along the rows of trees. The lime treatments were applied one month prior to transplanting. Other fertilizers were applied immediately prior to transplanting. All fertilizers were cultivated into a depth of 10 cm by raking.

Experimental Design and Plot Management

Plants were established in rows 150 cm apart, with intra-row spacing of 50 cm. Plots comprised four rows of six plants each. The resulting plots measured 6 x 3 m. The three species were replicated three times for each of six fertilizer treatments in a randomized block design. The fertilizer treatments (referred to in Table 2) were:

(F1) no fertilizer applied,
(F2) P,
(F3) P + Ca (nutr),
(F4) P + Ca (nutr) + TE,
(F5) P + Ca (nutr) + Ca (lime), and
(F6) P + Ca (nutr) + Ca (lime) + TE

These fertilizer treatments were chosen to examine the effects of various aspects of plant nutrition in acid soils (e.g., to separate the effects of calcium deficiency and low pH).

The seedlings at Sei Putih and Sembawa were transplanted in December 1986, whereas the Australian sites were not initiated until February 1987. At these latter two sites the seedlings were transplanted into a plastic mulch to minimize problems with weeds. At all sites the Leucaena plots were treated with insecticides to control potential psyllid problems.

The first harvest was made six months after transplanting. The center eight plants from each plot were cut to a height of 75 cm and separated into leaf (including small stems to 6 mm diameter) and wood components for dry weight determinations and subsequent chemical analyses. Only leaf yields are reported in this paper.

Results and Discussion

The intention of the initial harvest was to establish a more uniform experimental base from which to compare the subsequent regrowth over a number of cuttings. Nevertheless, data from this harvest are of interest in demonstrating some of the initial responses. These must be interpreted with care as the cutting height of 75 cm does not reflect the total amount of growth. Any plant not attaining this height would not contribute to the recorded yield. The method used tended to discriminate against Gliricidia, as a significant proportion of its leaf production was below the cutting height at this initial harvest. This problem should be overcome in subsequent harvests where regrowth above 75 cm is the criterion for comparison.

Utchee Creek

Data suggest the major nutrient deficiency at this site is phosphorus, although both Calliandra and Leucaena showed a possible response to calcium as a nutrient (Fig. 1). There were no obvious symptoms of nutritional deficiencies at this site.

Silkwood

Where no fertilizer treatments were applied, no plants grew above the cutting height of 75 cm. Even with added phosphorus, there was still no harvestable yield from Gliricidia, although both Calliandra and Leucaena showed a response to phosphorus. Leucaena and Gliricidia responded to added calcium as a nutrient, and indications of response to trace elements were evident in all three species (Fig. 2). Leaf discoloration was seen on most plants in the F5 treatment but not on the F6 (+ trace elements) plots.

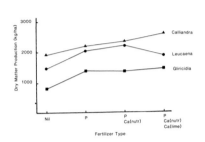

Fig. 1. Leaf production at Utchee Creek (Australia).

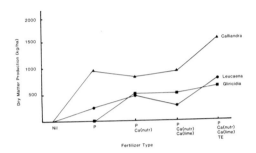

Fig.2. Leaf production at Silkwood (Australia).

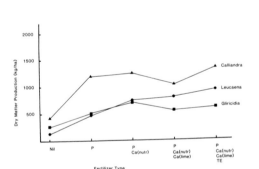

Fig. 3. Leaf production at Sei Putih (Indonesia).

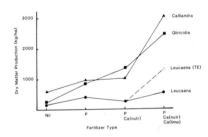

Fig. 4. Leaf production at Sembawa (Indonesia).

Unfortunately, at this time there is no information on the visual symptoms of mineral deficiencies in either Calliandra or Gliricidia. Calliandra had the most severe symptoms, the plants being stunted and the leaves yellow. The pinnae were tightly rolled and were brittle. A 0.5% solution of zinc sulfate was sprayed on plants in the border areas, resulting in alleviating the symptoms. With Gliricidia there was evidence of abscission of the second and third expanded leaves, suggesting the possibility of induced copper deficiency. Previous work on this soil has identified copper deficiency in pastures (J. Teitzel, personal communication). Samples have been taken for chemical analysis, both to try to confirm the suspected deficiencies and to investigate the possibility of an induced magnesium deficiency. Cool night temperatures in July (frequently below 12° C) caused some leaf fall in Gliricidia.

Sei Putih

All species showed a marked response to phosphorus. Calcium as a nutrient increased yield in Gliricidia and Leucaena. Trace elements further increased dry matter production. Calliandra and Gliricidia yields were depressed by lime, with a further response to trace elements. These data suggest that this site has a trace element deficiency that is exacerbated by lime addition (Fig. 3).

Sembawa

Calliandra and Gliricidia both performed relatively well at this site. The most significant response was to the addition of lime, with respective increases of 83% and 110% for Gliricidia and Calliandra (Fig. 4). Leucaena also showed a large response to lime after trace elements had been applied. These increases in yield are almost certainly due to an improvement in the soil pH effecting a reduction in aluminum saturation. However, the only lime source available in this region contained 10% magnesium, which could confound the interpretation.

Conclusions

Although the data reported above are only preliminary, the following conclusions can be drawn:

o Calliandra out-yielded the other two species at the first harvest.

o Except on extremely aluminum-dominated soils, Leucaena and Gliricidia can give reasonable production where possible calcium deficiencies have been overcome.

o Where lime is used to ameliorate problems of low soil pH, cognizance must be taken of any induced trace element deficiencies.

o Further study is needed to elucidate the differential nutrient responses of the three species involved and to follow their responses over several harvests.

Acknowledgments

The work reported here is part of a larger project, Multipurpose Shrub Legumes for Infertile Soils in the Tropics, funded by the Australian Centre for Agricultural Research, the Commonwealth Scientific Research and Industrial Organization, and the Government of Indonesia through the Agency for Agricultural Research and Development.

REFERENCES

Chadhokar, P.A. 1982. *Gliricidia maculata*: A promising legume fodder plant. *World Animal Review* 44:36-43.

Driessen, P.M., and M. Soepraptohardjo M. 1974. *Soils for agricultural expansion in Indonesia*. Bulletin No. 1, pp. 1-63. Bogor: Soil Research Institute.

Hill, G.D. 1971. *Leucaena leucocephala* for pastures in the tropics. *Herbage Abstracts* 41:111-119.

Murtha, G.G. 1986. *Soils of the Tully-Innisfail area, North Queensland*. Rep. No. 82. Canberra: CSIRO Aust. Div. Soils Div1.

Ranjaitan, M., D.A. Ivory, and R. Jessop. 1986. Regional evaluation of tree legume species and response to fertilizer application and rhizobium inoculation. *Forage research project annual report*, pp. 41-42. Indonesia: Balai Penelitian Ternak, Ciawi.

Performance of Nitrogen-Fixing MPTS
on Mountainous Wastelands in Low Rainfall Areas

L.L. Relwani, B.N. Lahane, and A.M. Gandhe

The Bharatiya Agro-Industries Foundation
Uruli-Kanchan, India

Nitrogen-fixing multipurpose tree species (MPTS) were evaluated on mountainous wastelands under rainfed, irrigated, and hand-watering systems with an annual rainfall of 350 to 400 mm over the experimental period. Under rainfed conditions, Leucaena leucocephala *achieved the best growth, followed by* Acacia tortilis *and* Acacia senegal. *With improved soil depth, moisture, and fertility, A. nilotica,* Pithecellobium dulce *and* Dalbergia sissoo *also performed well. Among the new introductions,* Acacia deamii *and* Sesbania formosa *were promising. Under protective irrigation,* Leucaena leucocephala *and* Sesbania grandiflora *were outstanding, closely followed by* Leucaena diversifolia, Gliricidia sepium, Dalbergia sissoo, *and* Erythrina indica. *In the hand-watering treatment,* Sesbania sesban, Leucaena leucocephala, A. nilotica var. *cupressiformis,* Gliricidia sepium, Parkinsonia aculeata *and* Albizia procera *performed well.* Leucaena leucocephala *showed consistently superior performance in all treatments. Prices of U.S.$ 40-50/ton wood are necessary to popularize growing of nitrogen-fixing MPTS by small farmers. One hundred metric tons of seeds and 1,600 metric tons of wood of* Leucaena leucocephala *have been sold to farmers and government and voluntary organizations.*

Introduction

The mountainous wastelands of the Deccan Plateau are characterized by low and erratic annual rainfall of 300-500 mm received in a short period of 3-4 months. The topography is rolling and undulating. The soils are shallow, deficient in plant nutrients, and poor in moisture-holding capacity. Agriculture is unstable and mostly not viable. Forests are disappearing due to population pressures and commercial wood extraction. Logic dictates that some of the lands should be planted with nitrogen-fixing multipurpose tree species (MPTS) to satisfy the needs of consumers, ensure high and sustainable profits, and improve soil fertility. An alternative appears to be combining woody perennials with herbaceous crops in agrosilviculture, silvopastoral, or agrosilvopastoral systems to guarantee productivity from one, two, or all three components.

The most important nitrogen-fixing MPTS adapted to such a harsh environment is *Acacia nilotica*. It is found scattered in small numbers in most of the cultivated and fallow lands, along field bunds, railway lines, and highways. The *vediana* variety is more xerophytic than the *telia* variety. The third variety, *cupressiformis*, has a tall main stem and broom-like, ascending thin side branches. *Telia* is the most important variety and can withstand dry summers, short periods of flooding, and saline conditions. The tree is an extremely valuable source of fuel, small timber, fodder, tannin, and honey. It reaches an average height of 10-15 m in 20 years (Anonymous 1980, Singh 1982).

The second most widely grown tree is *Prosopis juliflora*. Introduced in India over a hundred years ago for stabilizing dunes, it is now found growing rampantly in most of the barren lands, river banks, and edges of swampy areas. Goats eat the ripe pods and spread the seeds in their feces. Scarified seeds germinate easily in the monsoon season. The wood is hard, with a specific gravity of 0.7, and makes an excellent firewood and charcoal. It is generally cut in short rotations of 4-6 years, when it is 3-4 m in height. When widely spaced, it may grow up to 7-8 m and yield 30-40 tons of wood/ha. This tree has the greatest potential for solving quickly the fuel and feed problem of animals through aerial seeding of barren and eroded hil slopes.

The third type of small tree or bush is *Sesbania sesban*, which grows fast on better types of land and provides fodder and firewood in short rotations of about 2 years. Quite often it forms a hedge around sugarcane fields and is frequently cut for fodder. When intercropped with sugarcane, it is harvested along with the crop. There is a profuse and early formation of seed pods. Establishment is very fast, but growth then slows. *Sesbania grandiflora* is another popular tree planted for flowers and pods as vegetables, fodder for cattle and goats, and firewood for cooking. The tree grows luxuriantly in humid tropics with an annual rainfall of 1,000-1,200 mm. In drier areas, it is generally planted around farmers' cottages. Surplus flowers and pods are sold in vegetable markets. The tree is very attractive and is preferred for shade and ornamentation.

Parkinsonia aculeata grows very fast and survives extreme drought even in infertile, grave soils. It also is tolerant to salinity. Of the 20 tree species raised on the campus of the Bharatiya Agro-Industries Foundation (BAIF) in 1985, a drought year, only *Parkinsonia aculeata* survived. Apart from its use as fuel (specific gravity 0.6), it is grown for animal feed, particularly seeds and pods, erosion control, and beautification. The tree is not popular, however, because of its extreme thorniness.

Derris indica is grown on a small scale for fuel and oil for lamps in villages. Its other uses are fodder, fiber, timber for making cabinets, and shade and beautification. *Pithecellobium dulce*, *Dalbergia sissoo*, *Acacia auriculiformis*, *Acacia catechu*, *Samanea saman*, *Gliricidia sepium*, *Albizia procera*, and *Albizia lebbek* are some of the other trees grown in small numbers. *Casuarina equisetifolia* is suited to coastal areas. *Cajanus cajan* is sown for pulses, and the stems are used for fuel.

The most outstanding tree among the new introductions and all the species already mentioned is *Leucaena leucocephala*, due to its multiple uses as fodder, firewood, poles, posts, timber, organic manure, windbreak, shade, and ornamentation. It is now extensively grown on the BAIF campus for all these purposes. Only 10 years ago, the site was a barren, mountainous wasteland, but today it is lush green. There has been a tremendous public demand for seed of the K8 variety. We have sold about 100 tons of quality seeds of this Hawaiian Giant throughout the country, with excellent reports about its success. A wood depot has been established on the campus to supply firewood, poles, posts, and raw material to paper factories for newsprint. More than 1,600 tons of wood have

been marketed, at an average price of $US 30/ton. The psyllid pest has not yet entered India.

Indian foresters have had differing views about the suitability of various nitrogen-fixing multipurpose trees in wastelands. Qureshi (1986) recommends *Prosopis juliflora*, *Derris indica*, and *Albizia lebbek* for low- to medium-rainfall areas. Nimbkar and Zende (1986) recorded highest growth rates by *Leucaena leucocephala* compared to *Prosopis juliflora*, *Dalbergia sissoo*, *Acacia albida*, *Acacia nilotica* var. *cupressiformis*, *Albizia lebbek*, and *Acacia nilotica* var. *indica*. On low-rainfall, mountainous wastelands, Relwani et al. (1986) observed that *Leucaena leucocephala* grew best, followed by *Casuarina equisetifolia*, *Erythrina indica*, *Sesbania grandiflora*, and *Acacia auriculiformis*. Dwivedi (1980), Tiwari (1983), Dass and Shankar Narayan (1986), and Vaishnav (1986) have all strongly favored planting *Prosopis juliflora* for fast growth, hardiness, tolerance to salinity, drought and periodic flooding, profuse coppicing, immunity against grazing, palatable and nutritious pods, good quality firewood and charcoal, shade tolerance, and an excellent pioneer for site improvement. They consider it as an excellent species for eroded, degraded soils and as protection to agricultural lands against hot winds and browsing animals.

The results of the different trials on *Leucaena leucocephala* conducted at the BAIF Central Research Campus are detailed in the next section.

Forage Production and Effects on Animal Health

Hawaiian Giants K8, K29, K67, and K72 yielded an average of 11.1 tons green, 2.75 tons dry matter, and 0.66 tons crude protein/ha/cutting over 36 cuttings in 54 months with once-a-month irrigation in the dry season. They were about equal to one another and significantly superior to Hawaiian common and K341 (an improved Hawaiian common) with an average yield of 7.6 tons green, 1.96 tons dry matter, and 0.49 tons crude protein/ha/cutting (Relwani et al. 1982a). In the spacing management trial with K8, consisting of 1.0, 1.5, and 2.0 m inter-row and 0.1, 0.2, and 0.3 m intra-row spacings, the best combination was 1.0 x 0.1 m (100,000 plants/ha) with an average yield of 8.86 tons green, 2.29 tons dry matter, and 0.51 tons crude protein/ha/cutting (Relwani et al. 1982b).

Table 1. Growth observations of Hawaiian Giant K8 using different management practices.

Irrigation	Spacing (m)	Age (yrs)	Height (m)	Dbh (cm)	Vol ha/yr (m³)	Wt/yr t/ha*
Rainfed (400 mm/yr)	3 x 1	9.75	15.6	10.4	28.6	15.5
2 liters water/tree 4 times/yr (in first 2 yrs only)	2.75 x 2	7.75	13.7	9.9	15.8	8.6
Protective light irrigations once every 5-6 wks in dry season	1 x 1	3.50	10.5	8.0	96.0	51.9
Protective light irrigations once every 7-8 wks in dry season and interplanted w/	1 x 1	8.00	12.4	7.9	48.7	26.3
Guinea grass	3 x 3	8.00	15.1	14.4	21.8	11.8

Volume = height x dbh x 0.5 x plant population/ha.

Weight = volume x 0.54.

* Air-dried weight, about 15% moisture.

The best cutting height and frequency were 115 cm and 60 days (Relwani 1987). In a polypot experiment, the combination of NGR-8 and CB-81 Rhizobia was superior to NGR-8 on the green and dry matter yields and nitrogen content of the forage (Shinde and Relwani 1982). The amino acid mimosine poses no problem in the feeding of Leucaena forage to cattle as no toxicity effects were observed even in long-term feeding trials. Semen characteristics of bulls were also not adversely affected. Even the effect of feeding Leucaena seeds in limited quantities did not show any toxicity symptoms (Sobale et al. 1978; Rangnekar et al. 1983; Badve et al. 1985; and Waghmare, Joshi, and Rangnekar 1985). Feeding Leucaena forage at 33% of the total ration is the recommended practice.

Wood Production

Wood production trials using various management practices have been conducted in single plots, whose measurements range from 0.2 to 0.5 ha in different years in the past decade (Table 1.) Although statistically inconclusive, a few observations deserve mention. Uruli-Kanchan is located at 18.5° N latitude, 78.8° E longitude, and 560 m altitude, with an average annual rainfall of 400 mm from mid-June to mid-September. The maximum and minimum temperatures are 40-42° C in May-June and 7-8° C in January. Shallow soils, which overlay basaltic rock, have an average pH range of 7-8.

Table 1 shows that yields were lower in trial B than in trial A, in spite of additional watering, due to a lower plant population. The highest production was obtained from trial C, which had a high population and frequent irrigation. In trial D, the yields from 1 x 1 m spacing were lower than in C due to competition with Guinea grass and a longer irrigation interval. The spacing of 3 x 3 m was too wide and resulted in low yields (Relwani 1987).

Economics of Wood Production

Relwani and Hegde (1986a) estimated net profits of US $586, $405, $279, and $194 (US $1 = Rs 13) ha/year in areas with annual rainfalls of 1,250, 1,000, 750 and 500 mm, respectively, when sold as fuel at $25/ton. Profits increased to $1,739, $784.5, $442, and $292 when sold partially for poles and the rest for fuelwood. Under irrigated conditions, they reported net profits of $702 and $1,815/ha/yr when marketed as fuel or as poles and fuelwood, respectively (1986b). With the escalation in costs of wood production, producers must obtain $40-$50 per ton, depending on size and quality of wood. This is the major catalyst for popularizing large-scale tree plantations on wastelands and in agroforestry systems.

Tree Species Performance in Current Trials

Several tree species are being evaluated under dry land, protective irrigation, and hand-watering systems. Trees were planted on August 29, 1982, and observations were recorded on October 5, 1987. The growth data for the top, middle, and bottom sections of a slope under dry land conditions are given in Table 2. In each section, there were four trees of Leucaena (K8), alternating with an equal number of trees of other species, spaced 2 m apart.

There is a distinct increase in growth rates from the top, through the middle, to the bottom slopes. In the upper section, which is almost devoid of soil, *Acacia auriculiformis, Acacia nilotica, Casuarina equisetifolia, Dalbergia sissoo*, and *Derris indica* died from severe drought in 1986. Among the remaining species, Acacia *tortilis* grew relatively better than *A. senegal*, while *Pithecellobium dulce* performed the poorest, with only a 25% survival rate. In the middle section, *Acacia auriculiformis* died and *Casuarina equisetifolia, Dalbergia sissoo, Pithecellobium dulce*, and *Derris indica* had a survival rate of only up to 25%. Among the remaining species, the performance of *A. nilotica* was better than that of *A. senegal* or *A. tortilis*. At the bottom of the slope, the survival of *A. auriculiformis* and *Casuarina equisetifolia* was only 25%, *Dalbergia sissoo*, 50%; *A. nilotica* and *Derris indica*, 75%; and *A. senegal, Acacia tortilis*, and *Pithecellobium dulce*, 100%. *A. nilotica* showed higher growth rates, followed by *Dalbergia sissoo* and *Pithecellobium dulce*. *Casuarina equisetifolia*,

in spite of its higher growth potential, performed poorly. Taking the four parameters together, none of the species approached the outstanding performance of *Leucaena leucocephala*. It responded to the fertility gradient in a way similar to other species, with progressive improvement in growth and survival from the top to the bottom sections of the slope.

In another trial on degraded mountainous soils (Table 3), some new species were tested on two sites to evaluate their performance under dry land conditions. The trees were planted on October 14, 1985, on site 1 and November 4, 1985, on site 2. Observations were recorded on September 17, 1987. The trial was conducted as part of the Oxford Forestry Institute's international trials of dry zone species.

On these barren soils, *Leucaena leucocephala* and *Acacia deamii* on site 1 and *Sesbania formosa, Leucaena leucocephala, Acacia deamii*, and *Pithecellobium dulce* on site 2 performed best. Leucaena was the only species to be grazed by hares shortly after transplanting, but the plants recovered quickly and grew rapidly. *Acacia deamii* grew fast but later turned bushy. It is also thorny and can serve as a good protective fence. *Sesbania formosa* does not tolerate extended drought, and *Leucaena shannoni* and *Derris indica* established poorly.

In the third trial, 28 species, including 16 nitrogen-fixing species, were evaluated under protective irrigation with the idea of reducing mortality due to summer heat and drought. Trees were planted in July 1985, watered twice in 1985 and 1986, and once in 1987. Observations were recorded in September 1987.

Under protective irrigations, *Leucaena leucocephala* and *Sesbania grandiflora* performed best (Table 4). *Casuarina equisetifolia* established well but grew poorly and had high mortality in the hot and dry summer season. *Albizia lebbek* was severely affected by psyllids (order *Homoptera*, family *Psyllidae*) and its foliage turned yellow. Malathion sprayings proved ineffective, but two sprayings of Nuvacron (0.05%) at weekly intervals controlled the pest (Hegde and Relwani 1987). *Pithecellobium dulce* and *Prosopis juliflora* were slow starters and were over-run by weeds.

Table 2. Performance of five-year-old nitrogen-fixing trees on sections of slopes of mountainous wastelands.

Section of slope	Height (m)	Basal Diameter (cm)	Dbh (cm)	Survival (%)
Top				
Acacia auriculiformis*	--	--	--	0
A. nilotica	--	--	--	0
A. senegal	.99	1.9	0.3	100
A. tortilis	1.55	2.9	1.7	100
Casuarina equisetifolia	--	--	--	0
Dalbergia sissoo	--	--	--	0
Pithecellobium dulce	.87	1.8	--	25
Derris indica	--	--	--	0
Average for Leucaena leucocephala	4.91	7.0	4.9	78
Middle				
A. auriculiformis	--	--	--	0
A. nilotica	3.30	4.2	1.9	75
A. senegal	1.73	2.9	1.5	100
A. tortilis	1.70	1.9	1.0	100
Casuarina equisetifolia	7.12	7.7	5.5	25
Dalbergia sissoo	1.39	2.5	1.1	25
Pithecellobium dulce	1.55	3.5	0.5	25
Derris indica	.23	0.4	--	25
Average for Leucaena leucocephala	5.73	7.2	4.8	94
Bottom				
A. auriculiformis	3.63	5.2	3.5	25
A. nilotica	4.81	5.4	3.3	75
A. senegal	1.19	2.7	1.5	100
A. tortilis	2.41	3.1	1.9	100
Casuarina equisetifolia	5.57	5.2	3.5	25
Dalbergia sissoo	4.19	4.4	2.3	50
Pithecellobium dulce	3.02	4.4	2.0	100
Derris indica	1.30	3.3	1.5	75
Average for Leucaena leucocephala	8.66	10.0	7.4	100

* High mortality due to severe drought in 1986.

Table 3. Growth of two-year-old, dry-zone tree species.

Species	Average Height (m)	Average Basal Diameter (cm)	Survival (%)
Site 1			
Acacia auriculiformis	1.75	2.36	52.17
A. deamii*	2.29	3.86	95.45
A. farnesiana	1.35	1.60	78.26
A. pennatula	1.91	3.40	69.57
Albizia lebbek	1.67	3.28	86.96
Apoplanesia paniculata*	1.48	2.70	82.51
Enterolobium cyclocarpum	1.63	2.74	50.00
Myrospermum frutescens*	1.48	1.77	76.19
Leucaena leucocephala	2.32	3.30	95.24
Site 2			
Acacia auriculiformis	1.71	1.98	40.74
A. deamii*	2.43	3.46	96.30
A. farnesiana	1.52	1.93	100.00
Apoplanesia paniculata*	1.66	3.53	66.66
Derris indica	.73	2.01	28.00
Gliricidia sepium	1.14	1.85	80.95
Leucaena leucocephala	2.44	3.33	92.60
L. shannoni	1.87	1.40	8.33
Myrospermum frutescens*	.85	1.30	60.71
Parkinsonia aculeata	1.64	2.80	75.00
Pithecellobium dulce	2.09	3.23	92.86
Samanea saman	1.35	2.71	53.57
Sesbania formosa*	2.46	4.31	50.00

* Nitrogen fixation unconfirmed (Halliday 1984).

Table 4. Growth of multipurpose, nitrogen-fixing tree species with protective irrigation.

Species	Average Height (m)	Average Basal Diameter (cm)	Average Dbh (cm)	Survival (%)
Acacia auriculiformis	3.38	3.76	1.94	85.30
A. nilotica				
var. cupressiformis	2.05	2.61	0.83	100.00
Albizia lebbek	1.88	3.01	0.85	70.31
Casuarina				
equisetifolia	3.42	3.75	1.73	26.56
Dalbergia sissoo	3.32	3.98	2.38	93.75
Derris indica	2.09	3.03	0.99	85.94
Enterolobium				
cyclocarpum	1.94	4.20	1.31	71.88
Erythrina indica	3.15	4.40	3.21	90.63
Gliricidia sepium	4.13	3.66	2.73	100.00
Leucaena diversifolia	4.98	5.51	3.25	93.75
L. leucocephala	7.26	7.40	5.24	100.00
Parkinsonia aculeata	4.49	5.43	3.30	96.88
Pithecellobium dulce	3.26	4.31	2.74	39.06
Prosopis juliflora	1.10	1.93	--	60.94
Samanea saman	2.99	4.40	1.92	96.88
Sesbania grandiflora	6.98	8.88	5.72	100.00

In the fourth trial, an augmented block design with four replications, measured quantities of water were applied to trees, depending on the seasonal requirements for establishment and growth. Trees were planted in November 1986 and received two liters water each per week until January 1987. Each tree received four liters water per week from February until June, the dry summer, and from July to August, a period that experienced intermittent light showers. Watering then stopped.

Among the main plot species, *Sesbania sesban* started fast, branched, and produced seed pods profusely in the first six months (Table 5). Later on, its growth rate decreased considerably. *Leucaena leucocephala* grew slowly in the first five months but vigorously thereafter. Among the remaining species, *Acacia nilotica* var. *cupressiformis* maintained steady and satisfactory growth throughout the entire period.

In the subplots of the fourth replication, *Gliricidia sepium* and *Parkinsonia aculeata* established quickly and grew faster than the other species. *Albizia procera* was the next best but tended to bend in high winds. *Derris indica* grew slowly.

Table 5. Growth of multipurpose, nitrogen-fixing tree species with hand waterings.

Species	Average height (m)	Average canopy (cm)	Average basal diameter (cm)	Average Dbh (cm)
Main treatments				
Acacia nilotica				
var. cupressiformis	1.84	131.29	3.11	1.10
Albizia lebbek	1.56	81.93	2.64	1.00
Leucaena leucocephala	2.77	118.89	3.12	1.74
Pithecellobium dulce	1.66	152.78	2.42	0.89
Prosopis juliflora	1.13	326.96	2.57	1.32
Sesbania sesban	2.74	189.92	4.24	1.41
F. test	Significant	Significant	Not Significant	Not Significant
S.Em.+-	21.09	20.45	0.41	0.26
C.D. at 5%	62.04	60.16	--	--
Subtreatments*				
Acacia nilotica	1.21	92.22	1.80	--
Albizia procera	1.56	115.33	3.93	1.74
Casuarina equisetifolia	1.51	127.72	1.63	0.38
Dalbergia sissoo	1.50	88.72	1.90	0.58
Derris indica	0.48	37.17	0.89	--
Gliricidia sepium	2.47	136.83	3.59	1.52
Parkinsonia aculeata	2.21	254.67	3.57	1.00

* date of planting, 1/21/87.

REFERENCES

Anonymous. 1980. *Firewood crops*. Washington, D.C.: National Academy of Sciences.

Badve, V.C., A.L. Joshi, D.V. Rangnekar, and B.S. Waghmare. 1985. Mimosine metabolism in cattle fed *Leucaena leucocephala*. *Leucaena Res. Rep.* 6:22.

Dass, H.C. and K.A. Shankarnarayan. 1986. *Suitability of different Prosopis in waste lands development*. Bhavnagar: Chandan Printary.

Dwivedi, A.F. 1980. *Forestry in India*. Dehra Dun: Jugal Kishore & Co.

Halliday, J. 1984. Register of nodulation reports for leguminous trees and other arboreal genera with nitrogen fixing members. *Nitrogen Fixing Tree Res. Rep.* 2:38-45.

Hegde, N.G. and L.L. Relwani. 1987. Psyllids attack *Albizia lebbek* (Linn.) Benth. in India. *Common. For. Rev.* 66(2):193-94.

Nimbkar, N. and N. Zende. 1986. Comparative growth of tree species on waste lands under limited irrigation. In *The greening of waste lands. Proceedings of the national workshop on utilization of waste lands for bio-energy*, eds. N. G. Hegde and P.D. Abhyankar, pp. 122-27. Pune: BAIF.

Qureshi, I.M. 1986. Selection of species for waste lands and their protection. In *The greening of waste lands. Proceedings of the national workshop on utilization of waste lands for bio-energy*, eds. N.G. Hegde and P.D. Abhyankar, pp. 102-4. Pune: BAIF.

Rangnekar, D.V., M.R. Bhosrekar, A.L. Joshi, S.T. Kharat, B.N. Sobale, and V.C. Badve. 1983. Studies on growth performance and semen characteristics of bulls fed unconventional fodder (*Leucaena leucocephala* and *Desmanthus virgatus*). *Trop. Agric. Trinidad* 6(4):294-6.

Relwani, L.L. 1987. *Performance of Leucaena leucocephala*. New Delhi: Tata Energy Research Institute.

Relwani, L.L., S.S. Deshmukh, R.V. Nakat, and D.Y. Khandale. 1982a. Varietal trial on Leucaena cultivars for forage production. *Leucaena Res. Rep.* 3:39.

_____ 1982b. Effect of spacing management on the yield and quality of forage of K8 cultivar. *Leucaena Res. Rep.* 3:40.

Relwani, L.L. and N.G. Hegde. 1986a. Afforestation and waste lands developments. *BAIF Journal* 6(4):16.

_____. 1986b. Economics of subabul plantation. In *The greening of waste lands. Proceedings of the national workshop on utilization of waste lands for bio-energy*, eds. N.G. Hegde and P.D. Abhyankar, pp. 118-21. Pune: BAIF.

Relwani, L.L., B. N. Lahane, and D.Y. Khandale. 1986. Performance of different tree species on arid mountainous waste lands. In *The greening of waste lands. Proceedings of the national workshop on utilization of waste lands for bio-energy*, eds. N.G. Hegde and P.D. Abhyankar, pp. 114-16. Pune: BAIF.

Shinde, D.B. and L.L. Relwani. 1982. Effect of Rhizobium cultures on the nodulation, yield and nitrogen content of subabul. *The BAIF Jour.* 3(1):26.

Singh, R.V. 1982. *Fodder trees of India*. New Delhi: Mohan Primlani.

Sobale, B.N., S.T. Kharat, V.L. Prasad, A.D. Joshi, D.V. Rangnekar, and S.S. Deshmukh. 1978. Nutritive value of *Leucaena leucocephala* for growing calves. *Trop. Anim. Health & Prod.* 10:237.

Tiwari, K.M. 1983. Forestry for the rural poor. *The Indian Forester* 109:1-6.

Vaishnav, M.N. 1986. Prosopis juliflora: *Our saviour. Role of Prosopis in waste land development*. Bhavnagar: Chandan Printary.

Waghmare, B.S., A.L. Joshi, and D.V. Rangnekar. 1985. The nutritive value of Leucaena seeds for cattle. *Leucaena Res. Rep.* 6:56.

Nitrogen-Fixing Trees as Multipurpose Species for Soil Conservation

H.D. Tacio, H.R. Watson, and Warlito A. Laquihon

Mindanao Baptist Rural Life Center
Davao City, Philippines

Soil erosion due to deforestation and heavy rains presents an extremely serious problem in many parts of Southeast Asia, particularly in the upland Philippines. The Mindanao Baptist Rural Life Center has developed and spread an agroforestry scheme called Sloping Agricultural Land Technology (SALT) to help control erosion and increase crop yields. SALT is an integrated, diversified hillside farming system that uses thickly seeded rows of nitrogen-fixing trees, such as Leucaena leucocephala, planted in contour hedgerows. The trees serve as soil binders, producers of green leaf manure, and moisture reservoirs. SALT also includes annual and perennial diversified food crops grown in the spaces between the hedgerows. The SALT model has been tested both in demonstration plots and farmers' fields, and has proven to be appropriate for use by typical hilly-land farmers. The system can reduce soil erosion and restore moderately degraded hilly lands to a profitable farming system.

Soil erosion has been recognized as perhaps the most critical factor causing the instability and non-sustainability of upland production systems. Soil conservation measures include terracing, mulching, contour strip-cropping, multiple-cropping, cover-cropping, reforestation, and agroforestry. Farmers can choose any one of these methods or a combination to suit their needs and capabilities. However, for a farm size of 1-2 ha, the Mindanao Baptist Rural Life Center (MBRLC) recommends an agroforestry system called Sloping Agricultural Land Technology (SALT).

SALT is a package technology of soil conservation and food production that integrates several soil conservation measures in one setting. Basically, SALT is a method of growing field and permanent crops in bands 4-6 m wide between contoured rows of nitrogen-fixing trees. The nitrogen-fixing trees (NFTs) are thickly planted in double rows to form hedgerows. When a hedge is 1.5-2 m tall, it is cut back to a height of 40 cm and the cuttings are placed in the strips between the hedgerows, also called alleys, to serve as organic fertilizer.

With rows of permanent crops, such as coffee, cacao, and citrus dispersed throughout the farm plot, SALT can be considered an agroforestry system. The strips not occupied by permanent crops are planted alternately to cereals (corn, upland rice, sorghum, etc.) or other crops (sweet potato, melon, pineapple, castor bean) and legumes (mung bean, soybean, peanut). This cyclical cropping provides the farmer several harvests throughout the year. The average income to one family from a hectare of SALT farm, following instructions carefully, is approximately P1,000 per month ($US1 = P20.5).

On a SALT farm, a farmer can grow varieties of crops familiar to him. SALT can be adapted to incorporate new or traditional farming techniques. If farmers leave land fallow, the NFTs will continue to grow and may later be harvested for firewood or charcoal (Fig. 1).

SALT is a simple, applicable, low-cost, and timely method of farming uplands. It is a technology developed for farmers with few tools, little capital, and little training in agriculture. Contour lines are determined by using an A-frame transit that any farmer can learn to make and use.

Background

SALT was developed on a marginal site in Kinuskusan, Bansalan, Davao del Sur, when MBRLC started to employ contour terraces on their sloping lands in 1971. Dialogues with local upland farmers acquainted the Center's staff with farm problems and needs, and gave them the impetus to work out a relevant and appropriate upland farming system.

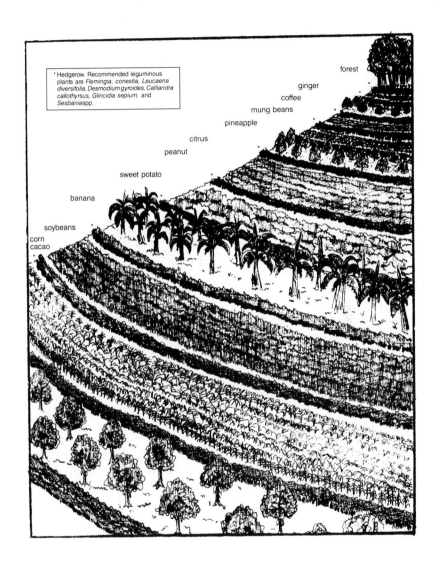

* Hedgerow. Recommended leguminous plants are *Flemingia, conestia, Leucaena diversifolia, Desmodium gyroides, Calliandra callothyrsus, Gliricidia sepium,* and *Sesbaniaspp.*

forest

ginger

coffee

mung beans

pineapple

citrus

peanut

sweet potato

banana

soybeans

corn
cacao

Fig. 1. The Sloping Agricultural Land Technology (SALT) model.

After testing different intercropping schemes and observing Leucaena-based farming systems, both in Hawaii and at the Center, the SALT prototype was finalized in 1978. During the development stage, some guidelines were considered essential. Scientists felt that the system should: (1) adequately control soil erosion, (2) help restore soil structure and fertility, (3) be efficient in food crop production, (4) be applicable to at least 50% of hillside farms, (5) be easily duplicated by upland farmers using local resources and preferably without making loans, (6) be culturally acceptable, (7) have the small family as the focus and food production as the top priority (fruit trees, forest, and others are secondary priority), (8) be workable in as short a time as possible, (9) require minimal labor, and (10) be economically feasible and ecologically sound.

In 1978, a one-hectare test site was selected at MBRLC. Typical of the surrounding farms, its slope was greater than 15 degrees and it had been farmed for at least five years. Its soils also were similar to those of most farms in the area. Contour lines were established and hedgerows of *Leucaena leucocephala*, commonly called Ipil-ipil, were planted. Alternate strips between the hedgerows were planted to maize, the traditional upland crop, leaving every other strip uncultivated to help control soil loss until the Ipil-ipil hedges were large enough to hold the soil.

Observations

The first corn crop was harvested in late 1978. Corn yields of the model and local farms were compared. Labor inputs and soil loss were also monitored (Table 1).

As indicated in Table 1, SALT requires more labor than conventional farm methods, but the increase in yields compensates for this additional labor. Furthermore, the use of 100% of available annual work hours for a SALT farm can be viewed as increased employment for farmers. The tools needed to cultivate the SALT farm are the same as those already used by the local farmer, i.e., carabao, plow, harrow, and a long knife for cutting grasses. The only tools introduced into the SALT system were the A-frames and hoes.

Soil loss in alleys was measured simply by placing marked stakes along contour lines at the upper and lower sides of alleys. The loss from the upper side of the alley and the accumulated soil on the lower side were measured. The amount of accumulated soil on the lower side was subtracted from the amount lost on the upper side; the difference was the amount of soil lost in the alley. On a local farm without SALT, erosion was determined by measuring topsoil depth, and in some cases, by comparing the amount of soil retained around coconut trees to the rest of the field.

Cost and Return Analysis

Table 2 shows the gross income, total expenses, and net incomes for the years 1980 through July 1987. The average annual income for upland farmers in the area is about P4,000 with most farmers planting more than 1 ha of land. A farmer can triple his farm income by adopting the SALT system.

Modifications

A SALT project that follows the MBRLC recommendations will have about 20% of its area planted to NFTs, 25% to permanent crops, and 55% to non-permanent crops (Fig. 1). Several variations, however, have been observed:

o Row crop SALT system. The farmer only plants annual crops such as corn, beans, or other row crops in the alleys. This system may be more economical for the farmer in the short run, but the recommended practice is that permanent crops be planted in every third or fourth strip or alley. Some farmers say that since they do not own the land, they do not want to plant permanent crops.

o Permanent crop SALT system. The farmer plants all the strips or alleys to perennials, such as banana, coffee, or fruit trees.

o Almost all SALT projects represent some variation of the Demonstration SALT at MBRLC.

Recent Developments

In October 1985, the Demonstration SALT, which included *Leucaena leucocephala* hedgerows, was attacked by a psyllid or jumping louse (*Heteropsylla cubana* Crawford). With this infestation, the MBRLC began intensive testing the

Table 1. Comparison of SALT model with a local farming system (1 ha).

Item	SALT Model	Local Farm
Labor (1 yr)	100% of work hrs	50% of work hrs
Corn yields (2 crops per yr)	2,000 kg/crop[1]	500 kg/crop[2]
Soil loss	Slight	Severe

[1] Only leaves from Leucaena in hedgerows were used for fertilizer. Leucaena herbage yields per ha/yr are 36,080 k of green leaves and stems, with the following NPK equivalents: 258.5, 120.2, and 90.1.

[2] The local farmer did not use commercial fertilizer or Leucaena green manure.

Table 2. Cost and return analysis of the SALT demonstration plot (1980-1987)[a]

Year	Gross Income	Total Expenses	Net Income	Net Income/ Month
1980	5,693	1,117	4,575	381[b]
1981	3,055	583	2,472	206[c]
1982	9,007	1,833	7,174	597
1983	6,471	1,228	5,242	436[d]
1984	14,287	1,741	12,545	1,045[e]
1985	15,559	1,858	13,701	1,141
1986	13,294	1,710	11,584	965[f]
1987	8,137	2,274	5,863	837[g]

[a] US$1 = Pesos 20.5.

[b] Includes seeds, insecticides, and fertilizer. No labor expense is included because the farmer uses his own labor.

[c] Permanent crops were not yet producing.

[d] A 6-month drought occurred in Mindanao in 1983.

[e] Permanent crops began producing.

[f] Psyllid infestation of Ipil-ipil was at its height.

[g] No Ipil-ipil available. Commercial fertilizers used. Leucaena hedgerows replaced by Flemingia. Covers 7 months period (January-July).

soil conservation capacity of several other NFTs that the Center was already studying.

The most promising species are *Flemingia congesta, Acacia villosa, Leucaena diversifolia, Gliricidia sepium,* and *Desmodium gyroides.* Other species under consideration and testing include *Calliandra calothyrsus, C. tetragona, Sesbania sesban, S. bispinosa, S. aculeata, Cassia confusa, C. fistula, C. siamea, C. spectabilis, Desmodium rotundifolia, D. distortum, D. discolor, D. rensonii, Acacia auriculiformis, Desmanthus virgatus,* and *Saba adenomthera.* Some of these species also are used as forage for goats and other ruminants.

The Animal SALT integration project, started in 1984, is doing well. The NFTs control soil erosion and their leaves are used as forage. Animal manure is recycled back into the contour strips.

The SALT System of Farming

SALT is an appropriate improvement over existing technologies. It is a simple, effective way of farming uplands without losing topsoil to erosion. It consists of the following 10 basic steps:

1. Making the A-Frame. The A-frame is a simple device for laying out contour lines across the slope. It is made of a carpenter's level and three wooden or bamboo poles nailed or tied together in the shape of a capital letter A with a base about 90 cm wide (Fig. 2). The carpenter's level is mounted on the crossbar.

2. Determining the contour lines. One leg of the A-frame is planted on the ground. The other leg is swung until the carpenter's level shows that both legs are touching the ground on the same level. A helper drives a stake beside the frame's rear (first) leg. The process is repeated across the field. Each contour line should be spaced 4-6 m apart regardless of the slope.

3. Cultivating the contour lines. One-meter strips along contour lines should be plowed and harrowed to prepare for planting. The stakes serve as guides during plowing.

4. Planting seeds of NFTs. On each prepared contour line, two furrows 50 cm apart should

be laid out. Four to six seeds are planted per hill, with a distance of 12 cm between hills. The seeds should be covered firmly with soil. When fully grown, the hedgerows bank the soil and serve as fertilizer.

5. Planting permanent field crops. Field crops are planted in the alleys between the thick rows of NFTs. Permanent crops may be planted at the same time the hedgerows are planted, but only the planting spots should be cleared and dug. When the seeds germinate, only ring-weeding should be employed. Full cultivation can start when the NFTs have grown to the point where they can hold the soil.

Permanent crops should be planted in one out of every three strips. Coffee, banana, citrus, cacao, and others of the same height are good examples of permanent crops that may be planted by an upland farmer. Tall crops should be planted at the bottom of the hill. Generally, short crops should be planted at the top.

6. Cultivating alternate strips. The soil can be cultivated even before the NFTs are fully grown (about two meters in height). Cultivation should be done on alternate strips (strips 2, 5, 9, and so on). The uncultivated strips collect the soil eroded by rainfall from the higher cultivated strips. When the NFTs are fully grown, all of the strips can be cultivated simultaneously.

7. Planting short-term field crops. Short- and medium-term crops should be planted between the strips of permanent crops as sources of food and income while the farmer waits for the permanent crops to bear fruit. Suggested crops for these strips include pineapple, ginger, castor bean, sweet potato, peanut, sorghum, corn, and upland rice. To avoid shading among crops, short species should be planted away from tall ones.

8. Trimming of NFTs. Once a month, the faster-growing NFTs like *Leucaena leucocephala* and *Flemingia congesta* should be cut back to a height of 40 cm. The cuttings should be piled at the base of the crops to serve as mulch and later as organic

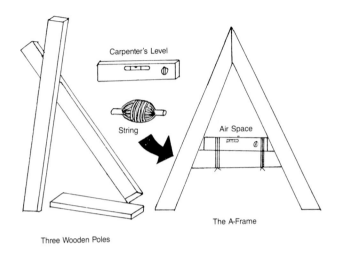

Carpenter's Level

String

Three Wooden Poles

Air Space

The A-Frame

Fig. 2. The A-frame used to determine contour lines.

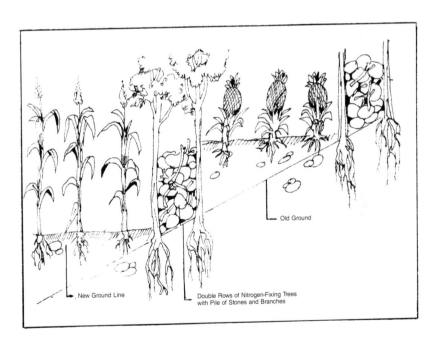

Old Ground

New Ground Line

Double Rows of Nitrogen-Fixing Trees
with Pile of Stones and Branches

Fig. 3. Straw, stalks, twig, branches, and rocks are piled at the base of hedgerows to control erosion.

fertilizer. When cutting the hedges, a sharp bolo, or machete, should be used.

9. Managing the system. The non-permanent crops should be rotated to maintain productivity, fertility, and good soil formation. One good way to do this is to plant cereal crops (upland rice, corn, and sorghum, among others), tubers (sweet potato, cassava) in strips where legumes (beans, pulses, and peas) were previously planted, and vice versa. Other crop management practices, such as weeding and insect control, should also be performed regularly.

10. Building green terraces. To enrich the soil and effectively control erosion, straw, stalks, twigs, branches, leaves, rocks, and stones should be piled at the base of the thick rows of NFTs (Fig. 3). With the passage of time, strong, permanent, and naturally green terraces will form to hold the soil in place.

By using SALT, the small upland farmer can conserve soil, reduce his purchases of commercial fertilizer, increase his yield and income and become generally self-sufficient. In this way, those living in marginal, hilly areas can break out of the common cycle of expensive monoculture, dependence on imported fertilizers and insecticides, and indebtedness to large landowners or banks.

Conclusions

SALT can adequately reduce soil erosion and restore moderately degraded hilly lands to a profitable farming system operated by a typical hilly-land farmer. The holding of topsoil by the hedgerows of NFTs and by permanent crops, the recycling of crop residues, the nutrients (NPK) furnished by the hedgerows, and minimum tillage account for the success of SALT. SALT is neither a perfect farming system nor a panacea. To establish a one-hectare SALT farm requires hard work, discipline, and time.

Recommendations

o For research and development:

-- Further studies on various NFT species for use as hedgerows on different soil types and pH ranges, and for compatibility with different intercropped annuals.

-- Tests to determine more accurately soil losses under SALT systems versus traditional and other systems of upland farming.

-- Studies to determine optimum cutting regimes for NFTs in hedgerows for use as organic fertilizer on crops.

-- Genetic work on the selection of tree cultivars suitable for SALT.

-- Feasibility tests of integrated animal and crop production.

-- Economic feasibility studies of SALT variations.

-- Continued search for better methods of farming upland soils.

-- Research on socioeconomic and other determinants of adopting hillside technologies such as SALT.

o For establishment:

-- An upland agroforestry research and training center should be established in Southeast Asia with a satellite center in each country. A center of this type can provide hope for securing the hills and forests from total degradation.

-- Regional training centers should be established in hilly areas with functional models of proven hillside farming and agroforestry systems.

-- National governments should create committees to assess the present conditions of their forests and land degradation in order to integrate conservation into all levels of society.

Backstopped by the MBRLC experience and in light of the recent study made by Laquihon (1987), a SALT strategy can now be formulated and recommended for trial by interested organizations and individuals. The SALT strategy requires four essential elements:

1. Simple technologies demonstrated in strategic locations; without good examples, nobody follows.

2. Holistic educational and training programs that emphasize that man is mandated to be steward of the earth.

3. Identification of factors inhibiting technology transfer in order to remove barriers to the adoption of SALT by farmers.

4. Credible, down-to-earth, and honest technicians to implement the approved programs. Sterling performance always speaks louder than electrifying but empty words.

REFERENCES

Laquihon, W.A. 1987. Some key determinants of SALT adoption in the Philippines. Ph.D. dissertation. Ateneo de Davao University/University of Mindanao Graduate Consortium, Philippines.

The Role of Nitrogen-Fixing Trees as MPTS for Small Farms

Kovith Yantasath

Thailand Institute of Scientific and Technological Research
Bangkok, Thailand

Many nitrogen-fixing trees are valuable multipurpose tree species (MPTS) that produce food, fodder, green manure, fuelwood, timber, and pulp and also control soil erosion. Research on these and other MPTS conducted by the Thailand Institute of Scientific and Technological Research (TISTR) on acid and non-acid soils identified some species as appropriate for small-farm use. Leucaena leucocephala, Cassia siamea (a non-nitrogen fixer), Gliricidia sepium and Sesbania grandiflora grew and coppiced well on non-acid soils and are considered suitable for animal fodder, fuelwood, and food for people. Acacia mangium, Acacia auriculiformis, Albizia falcataria, Casuarina equisetifolia, Casuarina junghuhniana and Eucalyptus camaldulensis adapted well in both acidic and non-acidic soils and have exceptional potential for fuelwood, charcoal, and other wood uses. These species should be planted densely as single or mixed stands or intercropped with agricultural crops in multipurpose plantings on small farms.

The definitions of multipurpose trees were discussed in 1983 at the Workshop on Multipurpose Tree Germplasm held at the National Academy of Sciences in Washington, D.C. Based on these discussions, multipurpose trees were defined as trees grown deliberately or kept and managed for preferably more than one intended use, usually economically motivated major products and/or services in any multipurpose land-use system, especially agroforestry systems (von Carlowitz and Burley 1984).

Under a systems approach rather than an individual tree approach, von Carlowitz (1984) described multipurpose trees as follows:

A multipurpose tree is a tree that clearly constitutes an essential component of an agroforestry system or other multipurpose land-use systems. Regardless of the number of its potential or actual uses, a multipurpose tree has to have the capacity to provide in its specific function(s) in the system a substantial and recognizable contribution to the sustainability of yields, to the increase of outputs and/or the reduction of inputs, and to the ecological stability of this system. Only a tree that is kept and maintained or introduced into an agroforestry system especially for one or more of these purposes qualifies as a multipurpose tree.

Nitrogen-fixing trees have been described as a subset within the category of multipurpose trees. They have the ability to fix atmospheric nitrogen in a form usable by individual plants, soil (through leaching from roots, leaf fall, and deliberate mulching), and domestic animals (through fodder). Nitrogen-fixation enriches the soil, foliage is a minor product, and wood is used for timber, poles, fuel, and living fences (Burley 1983).

Potential

The ability to fix nitrogen from the atmosphere makes these trees highly regarded for small-scale agroforestry in the tropics (Vergara 1982). Some outstanding examples of popular nitrogen-fixing tree species include *Leucaena leucocephala* and trees of the genera Acacia, Albizia, Alnus, Calliandra, Casuarina, Codariocalyx, Dalbergia, Erythrina, Gliricidia, Inga, Intsia, Parkia, Prosopis, Pithecellobium, Pterocarpus, Robinia, Samanea, and Sesbania (National Academy of Sciences 1979, Nitrogen Fixing Tree Association 1985).

Various nitrogen-fixing tree species may grow rapidly and are clearly adapted to particular stress environments. But these species may not be multipurpose in terms important to small farmers, unless their services and productivity for farmers can be more clearly understood, as is the case for *Leucaena leucocephala, Gliricidia sepium, Calliandra calothyrsus*, and some other species (MacDicken 1986). Although these may produce less biomass, they are probably much better suited to management and provision of important products and services, e.g., lopping for fodder, fuelwood, and soil improvement.

From the masterlist of 1,010 species, of which 600 are known to nodulate (Halliday and Nakao 1982), 44 nitrogen-fixing tree species are included in the "A" list of economically important species (Brewbaker and Styles 1982). These species are further divided by their primary uses (timber, fuelwood, fodder, food, soil improvement, medicinal and other uses) and their ecological zones (arid and semi-arid, tropics, humid tropics, highland tropics, and temperate). Another profile of selected nitrogen-fixing trees for small-farm planting, particularly their planting niches, has been reported (Macklin 1987). These guidelines can be used to select tree species most suitable for the environments and needs of different areas.

Choosing Trees for Small Farmers in Thailand

In Thailand, several nitrogen-fixing tree species (native and exotic) have been planted for multiple uses, including *Acacia catechu, Gliricidia sepium, Leucaena leucocephala, Parkia speciosa, Samanea saman,* and *Sesbania grandiflora* (Bhumibhamon 1986). Other nitrogen-fixing trees more or less known to possess MPTS characteristics include *Acacia auriculiformis, A. mangium, Albizia (Paraserianthes) falcataria, Casuarina equisetifolia* and *C. junghuhniana.* Recently, Leucaena and Casuarina have been planted with chili, corn, and cassava in an intercropping farming system in some southern provinces of Thailand.

Research on nitrogen-fixing tree species conducted by the Thailand Institute of Scientific and Technological Research on acid and non-acid soils in Thailand showed that *Leucaena leucocephala, Gliricidia sepium* and *Sesbania grandiflora,* which grew and coppiced well on non-acid soils and are suitable for animal feed, fuelwood or human consumption, are suitable MPTS for small-scale farmers (Yantasath 1987). *Acacia auriculiformis, A. mangium, Albizia (Paraserianthes) falcataria, Casuarina equisetifolia,* and *C. junghuhniana,* which adapted well in both acid and non-acid soils, are considered promising for fuelwood, charcoal, and other wood uses. However, these tree species are either newly introduced or have never been planted for multiple uses. Additional research on the potential of these species for intercrop farming/agroforestry is required to determine if they could be planted as

MPTS along farm borders, in crop fields, livestock systems, around the house, or in pure stands.

REFERENCES

Bhumibhamon, S. 1986. Survey methods to determine research priorities: A Thai case study. In *Forestry Networks. Proceedings of the first network workshop of the Forestry/Fuelwood Research and Development Project (F/FRED),* eds. N. Adams and R.K. Dixon, pp. 41-51. Arlington, Virginia: Winrock International.

Brewbaker, J.L. and B.T. Styles. 1982. *Economically important nitrogen fixing tree species.* NFTA 82-04. Waimanalo, Hawaii: Nitrogen Fixing Tree Association.

Burley, J. 1983. Global needs and problems of collection, storage and distribution of multipurpose tree germplasm. In *Multipurpose tree germplasm. Proceedings of a planning workshop to discuss international cooperation,* pp. 29-30. Nairobi, Kenya: International Council for Research in Agroforestry.

Halliday, J. and P.L. Nakao. 1982. *Masterlist of woody species under consideration as nitrogen fixing trees.* NFTA 82-03. Waimanalo, Hawaii: Nitrogen Fixing Tree Association.

Lyman, J. and J.L. Brewbaker. 1982. *Matrix of priority nitrogen-fixing tree species by use and ecology.* NFTA 82-05. Waimanalo, Hawaii: Nitrogen Fixing Tree Association.

MacDicken, K.G. 1986. Planning and implementing MPTS field trials. In *Forestry networks. Proceeding of the first network workshop of the Forestry/Fuelwood Research and Development Project (F/FRED),* eds. N. Adams and R.K. Dixon, pp. 25-31. Arlington, Virginia: Winrock International.

Macklin, B. 1987. Nitrogen fixing trees and their planting niches on small farms. Unpublished paper. Nitrogen Fixing Tree Association, Waimanalo, Hawaii.

National Academy of Sciences. 1979. *Tropical legumes: Resources for the future.* Washington, D.C.: National Academy Press.

Nitrogen Fixing Tree Association. 1985. *Annual report.* Waimanalo, Hawaii: NFTA.

Vergara, N.T. 1982. New direction in agroforestry: The potential of tropical legume trees. *Improving Agroforestry in the Asia-Pacific Tropics.* Honolulu: East West Center and United Nations University.

Von Carlowitz, P.G. 1984. Multipurpose trees and shrubs: Opportunities and limitations. The establishment of a multipurpose tree data base, March 1984. ICRAF Working paper No. 17. Nairobi, Kenya.

Von Carlowitz, P. and J. Burley. 1984. *Multipurpose tree germplasm.* Nairobi, Kenya: International Council for Research in Agroforestry.

Yantasath, K. 1987. Field trials of fast-growing nitrogen-fixing trees in Thailand. In *Australian acacias in developing countries,* ed. J.W. Turnbull, pp. 176-79. ACIAR Proc. No. 16. Canberra: Australian Centre for International Agricultural Research.

Table 1. Number of fodder trees owned in relation to farm size.

Farm size (ha)	No. of trees
0.5 or less	7.8
0.5 - 1.0	16.7
1.0 +	25.2
Overall average	15.1

Table 2. Number of fodder trees owned per household, by districts.

District	Percentage of Households in Each Class						
	< 10	10-20	20-30	30-40	40-50	50-60	60-70
Bhojpur	62.7	19.4	10.4	6.0	1.5	0	0
Dhankuta	58.0	19.4	9.7	6.5	3.2	0	0
Sankhuwasabha	68.4	21.0	5.3	0	0	0	0
Terhathum	44.8	24.1	17.1	3.5	3.5	0	3.5

Table 3. Tree ownership by region (number of plants).

Type	Eastern	Central	West	Far West	Hill Nepal
Fodder	13.5	9.4	16.7	8.6	12.1
Fuel/timber	12.8	9.2	17.8	7.4	11.9
Fruit	2.2	2.2	1.6	3.9	2.4
Bamboo plums	5.9	0.9	1.3	0.03	1.7
Total trees	33.8	21.7	37.4	20.0	28.1
Total seedlings	53.0	27.0	36.0	10.0	30.8

regions and lowest in the center and the far west (Table 3).

The Farming Systems Research and Development Division (FSR&D), Department of Agriculture, has been doing research on agricultural crops. Elements of the improved farming system it is developing will include use of more fruit trees and incorporation of a variety of multipurpose trees in contour tree strips. Because agriculture has a good potential in the area, FSR&D in cooperation with the Forestry Research Project has included forestry, horticulture and tree crops in the research program. *Albizia procera* was planted in one of trials in the summer of 1987. The survival is excellent and the seedlings are in good condition. We tried to convince farmers to plant *Alnus nepalensis*, since international interest in the species is high, but they were not interested because it is so plentiful in the farmland and public land there.

Nitrogen-Fixing Trees on Farmlands

Nitrogen-fixing trees can enrich the soil of farmland. Since the soil is poor in the hills, the role of NFT species is more important there than in the Terai. The hill people are mostly poor and cannot afford chemical fertilizer, which is difficult to obtain because of transportation problems in some parts of the country. The importance of legumes on the farmland had been recognized by Wyatt-Smith (1982) who concluded that cash cropping with legumes for fodder and soil improvement on terraced land should be undertaken. Nepal has about a dozen indigenous NFT species, most of which are multipurpose trees. *Dalbergia sissoo* and *Albizia lebbek* are found on boundaries of small farms in the terai. *Albizia chinensis* (Syn. *A. stipulata*), *A. procera*, *Alnus nepalensis*, and *Erythrina* species are found in the farmland in the middle hills. NFTs are rarely found in the farmland in the high hills.

Indigenous NFT species and their uses are described below:

Acacia catechu (Leguminosae: Mimosoideae): Known in Nepal as Khayer, its most important uses are for Katha, which is used with betel nut, and Cutch, a medicinal product. Other uses include poles in house construction, rice pestles, oil crushers and plows. The wood is hard and heavy, weighing

960 kg/m^3, and is excellent for firewood and charcoal. The leaves are quite good fodder. It has grown rapidly at the Adabhar trial site, and transplanting of stump-cuttings allows initial fast growth.

Albizia chinensis (Leguminosae: Mimosoideae): Its common names are Rato siris and Siran. Uses include timber for light furniture and planting. Due to its light heartwood, it is not good firewood. It is a reasonably good fodder with a high crude protein content. It has been used as a shade tree for tea. Isolated trees have grown very rapidly, with a mean diameter increment of up to 5 cm. In natural forest a mean annual diameter increment of 2.7 cm has been recorded.

A. julibrissin (Leguminosae: Mimosoideae): Also known in Nepal as Rato siris, the dark brown heartwood of this tree is used for furniture. Its green wood weighs about 700 kg/m^3, and it is not good firewood. It is used for fodder and bedding, and its pink flowers make it an attractive ornamental tree. Mean annual diameter increments of 1.0-1.6 cm have been recorded.

A. lebbek (Leguminosae: Mimosoideae): Commonly known as Siris and Kalo Siris, its heartwood is used for furniture and construction timber. It weighs about 680 kg/m^3 at 12% moisture content, and makes a good firewood (about 20,900 kj/kg). Leaves are reported to be a good fodder and also make a useful green manure. The tree is a good soil binder and therefore is often planted along embankments. On good sites, the trees should reach a height of 4 m in 4 years. It is a multipurpose tree.

A. procera (Leguminosae: Mimosoideae): Heartwood of Seto siris, as it is commonly known, is used for general construction, carts and furniture. It weighs about 640 kg/m^3 and is a very good firewood and charcoal. The leaves are fairly good cattle fodder. It has been included in trial plots at Chitripani and Adabhar but with rather poor results on these sites.

Alnus nepalensis (Betulaceae): Utis is used as timber for construction in unexposed conditions. The wood is very light (specific gravity of 0.32-.37), and its calorific value is relatively low, 18,230 kj/kg. Thus it is not good firewood, but it dries rapidly and burns easily. The mature leaves are eaten by sheep and goats but not cattle. It is a good soil binder and very important for bedding and compost. It grows fast in suitable conditions. In various trials in Pakhribas at an altitude of 1900 m, a height of about 2 m was reached 16 months after planting and between 3.5 and 4 m at 28 months after planting. At Kaskikot in the Phewa Tal Watershed near Pokhara, a 6-year-old plantation produced a mean annual yield of 4.3 t/ha. Alnus was mixed with other species at this site and eventually suppressed their growth.

Butea monosperma (Leguminosae: Papilionoideae): Timber of Palas is used mainly for construction. The leaves are lopped for buffalo fodder, the red gum and the seeds are used in medicine, and the leaves are also used for making plates.

Dalbergia sissoo (Leguminosae: Papilionoideae): Sissoo is a good and valuable timber used for building, furniture, cart wheels and tool handles. It is an excellent firewood and charcoal, weighing about 780 kg/m^3, and its heartwood has a calorific value of about 21,700 kj/kg. Its height at 18 months was 3.8 m when the plots were given full cultivation, compared with only 1.3 m when weeding was confined to a circle 50 cm in diameter around the plant. Table 4 shows height growth. Transplanting stump cuttings allows initial fast growth.

Erythrina arborescens (Leguminosae: Papilionoideae): Leaves of Thekikath and Phaledo are used for fodder, and it is an ornamental tree. It does not make a good hedge.

E. stricta (Leguminosae: Papilionoideae): Phaledo wood is soft and is used for making household utensils such as ladles, pots, scrubboards and sieve frames. Leaves are used for fodder.

E. suberosa (Leguminosae: Papilionoideae): Also known as Phaledo, its leaves are used for fodder, but its wood is not used for firewood. Growth is rapid, with a mean annual diameter increment of about 1.3 cm.

E. variegata (Leguminosae: Papilionoideae): Phaledo is a poor firewood, but it is used as a live fence and leaves are used as fodder.

Erythrina species are propagated through large cuttings, up to 2 m long and 5-8 cm in diameter, that are above grazing height and allow fast early growth.

Sesbania sesban (Leguminosae: Papilionoideae): It is good firewood, and the leaves are used as fodder. Growth is very fast for this multipurpose tree.

The regional distribution of indigenous NFT species is described in Table 5.

A taungya system in the Terai has been practiced since 1977/78. Tree intercropping may be new in the hill farming system, but planting trees on the boundaries or edges of the terraces is not. The agroforestry practices that are commonly found in the area are use of shrubs for live fences around farmland and pastures and use of strips of multipurpose trees and shrubs around sloping fields (Fonzen and Oberholzer 1984). No work has been done on management and yield of these trees on farms.

Problems

The main problems of planting trees in the farmland are:

o The hill farmers are not interested in planting trees in their khet land (irrigated land) because of shade effects on crops.

o Farmlands are very small, averaging 0.5 ha per household in the hills, and fields are scattered.

o Knowledge about how to plant trees is lacking.

o The desired species are not available in local nurseries.

Table 4. Height growth of *Dalbergia sissoo* at Sagarnath.

Age (yrs)	1	2	3	4	5	6	7	8	9
Height (m)	1	5	9	11	14	15	16	17	17

Table 5. Regional distribution of indigenous nitrogen-fixing trees.

Terai (to 1,000 m)	Lower Hills (to 1,500 m)	High Hills (to 3,000 m)
Acacia catechu	*Albizia julibrissin*	*Albizia julibrissin*
Albizia chinensis	*A. lebbek*	*Alnus nepalensis*
A. julibrissin	*Alnus nepalensis*	
A. lebbek	*Erythrina arborescens*	
A. procera		
Butea monosperma		
Dalbergia sissoo		
Erythrina suberosa		
E. variegata		
Sesbania sesban		

o Generally, farmland is far from houses, hence protection of young seedlings from cattle is a problem. Protection is also needed when the trees are ready for harvest because the trees might be stolen by other farmers due to shortages of firewood and small timber.

Conclusions

Since soil or land is poor in the hills, multipurpose nitrogen-fixing trees can play a more vital role in the hills than in the Terai. There are a few NFT species in the country, some of which are MPTS. Some are fast-growing, but usually not as fast as exotics such as Leucaena leucocephala. Growth data is not available for mature trees.

However, there is a wide distribution of MPTS in farmlands, mainly in non-irrigated land. More planting should be done in both non-irrigated and irrigated land in the Terai as well as in the hills. The greater planting of indigenous NFT species and some introduction of fast-growing nitrogen fixing multipurpose trees should be done. Research priority should be given to these indigenous NFT species, with respect to propagation, establishment, growth, yield (foliage and wood) and amount of nitrogen fixed.

REFERENCES

Abell, T.M.B. 1981. *Koshi hill area development programme, fuelwood and fodder collection within the Koshi hills.* Nepal: Koshi Hill Integrated Rural Development Project.

Brewbaker, James L. 1983. *Fodder and fuelwood N2-fixing trees in Nepal.* Nepal: Integrated Cereal Project, Department of Agriculture, Ministry of Food and Agriculture, .HMGN.

Campbell, J. Gabiel and Tara N. Bhattarai. 1983. *Private tree ownership and results of seedling distribution in the Community Forestry Project.* Lumle, Nepal: Workshop on Hill Forestry.

Fonzen, Peter E. and Erich Oberholzer. 1984. Use of multipurpose trees in hill farming systems in western Nepal. *Agroforestry Systems* 2:187-97.

Ginkel, Randolph Van. 1983. *Community forestry development in Tanahun Forestry Division 1982-83.* Nepal: Community Forestry Development Project.

Hawkins, T. 1982. *Community forestry development in Dhaulagiri Forest Division 1980-81.* Nepal: Community Forestry Development Project.

Nielsen, Erling M. 1986. *Community forestry development in Arghakhanchi District 1983-85.* Nepal: Community Forestry Development Project.

Stainton, J.D.A. 1972. *Forests of Nepal.* New York: Hafner Publishing Co.

Stewart, Martin. 1983. *Community forestry development in Acham Forest Division 1980-82.* Nepal: Community Forestry Development Project.

Wyattsmith, J. 1982. *The agricultural systems in the hills of Nepal.* Nepal: Agricultural Project Services Center.

Session IV: Fruit Trees and Other Woody Perennials

Chairman: Francis S.P. Ng

Discussants: K.G. Tejwani
Mohammad Zainul Abedin

Session IV Summary

Francis S.P. Ng

Forest Research Institute Malaysia
Kuala Lumpur

Fruit trees and other woody perennials can be important sources of income for farmers. In this session, we discuss ways in which they are integrated into farming systems in various parts of Asia.

Monoculture of fruit trees and woody perennials is practiced mainly in large landholdings; the products are intended chiefly for export. Polyculture is preferred by small farmers with holdings of five hectares or less, which constitute the major proportion of farmland in Asia.

Polyculture may involve as few as two or as many as 20 species grown on the same farm. Small farmers are said to prefer polyculture because it provides more security against failure or periodic price declines in any one crop. Trees and other perennials may also require less intensive day-to-day management. Ecologists favor polyculture for soil conservation, disease control, and optimal usage of soil and space. It also provides a more varied and nutritionally balanced diet.

Still, the socioeconomics of polyculture versus monoculture are poorly understood. Data from the Philippines are presented, showing that well-designed polycultures can give much better financial returns than can monocultures. But there are vast disparities of income among polycultures caused by variations in choice of species, spacing designs, regional climate and soils, and farmer skills.

These disparities pose a challenge for biological and social scientists. If we do not deal with them, we shall have trouble extending successful models from one place to another.

Among trees, monocots such as palms and bananas can combine well with most other crops of lower height. This is probably due to the architectural design of monocots; their shape and spatial requirements remain relatively fixed. A palm crown often may even decrease with height and age. Monocots may also be more compatible with intercrops because their roots do not develop massive thickening and do not spread far. The importance of coconuts, Arenga, and other palms in the rural economies of Asia is highlighted in several papers and posters.

With dicot trees, there are the usual problems connected with their habit of changing crown shape as they grow and respond to available space. In principle, dicot trees are much more amenable to pruning, shaping, and controlling than are monocots, which is why we can make a bonsai with a dicot tree but not with a monocot. But without planning and control, dicot trees can easily upset a farm design by outgrowing their original space. Most farmers know this, and confine such trees to the boundaries of their fields; but there are other alternatives such as alley cropping.

Many dicot fruit trees, such as *Artocarpus heterophyllus*, are also good timber trees with growth rates comparable to Albizia. However, management of such trees for fruit may conflict with management for timber production. Trees managed for fruit are usually kept short and stocky with wide crowns at wide spacings. Trees managed for timber are usually grown tall and straight at closer spacings. There is some rethinking on the latter practice, however.

Some foresters now believe timber trees should be grown short and stocky at wide spacings to produce a "single log tree." They argue that what is lost in height is gained in diameter. Still, it is possible in Southeast Asia to manage tall straight trees for timber and fruits simultaneously. The durian, for example, grows to 40 m as a timber tree. It is perhaps unique in that fruits drop when ripe so the tree need not be kept low for harvesting purposes.

The role of trees and other perennials as sources of herbal medicine is reviewed. There are many examples in China and India of farmers' incomes being augmented by the cultivation of medicinal plants, many of which occur in forest understorys. Such species would be ideal for intercropping under shade. In Southeast Asia, the integration of medicinal plants into smaller farming systems is still in its infancy; most herbal medicines are collected from the wild.

The integration of trees with crops on small farms has visibly improved the quality of life among farmers in the lowlands of Nepal and elsewhere in Asia compared to the situation 10-20 years ago when forests were indiscriminately cut. With the reintroduction of trees, soil erosion has been checked and the hydrological health of streams has been restored. These environmental lessons are spreading across Asia. We can perhaps look forward to a time when trees will regain much of the ground they have lost in recent decades. Soil and water are the most important components of a farming system, but they are highly vulnerable without the protection of trees.

The Status of Perennial Cropping Systems in the Philippines

R.R. Espino, J.B. Sangalang, and D.L. Evangelista

University of the Philippines at Los Banos

The status of cropping systems of selected perennial crops is discussed in relation to planting area, total production, farm-size distribution, and existing technology. Included also is a list of research and development needed in order to fully use existing technologies. The crops involved are coconut, banana, papaya, pineapple, coffee, and cacao.

Perennial crops constitute a major portion of the national economy of the Philippines. The land area devoted to these crops constitutes 37% of the total agricultural land. Many people depend on these industries. About one-fourth of the populace relies on the coconut industry alone for a livelihood. In terms of value of production, perennials contribute approximately 38% to our economy. The Philippines also supplies the world market with a considerable number of products from perennial crops, such as coconut, coffee, cacao, banana, abaca and others.

Despite the considerable contribution of perennial crops to the well-being of the Filipino people, it is lamentable to note that generally not many research efforts have been allocated to these commodities compared to annual crops. The actual farming practices for these crops are not yet completely understood. Hence, identification of existing technologies and possible ways to improve them are required. Farmers' reactions to new technologies of perennial crop farming systems also need invesigation.

The search for answers to the above propositions can be facilitated if farming systems are treated holistically, including the farmer, other crops planted, livestock, the environment, the market, and the socioeconomic forces that influence productivity (Sangalang 1987).

Crop and Technology Status

Coconut

The area devoted to the production of coconut (*Cocos nucifera* L.) increased by 15%, from 2.73 million ha in 1977 to 3.16 million ha in 1986. However, production decreased from 4.57 million metric tons (t) in 1980 to 2.92 t in 1984. This decrease can be attributed mainly to the low productivity of existing coconut plantations due to the predominance of old coconut trees (50-60 years old). Also, from 1980-1982, the price of copra was so low that farmers did nothing to improve yields. Since any treatment to coconut trees influences yield in 2-3 years, the lack of good cultural management practices in the plantations from 1980-1982 resulted in very low coconut production in 1983-1986.

Most of the coconut farms in the country are small holdings (5 ha or less). They comprise about 79% of the number of farms with mature trees and 75% of the farms with young trees. Medium-scale farms (5-25 ha) comprise 20.6% and 24.1% of all farms with mature and young trees, respectively. Large-scale coconut farms (25 ha and above) represent only about 0.7% of the total. In terms of land area, however, small farms comprise only about 34% of the total area devoted to coconut, and farms greater than 5 ha represent about 66% of the total coconut area (Cornista and Pahm 1987).

About 70% of coconut farms are monocropped. The remaining farms are intercropped in various ways. These can be classified as follows: coconut + annual crops; coconut + biennials; coconut + perennials; and multi-story cropping under coconuts. Distinct cropping systems can be observed in various provinces of the country

(Hilario 1986). For example, the coconut + annuals cropping pattern is common in Batangas (coconut + ginger or squash, bitter gourd, etc.), Sorsogon (coconut + cassava + upland rice), Ilocos region (coconut + garlic) and Bukidnon (coconut + tomato or corn or mungbean). The coconut + biennials (papaya or pineapple) cropping system is a familiar sight in Cavite, Laguna, Camarines Norte and Northern Samar. Among the perennial crops planted under coconuts are banana (in Quezon and Leyte), coffee (in Cavite, Quezon and Batangas), lanzones or langsat (in Laguna), abaca (in Camarines Sur and Albay), cacao (in Davao), and black pepper (in Batangas and Laguna).

Multi-story cropping systems under coconuts involve various systems with annuals, biennials and perennials, with perennials consisting of one or more species as the climax species. The annual and biennial crop species are planted to provide income while the perennial intercrops are at their vegetative or early reproductive phase of development. When the canopies of the perennial intercrops are about to overlap, planting of annuals and biennials ceases. Among the multi-story cropping technologies practiced by farmers in some areas in the Philippines are coconut + coffee + banana + fruit trees in Cavite, coconut + lanzones + black pepper + banana in Quezon and Laguna, and coconut + cacao/coffee + black pepper + banana in Batangas (FSSRI and PCA 1986).

Intercropping under coconut provides additional income. At the moment, it is not yet certain which crop combination would be most rewarding in different coconut areas in the Philippines. This depends on social, environmental, and economic factors, as well as crop varieties (Sangalang 1987). Pablo (1983) estimated the net benefits derived from various crop mixes (Table 1). It can be noted that coconut + string bean + bitter gourd + squash + papaya + cacao gave the highest net benefit of P80,587 ($US1 = P20.5) followed by coconut + bitter gourd + bottle gourd + squash + banana + cacao with P67,007 net. Coconut + rice + green corn + mungbean + soursop provided the lowest benefit, amounting to P3,607.

Banana is the most important fruit crop in the Philippines. It is widely grown throughout the country for both local and foreign markets. Based on the 1980 census, farm-size distribution was as follows: <1 ha, 19.3%; 1-4.99 ha, 63.28%; 5.0-9.99 ha, 12.5%; 10.0-24.99 ha, 3.9%; and >25 ha, 0.4%, with a total of 1,071,666 farms reported (NCSO 1985). This indicates that a big proportion of farm holdings were quite small and catered to the local market. For the export market, there are large farms (>50 ha) with highly efficient production systems and intensive management practices. Among the many cultivars grown, the most popular for local usage are Saba, Lakatan, Bungulan, and Latundan. Cavendish is for the export market.

From 1977-1986, banana production experienced fluctuations. Total production increased from 2.4 million to 4.0 million t from 1977-1980. The total value of the crop nearly doubled from P1.04 billion to P2.1 billion in the same period of time. There was a slight increase in total planting area (5.7%) devoted to this crop and an increase in the average yield per unit area from 8.1 t/ha to 12.5 t/ha. These increases can be attributed to the increased demand for fresh and dried banana fruits in the foreign markets. By 1980, total production had stabilized to approximately 4.0 million t/year until 1983 and then decreased an average of 2.4% per year until 1985, followed by an upward swing in 1986. This observed stabilization of the production can be attributed to the saturation of foreign markets. It is hoped that total export of bananas will increase in coming years through the opening of new markets and the use of Senorita, Latundan, Pitogo, and others cultivars.

Bananas can be a small- or large-scale monocrop. Most often, however, bananas are grown as a minor crop by small farmers either as an intercrop, a nurse crop, or backyard crop. In a survey conducted by Angeles and Ramirez (1983), a total of 18 cropping patterns were identified involving bananas, perennial, and annual crops (Table 2). The most common cropping combination observed was banana as an intercrop under coconut, which constituted 45% of the total number of farms surveyed, representing the four types of climatic patterns in the country. The preponderance of this cropping system is not surprising since there is quite a large hectarage devoted to coconut in the country. In addition, the low income derived from coconut and the availability of space between coconut rows makes planting of other crops possible. As a nurse crop, bananas are planted with perennials such as coffee, cacao, lanzones, rambutan, and others to

Table 1. Estimated accumulated net benefit per hectare from cropping patterns with coconut over a 10-year period.

Area/Cropping Pattern	Net Benefit (Pesos)
Batangas	
Coconut + papaya + bitter gourd + bottle gourd + sponge gourd + black pepper	25,027
Coconut + pineapple + taro + yam + cacao	60,680
Coconut + string bean + green corn + banana + lanzones	38,423
Cavite	
Coconut + taro + chayote + mung bean + papaya + jackfruit	28,995
Coconut + bitter gourd + bottle gourd + sponge gourd + squash + banana + cacao	67,007
Coconut + pineapple + mung bean + eggplant + yam + taro + sweet potato + guava	51,086
Laguna	
Coconut + pineapple + string bean + ginger + black pepper	20,732
Coconut + rice + green corn + mungo + soursop	3,607
Coconut + taro + sweet potato + yam + papaya + guava	22,614
Quezon	
Coconut + taro + sweet potato + yam + lanzones	25,038
Coconut + string bean + bitter gourd + squash + papaya + cacao	80,587
Coconut + pineapple + sponge gourd + bottle gourd + string bean + chico	50,261

Source: Pablo, 1983.

Table 2 . Cropping pattern using banana in various areas of the Philippines.

Cropping Pattern	Number of Farms	Percent
Coconut + banana	36.0	45.00
Banana alone	19.0	23.75
Coconut + banana + coffee	7.0	8.75
Banana + corn	3.0	3.75
Coconut + banana + cacao	2.0	2.50
Coconut + banana + coffee +		
calamondarin	1.0	1.25
Banana + coffee	1.0	1.25
Banana + pineapple	1.0	1.25
Banana + ginger	1.0	1.25
Banana + ginger + gabi	1.0	1.25
Banana + cassava	1.0	1.25
Coconut + banana + papaya	1.0	1.25
Coconut + banana + ipil-ipil	1.0	1.25
Coconut + banana + coffee +		
langsat + rambutan	1.0	1.25
Coconut + banana + vegetable	1.0	1.25
Banana + vegetables	1.0	1.25
Banana + other crops	1.0	1.25
Coconut + banana + coffee +		
cacao	1.0	1.25

Source: Angeles and Ramirez, 1983.

provide shade during the early growing period of the perennials.

Some of the intercropping schemes with banana are presented in Figs. 1, 2, and 3 (Angeles and Ramirez 1983). In terms of economic benefits, Table 3 presents the net income per hectare of different cropping systems involving banana. Coffee + banana + pineapple + papaya had the highest net income amounting to P76,362, and banana alone had the least (P5,711). In general, banana farms are managed following the traditional system of culture. Planting with other crops is unsystematic, management is labor intensive and involves deleafing, weeding, desuckering, and male bud removal (Angeles and Ramirez 1983).

Papaya

Papaya (*Carica papaya* L.) is another fruit in the country available year round in the market. Except for 1980, the area devoted to this crop gradually decreased from 7,650 ha in 1977 to 6,250 ha in 1986. This represents an average decrease of 140 ha per annum, which can be attributed to the outbreak of papaya ringspot virus in the main growing areas in the Southern Tagalog region. Infections reached as high as 100% in most farms, resulting in severe problems for many small farmers. In 1980, the farm-size distribution of papaya farms was as follows: <1.0 ha, 24.5%; 1.0-4.99 ha, 62%; 5.0-9.99 ha, 10%; 10.0-24.99 ha, 3.0%; and >25 ha, 0.5%. This was based on a total of 220,400 farms.

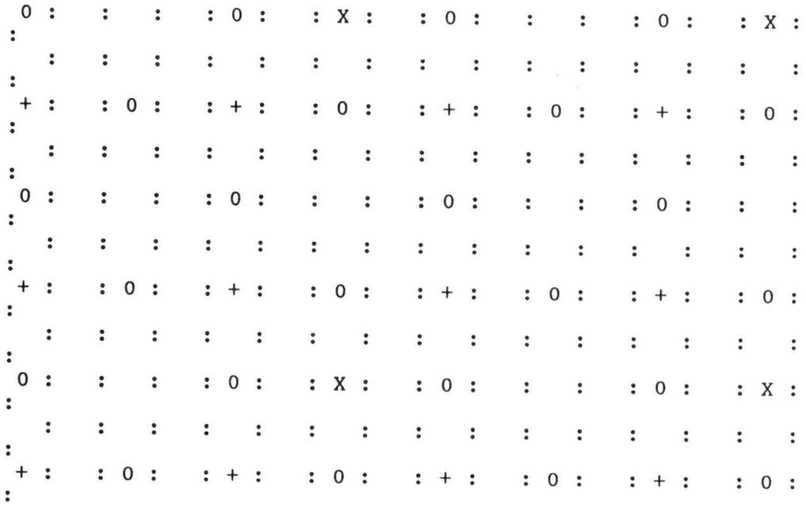

```
0 :    :    :    : 0 :    : X :    : 0 :    :    :    : 0 :    : X :
 :     :    :    :    :    :    :    :    :    :    :    :    :    :
 :
 + :    : 0 :    : + :    : 0 :    : + :    : 0 :    : + :    : 0 :
 :     :    :    :    :    :    :    :    :    :    :    :    :    :
 :
 0 :    :    :    : 0 :    :    :    : 0 :    :    :    : 0 :    :    :
 :     :    :    :    :    :    :    :    :    :    :    :    :    :
 :
 + :    : 0 :    : + :    : 0 :    : + :    : 0 :    : + :    : 0 :
 :     :    :    :    :    :    :    :    :    :    :    :    :    :
 :
 0 :    :    :    : 0 :    : X :    : 0 :    :    :    : 0 :    : X :
 :     :    :    :    :    :    :    :    :    :    :    :    :    :
 :
 + :    : 0 :    : + :    : 0 :    : + :    : 0 :    : + :    : 0 :
```

0 = coffee (3 x 3m), x = banana (6 x 6m), + = papaya (3 x 3m), : = pineapple (1.0 x 0.3m).

Figure 1. Coffee + banana + papaya + pineapple cropping system, Silang, Cavite.
Source: Angeles and Ramirez, 1983.

```
0 :   : 0 :   : 0 :   : 0
 :   :  :   :  :   :  :
 :   :  :   :  : + :  :
0 :   : 0 :   : 0 :   :
 : X :  :   :  :   :  :
0 :   : 0 :   : 0 :   : 0
 :   :  :   :  :   :  :
0 :   :  :   : + :   :  :
 :   :  :   :  :   :  :
0 :   : 0 :   : 0 :   : 0
 :   :  :   :  :   :  :
 : X :  :   :  :   :  :
0 :   : 0 :   : 0 :   : 0
 :   :  :   :  :   :  :
0 :   : 0 :   : 0 :   : 0
```

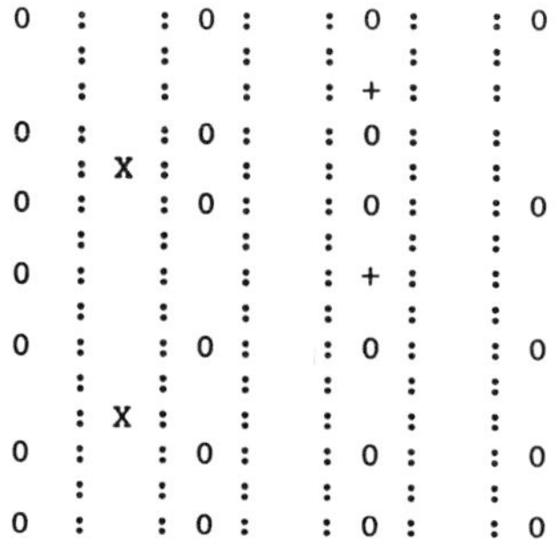

0 = coffee (2.5 x 3.0 m); x = banana (7.5 x 6.0m), + = papaya (7.5 x 6.0m), : = pineapple (1.0 x 0.15m).

Figure 2. Coffee + banana + papaya + pineapple cropping system, Tagaytay City.
Source: Angeles and Ramirez, 1983.

```
0   X   0   X   0   X   0
    X       X       X
0   X   0   X   0   X   0
    X       X       X
0   X   0   X   0   X   0
    X       X       X
0   X   0   X   0   X   0
```

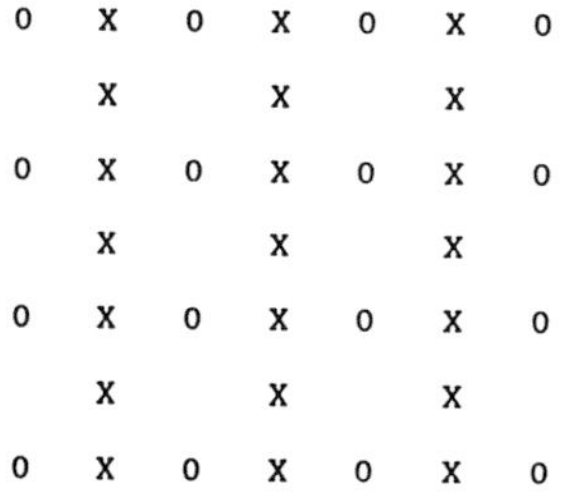

0 = coconut (7.0 x 7.0m), x = banana (3.5 x 7.0m).

Figure 3. Coconut + banana intercropping system.
Source: Angeles and Ramirez, 1983.

Table 3. Net income per hectare of various cropping systems using banana.

Cropping System	Net Income/ha
Coffee + banana	35,205.00
Papaya + pineapple + coffee + banana	67,699.10
Pineapple + banana	63,392.80
Coffee + banana + papaya	39,229.40
Coffee + banana + pineapple	15,035.00
Coffee + banana + pineapple + papaya	76,362.00
Coconut + banana	7,263.85
Banana alone	5,711.13

Source: Angeles and Ramirez, 1983.

Total production has ranged from 800,000-1,040,00 t annually. There are two distinct periods of increasing production, from 1979-1981 and 1983-1986. The former increase can be attributed to the expanding hectarage devoted to this crop, and the latter was due to higher production per unit area, with an average of 15.04 t/ha.

The main variety is Cavite Special. Its cultivation is either monoculture or in combination with other annuals or perennials. Most often, several papaya trees are grown in backyards with other perennial trees, such as coffee, cacao, and guava. In the Southern Tagalog region, especially Batangas and Cavite, papayas are planted under coconut with pineapple planted between rows of papaya. Aside from pineapple, one can find a multitude of crops being planted together with papaya in these areas, such as legumes and cucurbits.

Pineapple

Pineapple (*Ananas comosus* L.) is the second most important commercial fruit of the country. The biggest production areas are concentrated in Mindanao, where the Philippine Packing Corporation and Dole (Philippines), Inc. have extensive plantings and modern canneries.

From 1977 to 1984, the area planted with this crop increased from 36,130 to 63,070 ha, and then dropped in 1985 to 54,105 ha. The area began to increase again in 1986. Based on the 1980 census, farm-size distribution for pineapple farms was as follows: < 1.0 ha, 12.5%; 1.0-4.99 ha, 64.15%; 5.0-9.99 ha, 16.2%; 10.0-24.99 ha, 6.32%; and >25 ha, 0.82%, representing a total of 76,398 farms. The average yield is 25.73 t/ha.

Pineapple is generally planted as a monoculture. However, sizable plantings can be observed under coconuts in the Southern Tagalog and Bicol areas. These regions are the primary suppliers for the fresh fruit markets in Metro Manila and nearby population centers. The major varieties being grown are Smooth Cayenne and Queen. In addition, pineapple farms are managed following the traditional system of culture, i.e., weeding, fertilization, and flower induction using ethrel (an ethylene generating substance) or calcium carbide.

Cacao

Cacao (*Theobroma cacao* L.) production in the country has increased in the last 10 years (1977-1986) with the area of production growing from 4,350 to 15,330 ha. Most of the cacao farms in the country (70%) are considered small holdings (<5.0 ha). Twenty-eight percent of the farms are medium scale, ranging from 5-25 ha, while only about 2% are 25 ha or more. With the predominance of small farms, cacao is expected to be basically a component of a diversified farming system. Cacao trees are planted with other perennial crops, such as coffee, langsat, banana, and black pepper under coconuts in large and small plantations. Only in a few large plantations is cacao grown as a monocrop.

Coffee

There was a steady increase in the area of coffee (Coffea sp.) production from 76,180 ha in 1977 to 147,840 ha in 1986. In terms of total production, this has resulted in an average annual increase of 10.9%, except in 1982, when a sharp decline was experienced due to the drought of 1982-83. About 73% of the total coffee farms in the country are from 5 ha and under--basically small holdings. As such, most farms are managed intensively in mixed cropping systems. Medium-scale farms (5-25 ha) comprise around 27% of the total, and the remainder are large-scale farms. Coffee is mainly raised under coconut or in a highly diversified farming system. This is typified in the farms in Cavite, Batangas, and other Southern Tagalog provinces. In large corporate plantations in Mindanao, however, coffee is a monocrop.

Research and Development Status and Directions

Aware of the problems facing farmers, various government agencies have taken a major role in research and development of perennial cropping systems. Among these agencies are the Department of Agriculture, Farming Systems and Soil Resources Institute (FSSRI), Philippine Council for Agriculture, Forestry and Natural Resources Research and Development (PCARRD), Philippine Coconut Authority (PCA), Visayas State College of Agriculture (ViSCA), and the University of the Philippines, Los Banos (UPLB). Research undertakings involve the development of technology for perennial-based cropping system, screening for various annual and biennial crops under shade, integrating livestock with crop production, cropping patterns and soil management, pest management, and socioeconomic issues. In terms of funding, the bulk of the research in perennial cropping systems is devoted to coconut, cacao, and coffee. This is understandable since coconut occupies a large portion of agricultural lands in the country. Fruit-based cropping systems research has had little support and has usually been fragmented.

The following areas for research and development on perennial cropping systems are suggested:

Coconut-based Cropping System

o Development and piloting of technology transfer schemes for coconut-based cropping systems.

o Development of appropriate management systems in existing coconut-based cropping system in relation to production options, land conservation, integrated pest management, fertilization, irrigation, processing, and utilization.

o Integration of livestock into the cropping system.

o Development and piloting of village-level, coconut-based processing and utilization.

o Development of new varieties suited for various cropping system schemes.

o Determination of farm income of various cropping systems (impact assessment of introduced technology).

o Documentation and appraisal of other indigenous/existing cropping systems in relation to biophysical characteristics, crop combinations, cultural management practices, pest and disease controls, and socioeconomic aspects.

Fruit/Plantation Crops-based Cropping System

o Documentation and appraisal of existing cropping systems in relation to biophysical characteristics, crop combination, cultural management practices, pest and disease control, and socioeconomic impact.

o Development of appropriate management practices on existing cropping system in relation to production options, land conservation, integrated pest management systems, fertilization, irrigation practices, processing, and utilization.

o Development of technology transfer schemes for the fruit/plantation crops-based cropping system.

o Development and improvement of post-
harvest technology for fruits and plantation
crops.

o Development of new varieties suited for
various cropping systems schemes.

o Integration of livestock into the cropping
system.

o Determination of farm income of various
cropping systems (impact assessment of
introduced technology).

REFERENCES

Angeles, D.E. and C. Ramirez. 1983. *Survey and documentation of traditional banana growing practices in the Philippines*. IDRC-PCARRD Banana (Philippines) Project Report.

Cornista, L.B. and E. A. Pahm. 1987. *Direction of agrarian reform in the coconut sector*. Occasional Paper No. 17. Los Banos: Institute of Agrarian Studies, UPLB.

Farming Systems and Soil Resources Institute (FSSRI) and Philippine Coconut Authority (PCA). 1986. *Coconut-based farming system; Status and prospects*. FSSRI/PCA.

Hilario, F., ed. 1986. *Proceedings: Seminar workshop coconut-based farming system research and development thrust*. Cebu City: UPLB.

National Census and Statistics Office (NCSO). 1985. *1980 census of Agriculture*. National Economic and Development Authority.

Pablo, S.J. 1983. Multi-story cropping project under coconut. Paper presented during the Symposium in Coconut-Based Farming systems, June 1-3, 1983, at ViSCA, Leyte.

Sangalang, J.B. 1987. Intercropping under coconut. Lecture presented at the Training Course on Farming System Research and Development, April 1987, sponsored by FSSRI, UPLB.

Farm-level Management of *Artocarpus heterophyllus* Lam.

Eduardo O. Mangaoang and Romeo S. Raros

Department of Forestry
Visayas State College of Agriculture, Philippines

This study evaluated the growth performance of Artocarpus heterophyllus *grown as a permanent tree crop in an agroforestry farm, its response to varying slope, spacing and vegetative associations, and its effect on the growth of* Gliricidia sepium *hedgerows used as a primary ground cover. A Gliricidia-contoured agroforestry farm of 0.15 ha was planted with* Artocarpus heterophyllus *and food crops during the early stage of its development. Results show that its growth rate is comparable to well-known fast-growing tree species. At three years of age, it had produced a volume of 83.7 m^3. There were highly significant differences among slope effects on growth performance, but no pattern was identified. Wood growth was greater in stands where other trees were absent. Growth of Gliricidia hedgerows was reduced under an Artocarpus canopy.*

Origin and Distribution

Artocarpus heterophyllus Lam. is native to the rain forests of the Western Ghats of India and Malaysia. Arabs took it to the east coast of Africa, and eventually it arrived in the whole of tropical Africa and America. It is an important fruit crop in India, Burma, Sri Lanka, Malaysia, Indonesia, Thailand, Philippines, and many other tropical countries. It is now widely distributed throughout the Philippines, both cultivated and wild, at low and medium elevations. The tree belongs to the family Moraceae, which comprises about 55-67 genera and 900-1,000 species of mostly tropical herbs, shrubs, trees and sometimes vines, with 13 genera and about 150 species in the Philippines (Coronel 1983). The generic name was derived from the Greek words *Arto-carpus*, meaning breadfruit. The common name "jackfruit" came from the Indian name "jaka" or "tsjaka".

Botanical Description

Artocarpus is a small and medium-sized evergreen tree growing up to 15 m high and having a straight, cylindrical low-branched trunk that measures 0.30-0.50 m in diameter (FAO 1982). The tree species rarely reaches 20 m in height, and as a traditional fruit tree it possesses a dense, irregular or spreading crown. The bark is smooth and gray when young, becoming rough and grayish-brown and furrowed with age. It exudes a viscid and milky sap when injured. The leaves are alternate, stipulate, stiff and leathery, and elliptic-oblong in shape. They are dark green and shiny above and pale green and glabrous beneath, measuring 10-20 cm long and 3-12 cm wide (Coronel 1983).

Cultural Management

Traditionally grown as a fruit tree in orchards, Artocarpus is planted at a distance of 8-12 m between plants in a square planting design. In the Philippines, the tree species is usually grown as an intercrop with other fruit trees in the back yard. In India, it is used as an intercrop for mango and citrus and as a shade tree for coffee and black pepper. On small farms, the spaces between the Artocarpus are cultivated and planted to annual crops until the trees develop a closed canopy. Artocarpus start to bear fruit 3 to 8 or more years after planting, depending on the variety of the tree species used. Generally, Coronel (1983) observed that flowering takes place almost continuously throughout the year, so that fruits are visible in all stages of its development.

Economic Value

The fruit of *A. heterophyllus* is utilized at nearly all stages of its development. It is cooked as a vegetable when young and usually used as a dessert and flavoring ingredient when ripe. The seeds are also eaten when boiled or baked.

One of the most salient features of the tree is its strong, durable and beautifully marked timber, which polishes well. In India, the timber is used for furniture making, building materials, and the manufacture of musical instruments. In the Philippines, the wood is highly-valued for making guitars, ukuleles, and other musical instruments. The heartwood produces a yellow dye used for dying garments.

A. heterophyllus has been traditionally cultivated in the Philippines as a fruit tree in back yards and orchards. However, its potential as a perennial forest tree crop for wood production and soil conservation, as well as its suitability for small-farm use in the uplands, is still unknown. Many have said that Artocarpus could provide wood products, fruit, and a favorable environment for food crop production, especially in the critical uplands, if it received the best cultural conditions. Since little has been done in this regard, this study was undertaken.

Study Objectives

The study focused on evaluating the performance of *A. heterophyllus* grown as a permanent tree crop in an agroforest farm. The specific objectives were as follows:

o Evaluate wood volume growth of *A. heterophyllus* in comparison with selected known fast-growing tree species.

o Determine the growth response of Artocarpus to varying slopes and vegetative associations.

o Compare the growth characteristics of Artocarpus managed as a fruit and forest tree.

o Assess the growth and development of *Gliricidia sepium* hedgerows with and without an *A. heterophyllus* canopy.

Methodology

Volume Growth Measurement

A 0.15 ha farm with contoured hedgerows of *Gliricidia sepium* and varying slopes and vegetative associations was simultaneously planted with *A. heterophyllus* and food crops during the early stage of its development. Diameter at breast height (DBH), merchantable height (MH), and total height (TH) of the Artocarpus were measured two years after planting, and average wood volume/ha was calculated.

During the third year, growth measurement was taken to compare the performance of Artocarpus with the published performance of some known fast-growing hardwood tree species. Mean annual net volume increment was also calculated to determine its rate of growth when raised as a forest tree crop in an agroforest farm. Growth characteristics were also observed on Artocarpus managed as a fruit tree in orchards and as a perennial tree crop in upland farms.

Evaluation was also made to determine if different slopes and intercropping with other trees affected the growth of Artocarpus. A completely randomized statistical design (CRD) was used to test the significance of slope effects. An assessment also was made on the growth and development of the gliricidia hedgerows with and without an Artocarpus canopy. Measurements were in terms of mean stem diameter (cm) and crown density expressed in kilogram green weight.

Results and Discussion

Growth Performance

Planted along the contour and spaced at 2 x 3 m, *A. heterophyllus* can grow in three years an average DBH of 9.61 cm, MH of 6.92 m, a TH of 8.70 m, and an average wood volume of 83.67 m^3/ha with a minimum labor input for maintenance purposes. This growth compares very well with published reports on the growth of commercially known fast-growing tree species (Table 1).

Table 2 further indicates the innate capability of Artocarpus to produce a large wood volume in a few years. Published information indicates that the mean annual net volume increment/ha for *A. heterophyllus* ranges from 30-48 m^3; 30-40 m^3 for *Leucaena leucocephala*, and 25-40 m^3 for some other fast-growing trees.

Effects of Slope and Vegetative Association on Growth

There were highly significant differences among slope effects on the growth performance of Artocarpus, in terms of DBH, MH, and TH (Tables 3-5). The trees grew best at at a slope gradient of 32°, although no pattern was evident (Tables 6-8). The differences probably are due to some other unidentified factor. The key point is that Artocarpus spp. can grow well on steep slopes.

Table 1. Average growth data for selected three-year-old, fast-growing species.

Species	DBH (cm)	MH (m)	TH (m)	Volume (m³/ha)
Eucalyptus deglupta	9.60	4.85	6.78	55.54
Albizia falcataria	9.11	10.30	12.50	56.60
Leucaena leucocephala	5.09	5.72	7.35	52.78
Artocarpus heterophyllus				
2 x 3m spacing	9.61	6.92	8.70	83.67
2 x 4m spacing				62.75
3 x 3m spacing				55.78

Source: National Academy of Sciences, 1980.

Table 2. Mean annual net volume increment per hectare for selected, fast-growing tree species.

Species	Mean annual net volume increment (m³/ha)
Albizia falcataria	25-40
Gmelina arborea	25-40
Eucalyptus deglupta	25-40
Anthocephalus chinensis	25-40
Leucaena leucocephala	30-40
Artocarpus heterophyllus	30-48

Source: National Academy of Sciences, 1983.

Table 3. Analysis of variance for slope effects on dbh growth (cm) of A. *heterophyllus* three years after planting.

Source of variation	Degrees of freedom	Sum of square	Mean square	F-computed
Treatment	3	283.23	94.41	17.95*
Error	166	872.94	5.26	
Total	169	1,156.17		

* highly significant both at 1% and 5% levels.

Table 4. Analysis of variance for slope effects on merchantable height (MH) growth (m) of A. *heterophyllus* three years after planting.

Source of variation	Degrees of freedom	Sum of square	Mean sqaure	F-computed
Treatment	3	118.87	39.62	21.20*
Error	166	310.19	1.87	
Total	169	429.06		

* highly significant both at 1% and 5% levels.

Table 5. Analysis of variance for slope effects on total height (TH) growth (m) of A. *heterophyllus* three years after planting.

Source of variation	Degrees of freedom	Sum of square	Mean square	F-computed
Treatment	3	135.60	45.20	23.89*
Error	166	314.00	1.89	
Total	169	449.60		

* highly significant both at 1% and 5% levels.

Table 6. Mean dbh (cm) of *A. heterophyllus* at varying slope gradient three years after planting.

Slope gradient (degrees)	No. of observations	Total (cm)	Mean (cm)
26	38	334.53	8.80
30	31	190.13	6.13
32	75	683.60	9.12
35	26	168.77	6.49

Table 7. Mean merchantable height (m) of *A. heterophyllus* at varying slope gradients at three years of age.

Slope gradient (degrees)	No. of observations	Total (m)	Mean (m)
26	38	247.30	6.51
30	31	144.40	4.66
32	75	492.40	6.56
35	26	128.00	4.92

Table 8. Mean total height (m) of *A. heterophyllus* at varying slope gradient three years after planting.

Slope gradient (degrees)	No. of observations	Total (m)	Mean (m)
26	38	315.20	8.30
30	31	198.60	6.41
32	75	629.30	8.39
35	26	170.20	6.55

Table 9. Growth performance of A. *heterophyllus* related to other existing shade trees three years after planting.

Item	Associated with other trees	Not associated with other trees
Diameter at breat height (cm)	5.78	8.62
Merchantable height (m)	4.11	6.46
Total height (m)	5.89	8.14

Table 10. Growth and development of *Gliricidia sepium* in hedgerows under the canopy of A. *heterophyllus* and in the open at three years of age.

Item	Under canopy	Open canopy
Mean stem diameter (cm)	3.22	4.96
Crown density (kg green wt)	17.60	34.92

Wood growth/yield of the Artocarpus was greater in areas where other trees, such as *Pterocarpus indicus, Vitex parviflora,* and other overstory trees were absent (Table 9). Without other trees present, computed growth values of DBH, MH and TH were 8.62 cm, 6.46 cm, and 8.14 m, respectively. In areas where Artocarpus was grown with other trees, mean DBH, MH and values were 5.78 cm, 4.11 m, and 5.89 m. This result indicates that A. *heterophyllus* cannot grow favorably under shade or in association with bigger trees, and therefore is a suitable tree species for planting in open and denuded hilly lands.

Fruit and Forest Tree Crop

Fig. 1 shows some botanical differences in the growth and development of Artocarpus raised as a fruit tree in an orchard and as a forest tree crop in the hills of an agroforest farm. Observations indicate that as a forest crop, Artocarpus develops a long, straight and less-branched bole that is capable of self-pruning. The crown is narrow and light, a characteristic that facilitates fast growth. Branching is physiologically done in an orderly manner resulting in a nearly cone-shaped shoot form that is almost similar to pine tree species. Based on the age of the stand at three years, little can be said about its potential for fruiting compared to sites where the tree is raised as a fruit crop.

Effect of Artocarpus on Gliricidia hedgerows

The growth and development of Gliricidia hedgerows is better without an Artocarpus canopy (Table 10). Measured in terms of stem diameter and crown density expressed in kilogram green weight, mean values for Gliricidia hedgerows under the canopy were 3.22 cm and 17.60 kg, respectively. In the open, Gliricidia grew a mean stem diameter of 4.98 cm and produced a crown density equal to 34.42 kg in three years time over a 10-m long hedgerow.

147

(a)

(b)

Fig. 1. Botanical differences in the growth characteristics of A. *heterophyllus* as a (a) fruit and (b) forest tree.

Recommendations

With focus on the utilization of A. *heterophyllus* as a multipurpose forest tree in the upland farms, research must help develop appropriate farm management systems specific to given site conditions and based on existing traditional farming systems for a particular area or location. Efficient cultural operations, planting techniques, harvesting methods and other forest regulation strategies need to be determined for A. *heterophyllus* utilized as a perennial tree crop for wood and, secondarily, fruit production, in combination with various food crops in an agroforest farm (Mangaoang 1987). Research on crop diversification in steep slopes with A. *heterophyllus* as the perennial tree crop, simultaneously or sequentially planted with annual food crops, is also encouraged. Research also is needed to identify and test other potential tree species with multipurpose qualities for small-farm use. Further studies must also be done to develop useful and better methods for determining the economics of Artocarpus as a forest tree crop, primarily for wood production in an agroforest farm.

REFERENCES

Coronel, R.E. 1983. *Promising fruits of the Philippines*. Laguna, Philippines: College of Agriculture.

Food and Agriculture Organization of the United Nations. 1982. *Fruit-bearing forest trees*. FAO Forestry Paper 34. Rome.

Halos, S.C. 1983. Agroforestry: A new name for an old practice. *Scientia Filipinas*, Vol. 3, No. 1.

Mangaoang, E.O. 1987. Economic evaluation of selected upland farming systems. Unpublished MS thesis. University of the Philippines at Los Banos.

National Academy of Sciences. 1980. *Firewood crops: Shrubs and tree species for energy*, vol. 1. Washington, D.C.: National Academy Press.

_____. 1983. *Firewood crops: Shrubs and tree species for energy*, vol. 2. Washington, D.C.: National Academy Press.

Arenga pinnata: A Palm for Agroforestry

Gunawan Sumadi

Ministry of Forestry, Indonesia

Arenga pinnata *grows naturally in the forests of Indonesia and also is grown in monocultures and intercropped with food crops in agroforestry systems. The main products of this multipurpose tree are sugar, sago, saguer, and alcohol. The production and processing of these materials are increasing farmers' incomes and rural employment opportunities. Thus,* A. pinnata *is an excellent tree species for social forestry projects.*

Uses

One reason people like to plant *Arenga pinnata* is its year-round food production, especially in the dry season when other food is scarce. Young fruit is cooked as an appetizer and for use in drinks. Trees more than 15 years old produce sago in the inner part of the trunk, which people in some parts of Indonesia use as their staple food rather than rice. Sago also is used for cakes, noodles, and other dishes. In central Java, 100 households in one village have developed a home industry producing sago. They process 2,000 kg of sago from 20 trees every day. Each tree costs approximately Rp. 10,000-20,000 (US$1 = Rp. 1,641), and the sago from one tree will earn an income of Rp. 380,000. In North Sulawesi, one stem produces nearly 200 kg of sago, with solitary palms producing more than trees planted in groups. Since sago is obtained only by cutting trees, it is usually the last product obtained from them. In most cases, trees are cut for sago when they are more than 30 years old.

Sweet palm sap is another product that trees usually begin to produce 12-15 years after planting, although it can take as long as 35 years in rare cases. Based on production in North Sulawesi, fairly good trees will yield 35-40 liters palm sap every day, which contains 20% alcohol after one day. Saguer, a drink made from this sap, is used for traditional ceremonies in many parts of the country. Saguer also can be processed into palm sugar. In North Sulawesi, a 10 x 10 m spacing produced the highest yields. The estimated yield of various sweet palm sap products are listed in Table 1.

Table 1. Estimated potential income in Rupees from a one-hectare plantation (256 trees) of *Arenga pinnata*.

Yield/day	Unit Cost	Yield/ ha/day	Yield/ ha/day	Yield/ ha/50 days
Saguer palm sap (15 bottles)	50	750	192,000	9,600,000
Sugar (3 kg)	300	900	256,000	11,520,000
Alcohol (3 bottles)	350	1,050	268,000	13,440,000

Villagers also use other parts of the tree for various purposes. Leaves are 7-8 m long and used for decoration, making brooms, matting, baskets, roofing, and for fuelwood. The youngest leaves are sometimes used as cigarette paper. Leaf sheath fibers are used to make ropes, brooms, brushes, and concrete reinforcement. The outer part of the trunk is very hard and used for barrels, flooring, furniture, and tool handles. Other uses of the species include honey production and medicinal products from its roots.

Recommendations

Arenga pinnata is rarely attacked by pests or disease. Because it takes about 12 years to produce economic returns, this species is a long-term investment. A taungya system has been implemented to produce shorter-term income for villagers. *A. pinnata* is thus an ideal multipurpose tree for social forestry projects in Indonesia.

Impact of Multipurpose Trees on Small-Farm Systems of Nepal: A Case Study of Karmaiya Village

Madhav B. Karki

Institute of Forestry, Tribhuvan University
Pokhara, Nepal

The four-year impact of the Farm Forestry Project involving plantations of multipurpose trees on the farming systems of Karmaiya, a village in the central Terai of Nepal, was assessed. A 20% sample of households (34) was surveyed with questionnaires about on-farm plantations and the impact of such plantations on small-farm systems. Generally, on an average land holding of 1.08 ha, small farmers planted and maintained an average of 60 trees, of which Dalbergia sissoo *Roxb. ex DC. and* Leucaena leucocephala *(Lam.) de Wit constituted an average of 58.3 and 26.7% of the total, respectively. Ninety-one percent of the sample households reported no need to change their cropping system to accommodate trees on their farms; seventy-nine percent said that tree plantings allowed them better control of over-grazing and decreased dependency on forests. Farmers also noted better control of streams (82%) with no change in requirements for irrigation water (100%). Opinion was divided on the impact of on-farm trees on farm yields, with 53% saying that on-farm trees contributed to a decrease in farm yields. Nevertheless, most farmers (79%) reported that tree planting helped improve the general quality of life. Small farmers perceived greater benefit than harm from on-farm multipurpose trees.*

Nepal is predominantly an agricultural country. About 91% of the total population depends on agriculture and related activities for a living (CBS 1986). Forests cover about 38% percent of the total land area (LRMP 1986), but revenue from the forest has been continuously declining for the last 10 years. The production of major crops also has remained stagnant or even declined in the last two decades despite heavy investment in irrigation, fertilizers, and technical manpower training. The demand on the unmanaged forest and shrubland is far in excess of what it can now support on a sustained basis (Nield 1985). Wyatt-Smith (1982) estimated that about 3.5 ha of unmanaged forest was required to supply adequate fuelwood, fodder, and timber for one hectare of farmland in the western mid-elevation hills of Nepal. With the total land under "unmanaged" forests or shrublands at 6.3 million ha, an additional 4 million ha of forest lands are required (the total land under agriculture is 3 million ha). Therefore, the alternative is to grow more trees on farms. Since 88% of the farms in Nepal are characterized as small with under 1 ha per household (APROSC 1978), growing multipurpose trees on small farms is needed. Since hill farmers have traditionally been planting trees with crops on farmlands, this task should not be difficult to accomplish. The results of such an effort in a village in the Terai is reported in this paper.

The Study Area

Karmaiya Village Panchayat is located at the extreme northwest corner of Sarlahi District in the Central Region south of Kathmandu. The Panchayat is bisected by the east-west Mahendra Highway. Ward no. 5 of this Panchayat was selected to run pilot extension activities of the Farm Forestry Research and Development Project (FFP) jointly funded by Tribhuvan University, Nepal, and the International Development Research Centre (IDRC) of Canada. This village is situated in the Siwalik Hills at an altitude of about 200 m above sea level. The original vegetation of the area consisted of *Shorea robusta, Dalbergia sissoo, Acacia catechu, Adina cordifolia*, and *Terminalia* spp. This vegetation has been cleared for agriculture. Maize, rice, millet, wheat, and mustard are the major crops, and buffalo, cows, and goats are the major livestock. Due to population pressures, neighboring forests are being cleared for more agricultural land and forest products. As a result, the slopes are being eroded by heavy precipitation (2,000 mm per year) and silt suspended in the seasonal streams is deposited on farms by regular floods.

The village included 175 households of which 83% belonged to upper caste Brahmins and Chettris; 62% of the people here were literate. The average land holding of the sampled households was 1.09 ha. The estimated number of livestock units per household was 7.52 and the average family size was 6. Production per household was 1.8 ton cereals, 0.16 ton legumes, and 0.097 ton oilseeds.

FFP activities launched in 1983 included plantation of multipurpose trees. The villagers themselves took the initiative to plant trees. The project supplied suitable tree seedlings, and instructed them in planting methods and follow-up measures. Different planting patterns were developed with the help of the farmers and included border, block, stream-side, roadside, and mixed plantings. Planting was carried out every year from 1983. As the village lacked drinking water, a multipurpose water supply scheme was also implemented with active participation of the villagers. Besides Leucaena and Dalbergia, other trees planted were *Eucalyptus camaldulensis*, *Acacia catechu*, *Cassia siamea*, *Garuga pinnata*, *Bauhinia* spp., and *Albizzia* spp.

Farming System Model

The farming system has been defined as the combination of distinct functional enterprises, such as crop, livestock, and marketing activities (under a single farm unit), which interact because of the joint use of inputs they receive from the environment (Harwood 1979). Due to limiting factors of land, labor, cash, and technology, small farmers practice a complex farming system to maintain their subsistence living. Their fundamental strategy is to provide enough food for the family and forage for animals. A secondary strategy is to improve labor efficiency (Karki 1982). Each land use type has a specific importance. The small farmer's survival depends heavily on how well he combines inputs from various components to maximize production on his farm. System components such as grazing land, wasteland, and marginal land are also sources of negative effects that can be mitigated by strengthening positive effects of other components. This paper reports the results of a study on how farmers perceived the effects of tree planting on these various components.

A questionnaire survey was carried out in August, 1987. Thirty-four households (20% of the total) were randomly chosen as respondents. The questions were both open-ended and closed, and included structured and unstructured types. Questions were pre-tested on the study area, and post-sampling was carried out to verify the information collected. The data was compiled and discussed in the light of the benchmark information collected four years ago.

Results and Discussion

Each of the sampled households had an average of 60 multipurpose trees on their farms (Table 1). The most common were *Leucaena leucocephala* (an average of 16 per farm) and *Dalbergia sissoo* (34 per farm). These trees were mainly used for fuelwood, fodder, and the control of seasonal streams. Ninety-one percent of the sampled households listed Leuceana as the most preferred fodder species and 79% mentioned Dalbergia as the most popular fuelwood. The majority of households (82%) also reported that after planting multipurpose trees on farms, the stream flow was stabilized (Table 2). Similarly, 79% of the people felt that uncontrolled grazing practices prevailing in the village before tree planting began had been controlled due to the community concensus for tree planting. An overwhelming number (91%) of the households did not feel any need to change the cropping pattern to accommodate trees on their farms. However, opinion was divided on the question of whether there had been any change in crop production. About half (53%) felt that there had been an impact but the rest (47%) did not. The majority (79%) of the households said that there had been improvement in the general quality of village life from that of four years ago. All the farmers surveyed felt that there was not a greater demand for irrigation water, wood extraction from the forest was gradually decreasing, and that there was a greater supply of fodder and fuelwood from farms.

The factors that curtail food production in the small farms of the world are virtually unlimited in number and variety (Harwood 1979). Small farmers want stability and security and hesitate to take risks. Convincing them to add multipurpose trees as a new component to their farming system can be difficult. However, in the Karmaiya village, proper understanding of the farmers' felt needs and successful motivation resulted in high acceptance of tree planting. Initial concerns of the farmers regarding the so-called negative impact of the

Table 1. Number of trees planted by size of holdings in Karmaiya village of Nepal's Central Terai.

Size of Holding (ha)	Average Holding (ha)	Household Number	On-farm Trees		
			Dalbergia	Leucaena	Total
< 0.05	0.01	1	1	0	7
0.06-0.50	0.34	9	12	7	21
0.51-1.00	0.84	9	24	20	54
1.01-1.50	1.25	6	22	18	51
> 1.51	2.09	9	36	22	75

Table 2. Farmers' perceptions of impacts of tree planting project on their farming system and village life.

Impact	Response		
	Yes	No	No Opinion
Need to change cropping pattern	3 (9%)	31 (91%)	
Reduction in crop yield	18 (53%)	14 (41%)	2 (6%)
Stream/siltation control	28 (82%)	6 (18%)	
Wanton grazing control	27 (79%)	7 (21%)	
Greater need for irrigation	0 (0%)	34 (100%)	
Greater dependency on fuelwood from forests	0 (0%)	34 (100%)	
Greater supply of fuelwood/fodder from on-farm trees	34 (100%)	0 (0%)	
Impoved quality of life	27 (79%)	3 (9%)	4 (12%)

trees, such as the reduction of food crop production, were proven baseless. This resolution of their doubts contributed greater momentum to tree planting. Those who had not planted trees in the first year have also been successfully motivated by the results they have seen.

Leucaena and Dalbergia were the most accepted tree species for farm forestry. Both are fast-growing, multipurpose legumes that provide fodder and fuelwood while ameliorating soil infertility. *Dalbergia sissoo* is an indigenous tree whose merits and propagation techniques are common knowledge, but the benefits of Leucaena have been underscored by FFP demonstration plots.

The effect of trees on stream control was perhaps the most striking gain perceived by the small farmers. When the upper reaches of stream banks were planted with various trees, bamboos, and grasses, the dramatic decline in seasonal siltation by flash floods inspired the farmers to continue planting trees on farms.

Wanton grazing of livestock was often listed by the small farmers as the major disincentive against tree planting. However, with the increase in the plantation of trees on farms of each household, the problem of free grazing has been greatly reduced. Small farmers expressed no need to change their traditional cropping pattern, possibly because the intensity of on-farm tree plantation (60 trees per ha) was too small to induce significant negative impacts on food production. In fact, small farmers perceived an improvement in the general quality of life. One reason for this feeling of well-being was the availability of drinking water with the assistance of the FFP and the villagers. Other factors were the contribution of trees to the mitigation of stream erosion and siltation, easier fodder and fuelwood supply, and improvement of soil fertility.

The aforementioned farm forestry activities constitute a new development, possibly the first of its type, in the Terai region of Nepal. The results have been highly demonstrative and widely referred to by others working in the region. In only four years, several multipurpose trees (over 50 species) have been screened for adoption as useful components of the farming system. Moreover, these trees have already become productive. Farm forestry, an age-old tradition of Nepalese hill farmers, can be successfully adopted in the Terai region.

Conclusion

Nepal's farming system is highly dependent on inputs from trees. This critical forestry-farm linkage has been underscored in recent years by many prominent foresters and scientists (Mahat 1987, Nield 1985, and Wyatt-Smith 1982). It is generally accepted that current land under forests is insufficient to meet the demands of small farms for fuelwood, fodder, bedding material, green manure, and timber for construction and small tools. In this context, farm forestry or tree plantations on farms was considered necessary for the sustenance of Nepalese farmers and to conserve the already degraded forests of the country. To popularize the concept and practice of farm forestry, selection of appropriate multipurpose trees and the silviculture techniques to grow them are as important as community participation.

The example and experience of FFP in Karmaiya can be emulated by farmers and resource managers of small farms throughout Nepal. Although some tailoring may be necessary in selection of multipurpose trees and their silvicultural requirements for each village, certain general lessons are worth mentioning. First, farmers' aversion to risk must be recognized to avoid disturbing the fragile stability of the farming system. Second, after understanding the villagers' priorities through surveys, the most pressing priorities must be taken up concurrently with project activities to win people's trust and motivation. One must also appreciate that farmers are selective and careful in committing to tree planting on farms. Therefore, full acceptance of tree farming technology is followed by a trial period of their own with a few proven trees and practices. It is important to constantly monitor and evaluate farm trees and silvicultural techniques with the feedback from farmers to maintain the momentum of tree planting. Farm planting of trees is a complex operation requiring adequate understanding of socioeconomic conditions of the village as well as appreciation of the existing farming system. Finally, the increased use of multipurpose trees in small-farm systems of Nepal can improve and enhance the stability of traditional farming systems.

REFERENCES

Agricultural Projects Service Centre (APROSC). 1978. *Agrarian reform and rural development in Nepal*. Country Review Paper. Kathmandu, Nepal.

Central Bureau of Statistics (CBS). 1986. *Statistical pocket book*. National Planning Secretariat. Kathmandu, Nepal.

Harwood, R.R. 1979. *Small farm development: understanding improving farming systems in the humid tropics*. Boulder, Colorado: Westview Press.

Karki, M.B. 1982. An analytical approach to natural resources planning in Phewa Tal Watershed of Nepal. Master's thesis. Colorado State University, Fort Collins.

Land Resource Mapping Project (LRMP). 1986. *Forestry-land use report*. Mimeograph, Topgraphical Survey Branch. Kathmandu, Nepal.

Mahat, T.B.S. 1987. *Forestry-farm linkages*. Occasional Paper, No. 7. Kathmandu, Nepal: International Center for Integrated Mountain Development (ICIMOD).

Nield, R.S. 1985. *Fuelwood and fodder: problems and policy*. Working Paper. Water and Energy Commission Secretariat, Ministry of Water Resources, HMG/Nepal.

Wyatt-Smith, J. 1982. *The agricultural system in the hills of Nepal–the ratio of agricultural to forest land and the problem of animal fodder*. Occasional Paper No. 1. APROSC: Kathmandu, Nepal.

Medicinal Uses of Selected Fruit Trees and Woody Perennials

Saturnina Halos

National Science Research Institute,
University of the Philippines, Diliman, Quezon City

Herbal medicine practice is indigenous to South and Southeast Asia and is useful to farmers living in locations far from medical services. The most extensive study on medicinal plants in the Philippines lists 838 species with 262 tree species, including 31 fruit tree species. This study is derived mostly from surveys/reports on their efficacy by actual users and on their chemical composition. Later studies describe biological tests to detect anticancer, antimicrobial and mutagenic activities. Nine of 35 tree species tested had positive anticancer activity. Three of these are common fruit trees (Annona squamosa, Persea americana and Carica papaya). One is nitrogen-fixing, (Casuarina equisetifolia), and two are leguminous (Piliostigma malabaricum and Caesalpinia pulcherrima). Voacanga globosa also has been added to the list of trees with anticancer substances, based on a report on actual remission and inhibition of induced cell division. Exceptionally high antibacterial properties have been recorded for several tree species. Of note are the fruit trees Psidium guajava, the extract of which is active against dental pathogens, and Syzygium cumini, which is effective against the staph skin pathogen. The oleoresin of Canarium luzonicum has been identified in studies as mutagenic.

Medicine from Plants: Acceptability and Practice

Herbal medical practice is part of the culture of South and Southeast Asia. Countries in the region often have common uses for the same species of plants, but the art of healing appears to differ. In the Philippines, the more popular herbal healers are also more known as spiritual healers in that their powers are claimed to come directly from divine intervention: prayers are said during the healing session, after which prescriptions for herbs are given. Knowledge of the medical value and dosage formulation is obtained by the healer through a dream or trance. The healer can recognize the plant on sight although he or she may not even know its name. Healers are often as young as 13, have had no access to published literature, and often are not well educated.

Consequently, the bias against herbal healing has been well-entrenched among the formally schooled under strong Western influence, despite the fact that some people are healed. In the 1950s and 1960s, herbal healing was illegal. School children were taught that herbal and spiritual healers were charlatans who deserved to go to jail. This attitude began to change in the late 1970s after a government team went to the People's Republic of China to study herbal medicine and a national research program on medicinal plants was organized. "Herbolarios" (herbal healers and "hilots," or midwives) have now finally attained their place in primary health care through training that has integrated them into the system. Herbal medicines are now sold openly and the government is experimenting with syrup and tablet formulations of the well-studied plants. As of 1985, four species were producing medical syrups and tablets.

Medicinal Value of Selected Trees

Trees of medical value fall into three categories: those identified in the Philippine National Formulary as clinically tested for toxicity and dosage effectivity, those whose efficiency has been attested through a survey of herbal healers and from reports in medical literature, and those identified through phytochemical and biological screenings. Table 1 enumerates the formulary-listed species. Understandably, due to the short period of time so far involved in research and clinical trials, the list is brief: only 86 entries, which is about 10% of the first compilation of information on Philippine medicinal plants which listed 838 plants, 31 of which are common fruit trees (Quisumbing 1950). A later survey identified a total of 1,217 plants (Quintana 1979).

Table 1. Food trees with medicinal properties.

Species	Edible part	Therapeutic properties	Other claimed properties	Recent lab findings
Anacardium occidentale	fruit seed	bark and leaf infusion relieves toothache, sore gums, and throat.	antiscorbutic, anaesthetic, antidiabetic, antidiarrheal, insecticidal, diuretic.	
Canarium luzonicum (Blume) A Gray and C. ovatum Eng.	seed	seed for laxative; oleoresin for arthritis, boils, and abcesses.		oleoresin mutagenic
Citrus aurantifolia (Christm) Swingle C. decumana L. C. nobilis Lour.	fruit	leaves and rind as inhaler to relieve nausea and fainting.	C. aurantifolia — antihelminthic	C. decumana — antimutagenic 5 C. nobilis — antimicrobial 7
Citrus aurantium L.	fruit	rind decoction for gas, pain relief.	carminative, astringent, depurative.	
Citrus microcarpa Bunge	fruit	juice for cough and sore throat, skin bleach.	antiscabious, carminative, refrigerant.	
Mangifera indica L.	fruit	leaf infusion for cough; bark or kernel decoction for diarrhea.	antirheumatic, antiscabious antiscorbutic, antisyphilitic, laxative, diuretic	
Moringa oleifera Lam.	leaves fruit flower	cooked leaves and fruit as laxative; powdered, roasted seed for rheumatism; cooked leaves to promote lactation.	antivenomous, diuretic, anti-anemia, antihelminthic	antimicrobial 8 seed extract-mutagenic
Persea americana Mill.	fruit	seed or bark powder for rheumatism and neuralgia.	antihelminthic, emenagogue, resolvent, anti-inflammatory.	antimicrobial 8 antimutagenic 6
Psidium guajava L.	fruit	leaf decoction for diarrhea, skin ulcers, and vaginal wash after child-birth; leaves chewed for toothache.	astringent, antirheumatic, antiscabious, antispasmodic, tonic.	antimicrobial 8 antimutagenic 5
Pithecellobium dulce Roxb. Benth	fruit	bark decoction for loose bowel movement.		antimicrobial 8 antimutagenc 5
Syzigium cuminii (L.) Skeels	fruit	fruit for diarrhea, bark decoction for gum inflammation, gingivitis, or astringent wash.	antidiabetic, diuretic for tonsillitis.	antimicrobial 8 antimutagenic 5
Tamarindus indica L.	fruit	fruit preparation for antipyretic and laxative; leaf decoction for aromatic.	antihelminthic, anti-asthmatic antiscorbutic antirheumatic febrifuge.	

Source: Philippine National Formulary, 1980.

The Formulary also recommends a shorter list of therapeutic properties compared with the list obtained from the surveys with herbal healers. This indicates that for some therapeutic properties clinical tests have not yet been made or that tests have failed to detect such a property. In the latter case, either the tests are unsuitable or the plants really do not have such therapeutic properties. However, laboratory tests are also indicative of the efficacy of a plant. For example, when a plant used by an herbal healer to cure dysentery is found to contain antimicrobial properties, then one can be more confident of its capacity to cure such ailments.

Of the food tree species listed in Table 1, all bear fruits that are eaten raw except *Moringa oleifera*, which is cooked as vegetable, and Canarium, which has a seed kernel eaten raw or candied. Two are traditional and very good firewood species, *Psidium guajava* and *Pithecellobium dulce*. Of the two, *P. guajava* is early maturing, with some varieties yielding fruit 18 months after planting. It is also grown in plantations to support the fresh fruit market as well as fruit juice concentrate factories. Both species prefer moist areas. The Canarium species produce Manila elemi, a commodity in the international market of gums and resins. The wood of *Tamarindus indica* is preferred for chopping boards. The fruit of *Syzygium cumini* can be processed into an excellent wine. In the Philippines, this tree grows wild on dry mountain slopes. The other qualities of *Artocarpus heterophyllus* have been described in a paper by Mangaoang and Raros (this volume). *Moringa oleifera* is a useful tree on small farms. Its leaves are a very good source of dietary iron and, along with the fruits and flowers, are eaten cooked as vegetables. It is a very popular backyard plant, one of a few with many claimed therapeutic properties.

The only species in this list with a good timber value is *Mangifera indica*. More importantly, it fruits profusely and its fruit (Carabao variety) has a very good market. It matures late, about five years for seed-derived plants and about three years for grafted plants, but produces fruit for decades. Some farmers consider it a good hedge for old age. The fruit oil of *Persea americana* is used in cosmetics. Of the citrus species included, *C. macrocarpa* and *C. decumanus* are also plantation crops in the Philippines. *C. macrocarpa* is a common backyard plant and is an ingredient in a variety of Filipino dishes. *Anacardium occidentale* is a reforestation crop in the Philippines. Its shell is known to yield a high-value fine industrial oil.

Table 2 lists food trees, whose medical efficacy based on Western standards is not yet proven but which are used by herbal healers (Quisumbing 1950, Quintana 1979). Of these, *Bixa orellana* is most popularly used (Quintana 1979). *B. orellana* yields a dye from its seed coat that gives food a yellow-orange color. It is a good source of additional income since its seeds store well and command a good price. It is also medically interesting, its extract having been found to exhibit anticancer activities. *Averrhoa bilimbi* L. contains an antifertility factor, which is needed to help reduce the high population growth rate in the Philippines. *Annona squamosa* produces an anticancer factor and a popular fruit.

Piliostigma malabaricum is not extensively used for food, but yields an anticancer factor. It is a good fuelwood species and thrives in poor, dry areas. It is also fire-resistant. *Sandoricum koetjape* produces a popular fruit and a fine wood suitable for balusters and wood carvings. Its fruit is used to make a refreshing, tasty drink, a remedy for common colds that in our experience works.

Table 3 provides a list of tree species found to yield extracts with anticancer activity. I have focused here on cancer, a tricky and expensive disease. In the Philippines, one program of chemotherapy could cost as much as 400-500 cavans (50 kilos each) of palay.

Although *Ervatamia pandacaqui* has been the most studied among the plants enumerated here, *Voacanga globosa* leaves are the remedy prescribed by one healer I know. This same healer's patient was apparently cured of breast cancer with a concoction of leaves of this plant. The difficulty with cancer research, however, is the expense involved in testing a drug. Clinical trials must be preceded by extensive laboratory tests. Human test subjects either must have strong faith and courage to forego currently prescribed medications or they simply have no money for hospital treatment, as in the case mentioned above.

Vitex negundo is formulary listed as a cure for headache, wounds, ulcers, fever and coughs. Medical preparations in syrup and tablet forms are now available on a pilot scale. In China, its volatile

Table 2. Additional food trees with medicinal properties based on folk usage.

Species	Edible part	Therapeutic properties	Lab findings
Achras sapota L.	fruit	antidiarrheal, febrifuge, antipyretic, antidysenteric.	antimicrobial 8
Annona muricata	fruit	emetic, febrifuge, anti-spasmotic, antiscorbutic, diaphoretic.	
Annona squamosa L.	fruit	antihelminthic, antidyspeptic, astringent, tonic.	antimutagenic 6 anticancer 9
Annona reticulata L.	fruit	antihelminthic, antidysenteric, astringent, tonic.	
Antidesma bunius L. Spreng.	fruit	antisyphilitic, refrigerant, diaphoretic.	antimicrobial 8
Artocarpus altilis (Parkins) Fosb. and A. camansi Blanco	fruit	antidysenteric, vulnerary.	antimicrobial 8
A. heterophyllus Lam.	fruit	antivenomous, astringent, demulceant, febrifuge, laxative, refrigerant, vulnerary.	
Averrhoa bilimbi L.	fruit	antipyretic, antiscorbutic, antihemorrhagic, anti-syphilitic.	antifertility 3 antimutagenic 5
A. carambola L.	fruit	anti-asthmatic, anticolic, antiscorbutic, emenagogue.	
Bixa orellana L.	seeds	antipyretic, purgative, astringent, febrifuge, antidysenteric, diuretic.	antimicrobial 8 anticancer 9
Calocarpum sapota	fruit	antidandruff, remedy for heart afflictions, colds, renal colic, vomitive, sedative, antihelminthic.	
Chrysophyllum cainito L.	fruit	tonic, antidysenteric, antihelminthic, poultice, antidiabetic, antipyretic.	anticancer 9
Garcinia mangostana	fruit	antidiarrheal, astringent, antidysenteric, diuretic, febrifuge, stimulant.	antimutagenic 6
Lansium domesticum Corr.	fruit	antihelminthic, antipyretic, antispasmotic, treatment for women's diseases.	
Muntingia calabura	fruit	antispasmodic, emollient.	antimicrobial 8
Nephelium lappaceum	fruit	antidysenteric, astringent, febrifuge, carminative, narcotic, refrigerant.	

Table 2. Continued.

Species	Edible part	Therapeutic properties	Lab findings
Piliostigma malabaricum	leaves	antidysenteric, liver problems, headaches and fever.	anticancer 9
Sandoricum koetjape	fruit	antispasmodic, antidiarrheal, tonic, carminative, astringent.	
Sesbania grandiflora	flowers	antirheumatic, expectorant, diuretic, emetic, laxative.	
Spondias purpurea	fruit	astringent, antidiarrheal, antidysenteric.	antimicrobial 8 antimutagenic 6
Syzygium jambos	fruit	digestive, stimulant, antidysenteric, febrifuge, for sore eyes.	
Syzygium lalaccense	fruit	for thrush, antidysenteric, abortificant, diuretic, for amenorrhea.	

Table 3. Non-food tree species that yield extracts with anticancer properties.

Species	Remarks
Ervatamia pandacaqui Pich. (Tabernamontana pandacaqui Porr.)	Promising source of anticancer alkaloids.
Voacanga globosa (Blanco) Merr.	Of recent use by folk healers; one known case of complete remission from breast cancer; selective antimicrobial activity; prevents induced leukocyte division.
Vitex negundo L.	Formulary listed 12, extensively used to treat headaches, coughs, wounds, ulcers, and fevers; antibacterial properties 1.
Cinnamomum zeylanicum Blum.	Numerous folk medical uses, including antityphoid, antidyspeptic 10.
Lagerstroemia speciosa (L.) Pers.	Extensive use as contraceptive, for coughs, diabetes, diuretic, kidney trouble, wounds, antimutagenic 10, 11, 6.
Alstonia scholaris (L.) R.Br.	Antimicrobial 1, 8; antimalarial; antirheumatic; antidysenteric; and other uses 10.
Casuarina equisetifolia L. Forst.	Antiberiberi, antidysenteric, anticolic 10.
Erythrina variegata L. var. orientalis	Anti-asthmatic, narcotic expectorant 10.
Hopea plagata (Blanco) Merr.	
Hopea foxworthyii Elm.	
Albizia falcataria (L.) Fosberg	
Pinus merkusii Jungh de Vr.	
Callicarpa formosana Rolfe	
Alphitonia philippinensis Brard	
Goniothalamus amuyon (Blanco) Merr.	
Caesalpina pulcherrima (L.) Sw.	

Source: Halos and Catipon, unpublished.

oil is dispersed in soft elastic capsules for chronic bronchitis and as a sedative for asthma and coughs (de Castro 1980).

Other interesting trees are two dipterocarp timber species, *Hopea plagata* and *H. foxworthyii*, *Pinus merkusii*, and *Albizia falcataria*, a fast-growing leguminous agroforestry crop for paper pulp production.

Another species easily adaptable as an added income source in a small farm situation is *Cinnamomum zeylanicum*. Its leaves are harvested, air-dried, and sold as spice. It also reportedly cures a number of ailments. *Lagerstroemia speciosa* and *Alstonia scholaris* are also popularly used in medications for various ailments (Quintana 1979). Several therapeutic powers are also attributed to *Erythrina variegata*, a leguminous, fast-growing, and easily propagated species.

Conclusion

It is not the purpose of this paper to argue for extensive research in medicinal plants. It is, however, my aim to bring to your attention the medicinal possibilities offered by trees. For a while, multipurpose trees meant only nitrogen-fixing forage and fuelwood species. Such uses may or may not be attractive to a rural dweller. But it is often easy to talk to a farmer, and especially to his wife, about medicinal plants. The other issue that I would like us to focus on is this: shall we design forestry programs to revolve around the needs of an agroforest or farm family? Such programs would definitely differ from fuelwood programs and the like, for in the former we would focus on human needs of food, medicine, and shelter. In the latter, although we do talk of social forestry, we are more concerned with the production of the wood commodity itself.

REFERENCES

Beloy, F.B., V.A. Masilungan, R. M. de la Cruz, and E.V. Ramos. 1976. Investigation of some Philippine plants for antimicrobial substances. *The Philipp. J. Science* 105(4):205-211.

de Castro, N.F. 1980. Scientific travelogue--The People's Republic of China. In *Proc. Symp. Medicinal Plants*. Taguig, Manila: National Research Council.

Herrera, C.L., P.M. Cuasay, E.V. Ramos, E.P. Chavez, L.A. Dayap, and B.C. Rabang. 1986. Preliminary studies on the antifertility activity of *Anerrhoa bilimbi* L. *The Philipp. J. Science* 115(4): 307-316.

Land, T. 1987. Desert tree with large promise. *Development and Cooperation*, No.3, p. 24.

Lim-Sylianco, C.Y. 1986. Antimutagenic effects of expressions from twelve medicinal plants. *The Philipp. J. Science* 115 (1):23-30.

Lim-Sylianco, C.Y., J.A. Concha, A.P. Jocano, and C.M. Lim. 1986. Antimutagenic effects of eighteen Philippine medicinal plants. *The Philipp. J. Science* 115(4):293-298.

Manalo, J.B., V.Q. Coronel, and F.B. Beloy. 1982. Some local essential oils with antibacterial activities for pharmaceutical dosage forms. *NSTA Tech. J.* 7(3):69-71.

Masilungan, V.A., J. Maranon, V.A. Vicenta, N.C. Diokno, and P. de Leon. 1955. Screening of Philippine Higher Plants for antibacterial substances. *The Philipp. J. Science* 84(3):275-299.

Masilungan, V.A., R.N. Relova, and J.S. Raval. 1967. Screening of Philippine plants for anticancer activity. *The Philipp. J. Science* 96(4):393-398.

Quisumbing, E. 1950. *Medicinal plants of the Philippines*. Technical Bulletin No.16. Department of Agriculture and Natural Resources, p. 740.

Quintana, E.G. 1979. Survey of folk use of potential medicinal crops. In *Proc. PhilAAS 28th annual convention on medicinal plants in contemporary living*. Quezon City, Metro Manila, Philippines.

Venzon, E.L. 1979. Anticancer research at NIST. In *Proc. 28th annual convention PhilAAS*. Quezon City, Metro Manila, Philippines.

Philippine National Formulary. 1980. NSDB. Bicutan, Taguig, Metro Manila.

Zara, P. 1979. The integrated program on Philippine medicinal plants. In *Proc. 28th annual convention PhilAAS*. Quezon City, Metro Manila, Philippines.

Session V: Socioeconomic Considerations for MPTS Research

Chairman: Karim Oka

Discussants: Charit Tingsabadh
Charles B. Mehl

Session V Summary

Karim Oka and Charles B. Mehl

International Development Research Centre, India
Forestry/Fuelwood Research and Development (F/FRED) Project, Thailand, respectively

The papers presented in this section might best be divided into the following broad categories:

o Socioeconomic considerations in introducing or supporting the use of multipurpose trees by small farmers (equity issues [Malla and Fisher], nutrition [Robert], farmers' preferences and current use of trees [Khaleque]) and

o Socioeconomic factors affecting implementation of an agroforestry or farm forestry program (FAO pilot project in Thailand [Amyot], IDRC farm forestry project in Nepal [Dixit]).

The topics raised in the papers are diverse, reflecting the wide range of social, economic, and cultural conditions that affect villagers' use of trees.

Malla and Fisher raise serious questions about the distribution of benefits of farm forestry projects. Data from a preliminary survey indicate that programs that encourage tree planting on private lands can widen the gap between the rich and poor of a village. Those with larger land holdings, and hence more land to set aside for trees, are likely to benefit from farm forestry programs more than small-scale farmers. They suggest two strategies to ensure greater equity. First, tree planting can be encouraged on marginal lands (along trails, gullies, and agriculturally unproductive land), which even small-scale farmers own. Second, planting, protection, and utilization of common lands that are accessible to all farmers should be continued along with programs to encourage tree planting on private lands.

In his survey of rural Bangladesh, Khaleque found that tree planting is well-established at the farm level. Although fuelwood scarcity is acute, farmers prefer to plant tree species with primary end uses other than fuelwood. Fruit trees are preferred with timber-producing species sometimes planted.

Farmers always have in mind, however, that these trees will also produce enough residual fuelwood and fodder to satisfy their needs. Farmers clearly prefer multipurpose trees where nutrition and income take priority while remaining compatible with the satisfaction of subsistence fuel and fodder needs.

Nutrition as an important, yet often overlooked, purpose of multipurpose trees is the subject of Robert's paper. Malnutrition is endemic to large parts of Asia, especially in the uplands where land is only marginally productive for field crops and where tree planting programs are encouraged to counteract the degradation of forests. Promotion of commercial tree crops to increase farmer incomes will not alleviate malnutrition in many cases: the farmers will use their cash income to buy consumer goods rather than food. Robert argues that farmers' preferences can not always be the most important criterion for tree planting. Forest and agricultural extension programs should promote food trees, along with commercial tree crops, to supplement farm families' diets, especially where nutrition-deficiency diseases are evident. This suggests that one important use of multipurpose trees should be food.

How can farm and village forestry programs be promoted? Amyot studied a social forestry program among squatters in a reserved forest in Northeast Thailand. Lack of any rights to the land discouraged farmers from growing trees; a program to give farmers temporary and exclusive (usufructary) rights to use small parcels of land led some of them to take up agroforestry practices. Tree planting for apiculture, fruit tree plantation, and forest grazing of cattle were popular among the more innovative farmers. Forest plantation for other uses, including fuelwood and timber, were not as successful. As Amyot writes, "agroforestry approaches are quite alien to these maize farmers and the transition to a tree-farmer mentality does not come naturally." For most farmers, the uncertainty of long-term tenure, the severe restrictions placed by the government on the size of holdings farmers can use legally, the severe limits placed on rights to the land, and the lack of a market for many tree products contribute to farmers' reluctance to plant trees. The importance of the market cannot be underrated. Despite the relative popularity of apiculture and fruit production, the production levels reached the maximum consumption capacities at the local level within a few years. The products could not be sold profitably beyond the area in part because of poor market infrastructure in the area and increasing competition from places more accessible to markets. Land tenure and markets, then, are essential considerations for promotion of multipurpose trees for small farms.

In his paper on introducing multipurpose trees on the small farms of Nepal, Dixit points out the differences between the needs of farmers in the lowland plans and those living in the hills. Hill farmers have a critical need for fodder for their animals and fuel for their own consumption. Lowland farmers are more interested in income-generating trees, so they prefer to plant fruit or timber trees. Dixit also examines the constraints to growing trees as stated by lowland farmers. These are a shortage of land, lack of know-how in tree planting, lack of good-quality seedlings from appropriate species, manpower shortage at the time of tree planting (which coincides with the rice planting season), animal browsing, the long waiting period before tree maturity, and the problem of land tenure when absentee land owners do not make decisions and tenants cannot or will not make decisions on long-term investments like tree planting. Furthermore, most farmers are suspicious about exotic species unknown to them and prefer indigenous varieties or well-known and widely-used exotics.

Significantly, most papers point out the potential of cash income as a major criterion in farmers' selection of trees. In most cases, farmers will decide on their own to choose trees with multiple uses. Among the more successful tree planting schemes are those promoting fruit trees that can meet farmers' nutritional needs, provide some fodder and fuelwood, and improve income through sale of the produce. In the social forestry project in Thailand described by Amyot, propagation of fruit trees was quickly taken up by the farmers themselves, leaving project resources for other activities.

The conclusion reached by participants in the discussion was that farmers' needs and preferences must absolutely be taken into consideration. Their present farm tree management practices and their overall farming systems need to be well-understood before any intervention takes place. Constraints to tree planting need to be investigated and solutions identified to problems in seedling availability, land use management, time management, and production technology. Credit is also an aspect that can be identified and addressed. Tackling more complex problems, such as marketing and land ownership and distribution, requires a concerted effort by government and business in addition to the work of researchers, project planners, and implementers.

Manibhai Desai from The Bharatiya Agro Industries Foundation pointed out as a general conclusion that all the farm forestry programs and tree improvement schemes will have little more than a marginal effect if they do not address the root cause of the problem of rural poverty, which is the lack of productive employment. Malnutrition and backwardness can be linked directly to underemployment. Farm forestry or multipurpose tree schemes will be truly successful only if they help generate employment and thus, income.

The major issues discussed during this session were numerous and diverse: Land distribution and tenure, preferred end uses (fruit trees, apiculture, fodder), gaps in farmer welfare and needs, improved farm and tree management, markets for commercial tree products or surplus produce, patterns of household decision making, and the effects of government policy on farm forestry programs. These are by no means all the important socioeconomic issues that should be considered in a farm or village forestry program, but they represent some serious thought about how farmers use trees and how multipurpose tree research can be made more appropriate to farmers's needs.

Planting Trees on Private Farmland in Nepal: The Equity Aspect

Y.B. Malla and R.J. Fisher

Nepal-Australia Forestry Project
Kathmandu, Nepal

Equity aspects of private tree-planting programs in Nepal are discussed in a review of the literature and an analysis of data from a preliminary survey and field observations at two sites. In recent years, there has been increasing interest in planting trees on private land in Nepal, but it primarily benefits land-wealthy people. As there is little absolute landlessness in Nepal, shifting emphasis away from establishing private forests (on fairly large private plots) to planting on small pockets of private, non-agricultural marginal land (edges of fields, creek beds, etc.) may better benefit small farmers. However, analysis of field data suggests that the benefits to the poor are likely to be limited. It is possible, but not yet proven, that private planting may lead to reduced pressures on common forest resources and that this may increase access to these resources by the poorest farmers and landless people. Whatever the value of private planting, it must be seen as a supplement, and not an alternative, to plantings on common lands.

Tree Distribution and Program Participation

In terms of increased overall availability of forest resources, programs that encourage private planting have advantages. Ideally, all farmers would profit proportionately. While small landowners would benefit relatively less than large landholders, all would be better off in absolute terms. However, there are two problems with this scenario.

First, the extent of the discrepancy in relative benefit is very large due to the heavily skewed distribution of landholdings in Nepal. According to Wallace (1987), 13.6% of the population holds 62.8% of the agricultural land. Table 1 sets out land distribution figures based on the 1981-82 census. The absolute differences in landholding size is so large that the distribution of benefits overwhelmingly favors a small number of large-scale landholders, even if benefits are provided in proportion to landholding size.

Second, there is evidence that in practice, distribution of benefits is not proportional to landholding size, but instead heavily favors larger landholders. According to Gautam (1986), people owning more than one hectare of land represented only 2% of the population in a panchayat (local political and administrative unit) in Dolakha District, but they received 75% of all seedlings distributed. The 46% of the population that owned one-quarter of a hectare of land or less received no seedlings. Gautam does not attempt to account for the massively disproportionate distribution of benefits. Only 40 households within the panchayat received seedlings, so it is quite possible that the recipients are all influential locals aware of nursery activities. If so, increased extension to reach smaller farmers might help correct the skewed distribution.

Analysis of involvement in private planting programs from four panchayats within the Pakhribas Agriculture Centre's project area showed a much higher rate, with 59% of all households participating (Malla 1987). Again, there was a difference in the average land-holding sizes of participants and non-participants (3.45 ha versus 2.0 ha). Information on the average numbers of trees taken by participants within various landholding categories is not available.

Other studies are not particularly helpful in showing the distribution of seedlings or trees on private land by landholding size. Shrestha and Evans (1984) report an average landholding of 0.39 ha for households in Chautara (Sindhupalchok District) and an average of 17.01 trees per household. There is no information, however, on the relationship between tree numbers and landholding size. In any case, the number of trees per household seems extraordinarily low. In a study of patterns of ownership and use of farm fodder trees in three panchayats in the middle hills, Hawkins and Malla (1983) looked at the average numbers of fodder trees per household and at the average number of fodder trees per livestock unit, but did not relate tree numbers to landholding size.

The literature is not very helpful in showing present distribution of private trees according to the size of landholding or in showing whether larger or smaller landholders are more likely to respond to private planting programs. Nevertheless, it is clear that in the context of the highly skewed distribution of landholdings in Nepal, attention must be paid to the differing effects of private planting on land-rich and land-poor farmers. It also appears that the benefits of private planting programs may be diverted towards larger landholders. The evidence is not conclusive, but it raises serious questions.

Potential Benefits to Small Farmers

The above discussion outlines arguments and supporting evidence against emphasizing private planting. However, quite apart from the proportionate benefit model, there is one strong argument in favor of private planting for the benefit of smaller landholders. Chambers and Leach (1987) argue that trees on private land are an important resource for the rural poor, not merely because they provide increased income but because they act as a form of security in the case of unexpected events or irregular needs, such as wood for funeral fires or to replace houses after floods or fires.

It may also be possible to avoid the tendency of planting programs to favor larger landholders disproportionately. It has been pointed out that a great deal of private land is not used directly for agriculture (Malla 1987). As there is little absolute landlessness in Nepal, it is possible that shifting emphasis away from establishing private forests on large plots to planting on small pockets of private, non-agricultural, marginal land may enable planting to benefit the majority of small farmers.

Field Study

Aims and Methodology

To obtain more information on the relationship between land distribution, trees on private land, and planting on marginal non-agricultural lands, we carried out a preliminary study in two villages in the project area of the Nepal-Australia Forestry Project. As this was a preliminary study, we did not aim to establish firm conclusions, but rather to identify issues for later investigation and to provoke discussion.

The two villages selected were Pandegaon, about a half hour's drive east of Kathmandu in Kavre Palanchok District, and Buchakot, a half day's walk from Dhulikhel, the District headquarters of Kavre Palanchok District.

The objectives of the study were as follows:

o Determine who is most likely to benefit from the private planting program.

o Ascertain whether the program will widen or decrease the gap between rich and poor.

The approach taken centered around direct observation and informal discussion and questioning. Some quantitative information was collected using an informal checklist of questions. While we concentrate discussion in this paper on analysis of the quantitative data, we must stress that we have very serious reservations about overuse of the questionnaire approach and about the reliability of quantitative data based on survey questionnaires. Doubts about the validity of statistics obtained by survey questionnaires in the developing world have been raised frequently in recent years (Campbell, Shrestha, and Stone 1979; Hill 1984, 1986; Gilmour, King, and Fisher 1987; Fisher 1987). Among the factors that lead to unreliable data are suspicion about the motives of investigators and questions inappropriate to local patterns of classification (in this case, the classification of land). We are confident that suspicion about motives was minimized because the field investigator spent considerable time in each village and established good rapport with villagers.

A structured interview focused on individual households. One informant (usually the household head) was interviewed from each household. The sample was essentially a random survey. It should be noted that in the case of Pandegaon, all villagers were previously known to the field investigator.

The sample sizes were small, although they represent a significant proportion of households in each village (Table 2). The small sample sizes were intentional, given the trial nature of the survey.

The land types considered were as follows:

o   khet (irrigated fields suitable for rice cultivation)

o   bari (rainfed terraces on which maize and millet are grown)

o   marginal land and abandoned terraces

o   stream banks, gullies, and landslides

Reported figures on landholding size are reasonably accurate except for various types of marginal land. In these cases, figures are little more than "quantified guesses" and are unlikely to be reliable beyond a general order of magnitude. Figures on landholdings have been converted from the locally used measure (ropani) to hectares (20 ropanis = 1 ha). As a consequence of this, figures on average tree densities are derived from quite small absolute numbers of trees. For example, 20 trees on a half ropani of marginal land translates to a density of 800 trees/ha.

For the purpose of comparing the economic status of households, total reported landholdings have been divided into various categories based on the range of landholding sizes used by Wallace (1987). It is important to note that landholding size alone can be quite misleading as an indicator of land wealth. Agricultural land can be divided into two categories, khet and bari, with smaller areas of the former much more valuable than larger areas of the latter.

The Two Villages

Selection of villages was based on practical considerations, not random sampling. Given the difficulty of obtaining accurate information on sensitive matters like landholding size, villages were selected where the project was well-known and in which the field investigator had established rapport.

Differences in location have probably contributed to differences in socioeconomic conditions in the two villages. In Pandegaon, relatively near Kathmandu, many people are engaged in off-farm activities. Even some of the land-poor people claim to have considerable cash resources. One lady from an untouchable caste claimed to have enough money to purchase land but had not done so because none was for sale. Buchakot, on the other hand, is more typical of the subsistence-oriented villages of the middle hills. While agropastoralism is the major source of income in both villages, it is heavily supplemented by other sources of income in Pandegaon.

Patterns of land distribution differ between the villages (Table 3). The most significant difference is that land distribution in Pandegaon was relatively homogeneous. No household in our sample owned more than 1.5 ha. On the other hand, land ownership is more skewed in Buchakot. Of those surveyed, 53% own more than 1.5 ha, totalling 83% of all land held by households in Buchakot. While land distribution in Buchakot is uneven, there are more farmers in the larger landholding size-categories than described by Wallace for all of Nepal (Table 1).

Distribution of Trees on Private Land

Analysis of tree distribution on private land in the two villages suggests a number of patterns. There is also an absence of some patterns we expected to find.

Pattern 1. While sample sizes for each land-holding category are small, we expected to see tree numbers and density increase with the size of landholding. But no clear evidence of such a trend in densities exists for the landholding size-categories below 1.5 ha (Table 4).

In Pandegaon there are no clear trends relating to tree density between landholding categories. The same absence of pattern applies to the landholdings below 1.5 ha in Buchakot. On the other hand, there was a contrast between tree densities on landholdings above and below 1.5 ha. There is a very clear increase in the average tree numbers as landholding size increases (Table 5). This increase is massive for farms with more than 1.5 ha.

Why does a landholding size over 1.5 ha appear to be a take-off point above which much higher tree densities become possible? This occurs in Buchakot even though smaller farmers have lower tree densities than owners of similar landholdings in Pandegaon.

Table 1. Agricultural land distribution in Nepal, 1981-82.

Category	% Population	% Area
No land or livestock	15.1	0.0
No land, but own livestock	0.3	0.0
0-0.5 ha	42.5	6.6
0.5-1.0 ha	13.7	10.8
1.0-2.0 ha	14.7	19.8
over 2.0 ha	13.6	62.8
Total	100.0	100.0

Source: Wallace, 1987.

Table 2. Sample sizes in the two villages surveyed.

Village	No. households interviewed	Total households in villages
Pandegaon	16	31
Buchakot	17	40

Table 3. Land distribution patterns in the two villages surveyed.

Village	0	0-0.49	0.5-0.99	1.0-1.49	1.5 and over
Pandegaon					
No. households	1	4	9	2	--
% households	6.2	25	56.2	12.5	--
Buchakot					
No. households	--	3	2	3	9
% households	--	17.6	11.8	17.6	52.9
Combined					
No. households	1	7	11	5	9
% households	3	21.2	33.3	15.1	27.3

Table 4. Tree density by landholding size.

Size landholding (ha)	Tree density/ha		
	Pandegaon	Buchakot	Combined
0	--	--	--
0.-0.49	236	33	112
0.5-0.99	126	42	112
1.0-1.49	165	35	87
1.5 and over	--	225	225

Table 5. Average number of trees per farmer within each landholding category.*

Landholding size (ha)	Pandegaon	Buchakot	Combined
0-0.49	28 (4)	8 (3)	19 (7)
0.49-0.99	95 (9)	26 (2)	82 (11)
1.0-1.49	199 (2)	44 (3)	106 (5)
1.5 and over	--	671 (9)	671 (9)

* Nos. of farmers are in parentheses. Landless farmers are omitted.

As a tentative explanation, we suggest that farmers with landholdings above 1.5 ha are able to meet more than their subsistence requirements for grain and are thus in a position to diversify, spread risks, and gain opportunities for alternative income. The validity of this explanation should be explored by further study.

Pattern 2. If we look only at farms of less than 1.5 ha, the average tree density was higher in Pandegaon than in Buchakot for all landholding size-categories.

Why is there a difference in tree density between smaller landholders in Pandegaon and in Buchakot? Different qualities of land may be part of the explanation. The land at Buchakot may be of lower quality than that at Pandegaon. But this does not fully explain the differences since land-rich farmers at Buchakot own many more trees than poorer farmers in either village. A probable explanation is that Pandegaon is less dependent on farm sources for subsistence and income. Consequently, farmers are more able to spare small amounts of land for trees. One farmer in the 0.5-0.99 ha category at Pandegaon had considerable numbers of trees, including trees on various types of marginal land and on the edges of khet. He recognized that the shade from the trees was reducing crop production by an estimated 20%, but he felt the economic advantages of trees outweighed the losses. A number of *Choreospondias axillaris* (locally known as 'lopse') grown on his land illustrate his point. The fruit is extremely valuable as a cash crop.

Pattern 3. We anticipated a pattern that would show a relationship between tree density on marginal land and landholding-size categories. The results are mixed, probably because of the small sample size. Farmers with 1.5 ha or more have generally higher tree densities on marginal lands. The 0-0.49 ha category also has a high density of trees in stream beds, gullies, and land slides (Table 6), but all of the land and all of the trees in that category are owned by a single farmer and the density of 1,400 trees/ha is derived from 35 trees on 0.025 ha.

This suggests that marginal lands are underused by smaller farmers in both Pandegaon and Buchakot. It should be pointed out that the area of marginal land owned by farmers in the Pandegaon sample amounted to only 0.61 ha (6.2% of all land).

In Buchakot, 4.86 ha (14.9% of total private land in the sample) was marginal land, but only 0.78 ha of this is owned by farmers with less than 1.5 ha (Table 7). The potential for small farmers to gain from increased tree planting on marginal land is slight.

Analysis of the distribution of bari land among small farmers points to an area with potential to reach the land-poor. Much of the bari land (9 ha or 40% of all bari land) is owned by farmers with less than 1.5 ha (Table 8). The importance of bari land for subsistence farming might prevent poor farmers from substantially increasing tree planting on bari unless they have alternative sources of income. The often-observed phenomenon of trees on bari would appear to be more common among larger landholdings (Gilmour 1987).

Conclusions

A USAID report (Kernan, Bender, and Bhatt 1986) made a case for increased privatisation of common land for forest purposes, arguing that private ownership would contribute to better management of forest and other natural resources. Robinson (1986) acknowledges that there may be "advantages in the equitable privatization of some land," but goes on to say that "the need for any privatization to be equitable would be essential; small farmers often rely more on land other than their own for fodder and other foliage, compared to wealthier farmers."

This is an extremely important point. We believe that increased privatization, unless accompanied by extensive redistribution of land, could only decrease the access of poor farmers to forest products. They would not only be relatively worse off, as larger farmers become absolutely better off, but they could easily become absolutely worse off.

On the other hand, it may be that increased planting on private land (as opposed to privatization of common land) will reduce pressure on common land. This may benefit smaller landholders, but there is no guarantee that reduced pressure on common land will translate to increased access to resources by the poor.

The findings from Pandegaon and Buchakot leave little room for optimism about the potential for private planting programs to benefit small farmers. Equity in tree ownership is clearly impossible in the

Table 6. Trees per hectare by land type and landholding size.

Land type [a]	Landholding size			
	0-0.49	0.5-0.99	1.0-1.49	1.5 and over
Pandegaon				
1	0	23	0	--
2	167	175	258	--
3	0	265	0	--
4	1,400 [b]	244	460	--
Buchakot				
1	0	0	0	55
2	23	54	56	89
3	13	0	11	642
4	0	176	216	1,706
Combined				
1	0	21	0	55
2	109	157	135	89
3	13	193	11	642
4	1,400 [b]	220	355	1,706

[a] 1 = ket, 2 = bari, 3 = marginal land including abandoned terraces, 4 = stream banks, gullies, and landslides.
[b] single household.

Table 7. Land types available for whole sample.

Land type*	Pandegaon		Buchakot		Combined	
1	3.40	(34.8)	11.32	(34.6)	14.70	(34.6)
2	5.75	(58.9)	16.50	(50.5)	22.25	(52.4)
3	0.20	(2.0)	3.00	(9.2)	3.20	(7.5)
4	0.41	(4.2)	1.86	(5.7)	2.29	(5.4)
Total	9.76	(99.9)	32.68	(100.0)	24.46	(99.9)

* 1 = khet, 2 = bari, 3 = marginal land including abandoned terraces, and 4 = stream banks, gullies, and landslides.

Table 8. Distribution of bari land by landholding size for two villages.

Landholding size	Area (ha)	% of total bari
0-0.49	0.95	4
0.50-0.99	4.85	22
1.0-1.49	3.20	14
1.5 and over	13.25	59
Total	22.25	99

context of the dramatically skewed distribution of landholding in Nepal, and there is little room to increase the absolute numbers of trees owned by poorer farmers. Nevertheless, in the context of very small absolute numbers of trees, small increases may be of value.

Our figures are tentative and based on small samples. While they are consistent with other evidence from the literature, it is essential that much larger studies be carried out to confirm or deny our suspicions. Further, even if our observations appear generally true, there may be some locations where the land distribution patterns allow private planting programs to reach the land poor.

For such a program to work, it would be necessary to concentrate activities on small farmers. One element of this would be increased extension work aimed at making small farmers aware of the program. However, it is unlikely that awareness alone is a solution, since farmer-ignorance is almost certainly overrated and unjustly assumed to be an explanation for failure to take advantage of program benefits.

To prevent monopolization of benefits by relatively wealthy farmers, it might be worth thinking about a system of phased charges. Up to a certain number of seedlings, perhaps 200, could be provided free, and seedlings in excess of this maximum could be provided at a small fee.

Finally, while programs emphasizing private planting on existing private land may have their place, we are overwhelmingly in favor of a policy that treats private planting as part of a broad range of approaches. To maintain any equity of access to resources, private planting must supplement, not be an alternative to, continued planting, protection, and utilization of common lands.

REFERENCES

Campbell, J.G., R. Shrestha, and L. Stone. 1979. *The use and misuse of social science research in Nepal*. Kathmandu: Centre for Nepal and Asian Studies.

Chambers, Robert and Melissa Leach. 1987. *Trees to meet contingencies: Savings and security for the rural poor*. Institute of Development Studies, Discussion Paper 228.

Fisher, R.J. 1987. Guest editorial on social science in forestry. *Banko Janakari*: Vol. 1, No. 3.

Gautam, K.H. 1986. *Private planting: Forestry practices outside the forest by rural people*. Kathmandu: Winrock International Forestry Research Paper Series, No. 1.

Gilmour, D.A. 1987. *Not seeing the trees for the forest: A re-appraisal of the deforestation crisis in two hill districts of Nepal*. Kathmandu: Nepal-Australia Forestry Project.

Gilmour, D.A., G.C. King, and R.J. Fisher. 1987. Action research into socio-economic aspects of forest management. In *IUFRO Symposium: Role of forest research in solving socio-economic problems in the Himalaya Region*. Peshawar, Pakistan.

Hawkins, T. and R.B. Malla. 1983. *Farm fodder trees: Patterns of ownership and use*. Nepal Forestry Technical Bulletin 9.

Hill, Polly. 1984. The poor quality of official socio-economic statistics relating to the rural tropical world with special reference to South India. *Modern Asian Studies* 18:491-514.

_____. 1986. *Development economics on trial: The anthropological case for the prosecution*. Cambridge: Cambridge University Press.

Kernan, Henry, W.L. Bender, and Bal Ram Bhatt. 1986. *Report of the forestry private sector study: Sept. 28-Nov. 27, 1986*. Kathmandu: USAID Mission to Nepal.

Malla, Yam B. 1987. The case for placing more emphasis on private tree planting programs: A case study of Pakhribas Agricultural Centre's private tree planting program. Unpublished paper prepared for Winrock International, Kathmandu.

Robinson, Patrick J. 1986. The dependence of crop production on trees and forest land. In *Amelioration of soils by trees: A review of current concepts and practices*, eds. R.T. Prinsley and M.J. Swift, pp. 104-120. London: Commonwealth Science Council.

Shrestha, R.L.J. and D.B. Evans. 1984. The private profitability of livestock in a Nepalese hill farming community. *Agricultural Administration* 16:145-158.

Wallace, Michael B. 1987. *Forest degradation in Nepal: Institutional contexts and policy alternatives*. Kathmandu: Winrock International.

Let Them Eat Trees: Some Observations on Nutrition and Social Forestry

G. Lamar Robert

Research and Development Center, Payap University
Chiang Mai, Thailand

The purpose of this paper is to focus attention on the need for increased attention to health and nutrition in social forestry. A health/nutrition and agroforestry/sociological survey of northern Thailand hilltribe farmers indicated that a significant number suffered from various nutritional problems. It was found that many development activities that promote commercial tree crops to improve hill tribe incomes and nutrition often overlook food tree crops, reasoning that with increased income a farmer will purchase additional food. However, cash income is more likely to be spent on consumer goods than food. Additionally, due to limited household labor, commercial tree crop production can actually result in less food production and thus cause an overall reduction in nutrition. Food tree crops, therefore, should also be promoted. Careful attention to the current health and nutritional condition of the target population is needed, however. For example, promotion of citrus trees to provide vitamin C in areas where there is a sufficient supply from chili peppers is not appropriate, whereas promotion of papaya in an area deficient in vitamin A can significantly improve nutrition and health.

Causes of Malnutrition

The obvious question is why, in the midst of what would seem to be fertile mountain areas, is there such a serious malnutrition problem? The answer in many locations is rising population density, aggravated by rapid introduction of cash crops to subsistence farming areas.

Population density in the hills of Thailand, Nepal, and many other regions is increasing rapidly and is straining the carrying capacity of the land. Traditional swidden agriculture is no longer as viable as it once was. Fallow cycles have been shortened due to a shortage of land. An increasing number of farmers in Thailand must clear more forest to make room for increased rice production.

Many people can barely produce enough rice for their own consumption. They are increasingly less able to gather from the dwindling forests the foods on which they formerly depended for protein, vitamins, and minerals to balance the carbohydrates and minimally sufficient amounts of protein and B vitamins in rice. In some areas, shortages of rice or other carbohydrate staples occur during the late dry season, and people live on only one or two meals a day.

The forest used to be the hill tribesman's supermarket, but no longer. Forest destruction for agriculture, combined with hunting and gathering by the increased population, have caused a very serious decline in the availability of edible wild plants and animals.

In the past, virtually all hilltribe villages were located near adequate year-round sources of water. Increased population pressures have forced many villages to locate in areas where water supplies are insufficient for growing vegetables in the dry season, a situation that worsens chronic vitamin and mineral deficiencies.

Aggravating the problem of population growth is the rapid and large-scale introduction of commercial crops, including tree crops, into areas of subsistence production. The idea behind such introductions is that increased incomes will lead to a better diet. Unfortunately, many studies indicate that nutritional levels decline, and that improvements in the subsistence system are more likely to improve nutrition (McElroy and Townsend 1979; Foster and Anderson 1978; Schubert 1986; and Vyrheid, Wongcharoen, and Robert 1987).

When cash cropping is introduced on a large scale in a relatively short period of time, the indigenous population does not have time to assimilate the changes. Various social and other pressures encourage them to spend their increased income on non-food items, such as watches and

radios. In some cases, the range of foods needed for a balanced diet are not available for purchase or people do not know which foods to purchase to obtain a balanced diet.

A corollary for Thailand and other opium-growing areas is that, although alternative crops may be desirable, there is a danger that the vegetables usually grown with poppies also may be lost in the changeover (Krantz 1986).

How Multipurpose Trees Can Help

This paper is not intended as a comprehensive encyclopedia of nutritionally valuable tree crops. Nevertheless, a few typical examples of how trees can improve nutrition are provided as a guideline.

B-complex Vitamins

Dark green vegetables are high in many nutrients, including vitamin A, B-complex vitamins, iron, and calcium. Obviously, dark green vegetables are not trees. However, trees can help increase the supply of this valuable food group. Many villagers already have bamboo fences (various species) around their houses or gardens. Others have living fences of *Leucaena leucocephala, Jaropha curcas* or some other tree species. Any of these make very functional trellises for the climbing vine *Coccinia indica* and similar crops. These crops growing on the fence close to the home can be easily watered with leftover rinse water in areas where water is scarce or difficult to obtain. A Leucaena trellis also can provide firewood and fodder. Eaten by humans, leaves of this tree are a source of vitamin B1 and could help prevent beri beri. Unlike some livestock, humans generally do not have to be concerned with mimosine-related problems from ingesting too much Leucaena.

Vitamin A

Papaya (*Carica papaya*) is a tree crop easily grown in warmer climates where there is sufficient water. It is an excellent source of vitamin A, a vitamin often lacking. Other fruit tree crops high in vitamin A include mango (*Mangifera indica*), persimmons (*Diospyros* spp.) and apricots (*Prunus armeniaca*). These three crops can easily be dried for a good year-round source of Vitamin A. These trees are already growing in some areas of Northern Thailand (Krantz 1986). These fruit trees

also help control soil erosion, provide fuel, and can shade other crops, especially in the hot season. Pumpkin (*Cucurbita* spp.) is also a good source of Vitamin A. This crop will climb up larger trees or bamboo trellises.

Vitamin C

Often well-meaning projects promote production of tree crops for home consumption such as lemons (*Citrus limon*) and tamarind (*Tamarindus indica*), which mostly contain Vitamin C. In areas where chili peppers (*Capsicum* spp.) are a regular part of the diet, they normally provide sufficient Vitamin C without supplements from citrus or other similar tree crops.

Where this vitamin is needed, the number of crops containing Vitamin C is quite extensive and includes, in order of importance, guavas (*Psidium guajava*), limes (*Citrus aurantifolia*), papayas (*Carica papaya*), persimmons (*Diospyros* spp.), lychee (*Nephelium litchi*), pomeloes (*Citrus maxima*), mangoes (*Mangifera indica*), and jackfruit (*Artocarpus heterophyllus*) (Krantz 1986).

Oil

Although oil and fat are extremely important to the diet for energy and the utilization of Vitamin A, there is a significant deficiency of dietary oils and fats in several developing countries, particularly in Southeast Asia (RAPA 1987). The common coconut *Cocos nucifera* can be a valuable source of oil in areas where villagers cannot buy or obtain vegetable or animal oil. Other uses of the fruit are well-documented and range from crop mulch (from the husks) to drinking vessels made of the inner shell. This tree is limited to areas below about 900 m elevation. Other potential sources of oil-seeds which are or could be promoted include chestnuts (*Castanea sativa*), and walnuts (*Jreglams regia*) (Krantz 1986).

Animal Fodder

There are many tree crops, such as *Leucaena leucocephala*, that can serve as animal fodder as well as fuel. Legumes such as Leucaena also increase soil fertility. The only limitation is that a diet too high in concentration of Leucaena leaves can induce undesirable side effects in non-ruminant animals due to the presence of mimosine, a toxic compound (Robert 1982). Ruminants in many

areas, however, contain rumen bacteria that detoxify this compound and its by-products. If it is not present, it can be introduced into the animals.

Green Manure

In areas where livestock fodder is not in short supply, there are many tree crops that can serve as green manures. These crops can be grown in village woodlots near agricultural fields or on bunds or windbreaks along the edges of fields. Leguminous trees are particularly suitable for this application as their nitrogen-fixing capability further enhances soil quality.

Coffee and Tea

In many areas of the world, coffee (*Coffea arabica* and *Coffea robusta*) and tea (*Camellia simensia*) are frequently promoted as cash crops in highland areas. In addition, they can help prevent soil erosion and, in the case of coffee, provide shade to other crops. Local consumption can cause nutritional problems, however, as tea and coffee taken with a vegetarian meal can seriously affect iron absorption (Krantz 1986).

The trees listed above are only a few of the many species that can benefit the nutritional status of people in rural areas. Generally, the problem is not a lack of suitable trees, but a question of which tree is appropriate for which area. Answering this question is the topic of the next section.

Promoting Nutritionally Beneficial Trees

There are a series of procedures that, if implemented in order, can greatly increase the chances of a successful reforestation project. If these steps are followed, villages in the reforestation area will be more inclined to compete to see who can plant the most trees, rather than competing to see who can pull up the most seedlings newly planted by extension officials.

The first step is to determine the nutrition status of the population that will be affected. It is highly recommended that all reforestation projects include some form of orchard or other component that will be of direct value to the local population. The specific nutrients lacking in their diet can be identified by surveying for key indicators of specific types of malnutrition, such as those mentioned

earlier in this report. In addition, local health officials should be consulted regarding nutritional deficiencies they are aware of, and any on-going projects to alleviate those deficiencies.

If the population raises a substantial amount of livestock, then the residents of the area also should be surveyed regarding adequacy of fodder. Local livestock extension agents could help with the survey and provide general knowledge about the local fodder situation.

Another aspect to be considered in determining forest needs is that forest resources are not limited to trees and tree products. Forest resources also include animals, insects, other low-growing shrubs and herbs (including many medicinal plants) that can only grow in forest-like conditions. These may not prosper in an area replanted to a single species like Eucalyptus or Pine.

The next step is to determine which species of trees might provide the needed nutrients and/or fodder. Once this basic list of potentially promotable tree crops has been compiled, it should be reviewed for suitability in the target area.

It should be borne in mind that just because a tree can grow in a given area is not proof that the local population will actually use it. Determining the acceptability of the tree products to the local people is thus the next step. One way to increase the odds of acceptance is to promote trees already familiar to the local population. Some research on what tree crops are generally used by the local population can be very helpful. Two examples of the fruits of such research, listings of trees used by indigenous populations in Thailand and Nepal, are shown in Tables 1 and 2. Even in a rather limited area, different ethnic groups make use of different trees (Table 1).

Exotic trees or those not currently used by the villagers can also be successfully introduced, but a program of education is required. Villagers must first be taught why they should eat the products of a certain tree. When nearly all the residents of an area suffer malnutrition, they view it as a normal part of life. Thus, it is necessary to teach villagers the nature of their malnutrition, its causes and ramifications, and how the trees being promoted can help cure the problem. Involving health officials and welfare department officials can be very helpful in this activity. Simple as it may sound,

Table 1. Trees used for food by hilltribes in northern Thailand.

Species	Family	Tribes
Annona spp.	*Annonaceae*	Karen
Araucaria cunninghamii	*Araucariaceae*	Akha
Arenga pinnata	*Arecaceae*	Akha, Karen
Artocarpus heterophyllus	*Moraceae*	Akha, Karen, Lahu
Averrhoa carambola	*Oxalidaceae*	Akha, Karen, Lahu
Azadirachta indica	*Meliaceae*	Karen
Baccaurea ramiflora	*Euphorbiaceae*	Lahu
Bauhinia purpurea	*Caesalpinaceae*	Karen
Borassus flabellifer	*Aracaceae*	Lahu
Broussonetia papyrifera	*Moraceae*	Lahu
Canarium subulatum	*Burseraceae*	Akha, Karen
Caryota mitis	*Arecaceae*	Akha, Karen
Castanopsis spp.	*Fagaceae*	Akha, Karen, Lahu
Citrus aurantifolia	*Rutaceae*	Akha, Karen, Lahu
C. maxima	*Rutaceae*	Akha, Karen, Lahu
C. reticulatus	*Rutaceae*	Akha, Karen, Lahu
C. sinensis	*Rutaceae*	Akha, Karen, Lahu
Cocos nucifera	*Arecaceae*	Karen, Lahu
Corypha umbraculifer	*Arecaceae*	Akha, Lahu
Crateva magna	*Capparidaceae*	Akha, Karen, Lahu
Dillenia indica	*Dilleniaceae*	Akha, Lahu
Duabanga grandiflora	*Sonneratiaceae*	Akha
Ficus auriculata	*Moraceae*	Lahu
F. carica	*Moraceae*	Karen
F. lacor	*Moraceae*	Lahu
Garcinia spp.	*Guttiferae*	Akha
Garuga pinnata	*Burseraceae*	Akha
Grewia paniculata	*Tiliaceae*	Akha
Litchi chinensis	*Sapindaceae*	Lahu
Mammea siamensis	*Guttiferae*	Akha
Mangifera indica	*Anacardiaceae*	Akha, Karen, Lahu
Markhamia stipulata	*Bignoniaceae*	Akha, Karen, Lahu
Nephelium hypoleucum	*Sapindaceae*	Akha, Karen
Parinari anamense	*Rosaceae*	Akha
Phyllanthus emblica	*Euphorbiaceae*	Akha, Karen, Lahu
Protium serratum	*Burseraceae*	Lahu
Prunus persica	*Rosaceae*	Akha, Lahu
Psidium guajava	*Myrtaceae*	Akha, Karen, Lahu, Hmong
Radermachera ignea	*Bignoniaceae*	Akha
Shorea thorelii	*Dipterocarpacea*	Ahka
Sterculia foetida	*Sterculiaceae*	Akha
Tamarindus indica	*Caesalpinaceae*	Akha, Karen, Lahu
Turpinia pomifera	*Staphleaceae*	Akha
Zanthoxylum limonella	*Rutaceae*	Lahu

Source: Anderson, personal communication.

Table 2. Trees used for fodder in rural Nepal.

Botanical name	Local name
Melia azedarach Linn.	Bakaino
Dendrocalamus spp.	Bans
Buddleja asiatica	Bhimsen pati
Ficus nemoralis Wall.	Dudilo pati
Proman spp.	Ginderi
Leucaena leucocephala	Ipil ipil
Myrine semiserata	Kalikath
Machilus gamlei	Kaulo
Ficus lacor	Kavro
Ficus semicordata	Khaniyo
Litsea polyantha	Kumiro
Ficus roxburghii Wall.	Nimaro
Prunus cerasoides D. Don.	Paingyo
Shorea robusta Gaertn.	Sal
Grewia tiliafolia	Syalphusro
Bauhinia purpurea Linn.	Tanki
Cedrela tonna Roem.	Tooni

Adapted from Gautam, 1986.

picture posters and charts can be very beneficial in introducing new tree species or new uses for under-utilized species. Rather than just saying "don't destroy the forest," the posters can instead say "plant trees of the XYZ variety for eating, for fuel, and to keep topsoil in place."

The next step in introducing nutritionally beneficial trees is to obtain seeds. This can be difficult. Few major seed suppliers carry stocks of many of the trees appropriate for a given area. It may be necessary for forestry and agricultural extension workers to depend on the indigenous population to obtain seeds from the remaining wild stock. This process may be more time-consuming than simply purchasing a hundred kilograms of pine seeds, but it will provide villagers with beneficial trees and will involve them in the reforestation process from the outset.

Once seedlings are grown, they need to be distributed to villagers. If the above steps have been followed, the villagers will know what the trees are for and will be prepared to assist in the planting. There is one corollary to this final step. If trees are being distributed to individuals, as opposed to large-scale forest planting on public land, it can help to sell rather than give the young trees away. People tend to place more value on purchased rather than free goods (Gautam 1986).

Finally, after the trees have been planted, it is necessary to continue to monitor the nutritional status of the population. If some of the introduced species are not effective as nutritional supplements, one should find out why and rectify the situation through additional instruction of local residents or introduction of more acceptable tree species with similar nutritional benefits. It is not enough just to plant trees with nutrition-improving potential, report to higher authorities the number of trees planted, and then assume the new trees will improve the nutritional status of the population. It is essential that the well-being of the target population be monitored regularly. Following-up to ensure that potential benefits are realized is vital to a truly successful reforestation program.

Network on Tree Nutrition

The establishment of an Asian Food and Nutrition Network, involving institutions from selected countries of the region, was recommended by the FAO Consultation (RAPA 1987). A network designed specifically for exchange of information and experiences with nutritionally beneficial trees could be a very valuable addition to the proposed overall Nutrition Network for those engaged in social forestry activities.

REFERENCES

Anderson, Edward F. 1987. Whitan College, Walla Walla, Washington. Personal communication.

de la Cruz, R.E. and N.T. Vergara. 1987. *Protective and ameliorative roles of agroforestry: An overview. In Agroforesty in the humid tropics, its protective and ameliorative roles to enhance productivity and sustainability*, eds. Napoleon T. Vergara and Nicomedes D. Briones. Honolulu: Environment and Policy Institute, East-West Center.

Foster, George and Barbara C. Anderson. 1978. *Medical anthropology*. New York: Wiley and Sons.

Gautam, K.H. 1986. *Private planting: Forest practices outside the forest by rural people*. Forestry Research Paper Series No. 1. HMG-USAID-GTZ-IDRC-Ford-Winrock Project, Strengthening Institutional Capacity in the Food and Agricultural Sector in Nepal.

Krantz, Miriam. 1986. *Nutrition report: Thai-Norwegian Church Aid Highland Development Project*. Mimeograph. Bangkok.

McElroy, Ann and Patricia K. Townsend. 1979. *Medical anthropology in ecological perspective*. Belmont, Calif.: Wadsworth.

Pandey, Tulsi Ram. 1987. *The subsistance farmers and workers of Sunwal Village Panchayat, Nawal Parasi District*. Forestry Research Paper Series No. 12. HMG-USAID-GTZ-IDRC-Ford-Winrock Project, Strengthening Institutional Capacity in the Food and Agricultural Sector in Nepal.

RAPA 1987. Report of the regional consultation on the progress and problems of food production for nutritional adequacy. Unpublished. Bangkok: Regional Office for Asia and the Pacific (RAPA), Food and Agriculture Organization of the United Nations.

Robert, G. Lamar. 1982. *Economic returns to investment in control of Mimosa pigra in Thailand*. Chiang Mai, Thailand: Chiang Mai University.

Schubert, Bernd. 1986. *Proposals for farming systems-oriented crop research of Wawi Highland Agricultural Research Station in Northern Thailand*. Berlin: Center for Advanced Training in Agricultural Development, Technical University of Berlin.

Somnasaeng, Prapimporn, Phakkarat Ratkhaet, and Sumalee Rattanabanya. 1986. *Natural food resources in northeast Thailand, KKU-Ford Project: Socio-economic studies of the farmers in rainfed areas of Northeast Thailand*. Khon Kaen: Khon Kaen University.

Uprety, Laya Prasad. 1986. *Fodder situation: An ecological-anthropological study of Machhegaon, Nepal*. Forestry Research Paper Series No. 5. HMG-USAID-GTZ-IDRC-Ford-Winrock Project, Strengthening Institutional Capacity in the Food and Agricultural Sector in Nepal.

Vergara, Napoleon T. and Nicomedes D. Briones, eds. 1987. *Agroforesty in the humid tropics: Its protective and ameliorative roles to enhance productivity and sustainability*. Honolulu: Environment and Policy Institute, East-West Center.

Vryheid, Robert, Sriwan Wongcharoen, and G. Lamar Robert. 1987. Report on the results of a survey of the nutrition status in the project area of the Thai-German Highland Development Programme, Mae Hongson Province. Unpublished manuscript. Chiang Mai: Thai-German Highland Development Programme.

Agroforestry and Smallholder Financial Viability

Jacques Amyot

Chulalongkorn University
Bangkok, Thailand

This study of the exclusive usufructuary rights to use but not own examines the potential of agroforestry to help small farmers who have disturbed government forest land to achieve financial viability. The location is an FAO-assisted social forestry project initiated in 1979 in Northeast Thailand. The findings are derived from on-site visits, conversations with farmers and project staff, as well as project documentation and reports. Agroforestry practices promoted include private forest tree plantations, use of these trees for charcoal production, forest grazing of cattle, fruit tree plantations and beekeeping. As of 1986, marketing constraints and real or imagined legal constraints stopped many farmers from accepting the first two practices. Although not yet adopted by a large number of farmers, the success of individual farmers engaged in grazing cattle, planting fruit trees, and beekeeping demonstrates that these practices can yield substantial income. Whether all farmers in the project area will eventually follow the agroforestry route rather than the more tempting and currently more common practice of expanding holdings by illegal farming of forest lands remains to be seen.

Background

Khao Phu Luang is the site of a once degraded national reserved forest some 250 km northeast of Bangkok. The rehabilitation of this area as a forest and human habitat was the project's main objective. The area covers approximately 1,178 km^2 of rolling hill country with many streams flowing (at least at one time) in the valleys. As recently as 40 years ago, the whole area had a dense natural cover of dry evergreen and dipterocarp forest. It was largely uninhabited except for a scattering of isolated forest dwellers, who gathered forest produce as a livelihood, and a few farming communities along the edges of the reserve. By 1980, less than 10% of the natural and undisturbed forest remained, and it was forest land only in the legal sense. This was the result of successive waves of legal and illegal loggers, shifting cultivators, and commercial farmers and entrepreneurs planting cash crops. Many of the streams dried up with the destruction of the forest. There was a considerably larger resident population dispersed throughout the area, not concentrated in village communities. This population was poor, living in generally depressed conditions. Virtually no government health, education, or agricultural extension services reached them, and there were no roads. As encroachers on a national reserved forest, people were illegal squatters with no status and rights.

The Royal Forest Department (RFD) approached UNDP/FAO for assistance in dealing with these problems. It was agreed that a social forestry approach would be the most appropriate. Agreements were made for a preparatory phase of the project in March, 1979 and for an implementation phase in October, 1981. The latter phase, which was under UNDP/FAO technical assistance, was extended to September, 1986. A follow-up phase implemented solely by the RFD and other cooperating RTG agencies is still in progress.

The project document for the UNDP/FAO-assisted implementation phase stated that the project's four objectives were: (1) forest rehabilitation, (2) socioeconomic development, (3) project staff development, and (4) infrastructure development.

The immediate objective of the project was reforestation of 40% of the total area. Reforestation was intended to be harmonized with alleviation of poverty by resettling the people on the remaining 60% of the project area suitable for agriculture. People were expected to participate willingly in agroforestry activities on this land to complement cash earnings from agricultural production.

The STK Land Licensing Program

The land allocation activity of the social forestry project was carried out in terms of the government-sanctioned STK program. The program is named after the Thai acronym of the usufructuary certificate issued to qualified reserved forest land residents to legalize their status. The STK1 certificates authorize temporary use of the land for a specified number of years. The UNDP was assured that the STK1 would be followed eventually by the STK2, a permanent usufructuary certificate. By law, the STK certificate can cover only up to 15 rai (2.4 ha) of land, and conditions are imposed on the holder. This land can be transferred only by inheritance to direct descendants and cannot be rented, given to others, or sold. STK holders are required to report to the forestry authorities all illegal activities observed in their neighborhood or their occupier rights will be revoked. The project also had provisions for the allocation of an additional 10 rai (1.6 ha) per household on a communal basis, mainly for the establishment of fruit trees. The benefit expected from land allotment under the STK program was to give this population a feeling of security and a new sense of respectability.

In the experience of all land settlement programs in Thailand for rural poverty alleviation, the allocation of land alone to poor farmers is not usually sufficient to ensure success. Sufficient production support must be provided to make it possible for the farmers to generate adequate income from their farms. The project's challenge was to create a financially viable situation for farmers and ensure that the forestry protection and rehabilitation objectives of the project were met. This was simply not feasible on the basis of continued reliance on maize monocropping, especially with swidden techniques.

Several approaches to achieve financial viability used singly or in combination with others were possible. Theoretically, the simplest way was to expand each farmer's area of cultivation with 10 rai (1.6 ha) of the communal land. This would bring the total area of cultivation per farmer to 25 rai (4 ha), which most farmers would consider adequate. In practice this is very difficult to achieve. At the end of Phase II in September 1986, the majority of farmers cultivated or held considerably more than 25 rai, and many had less then 15 rai. A more efficient and better approach pursued by the project was to get the farmers to make more efficient use of available land by better farm management and to introduce cottage industries and part-time, off-farm work on tree plantations to supplement farm income.

The strategy employed to help farmers enhance their income earnings drew heavily, but not exclusively, on agroforestry. Another measure was promotion of crop diversification. Maize monocropping was widely practiced by local farmers. Efforts were made to introduce other crops grown in rotation to help maintain soil fertility and provide a sustained, reasonable income throughout the year.

Forest Tree-Farming Promotion

Agroforestry practices promoted by the project have included forest and fruit tree planting by the people for their own use, forest grazing of cattle, apiculture, and charcoal-making. Extension was provided mainly by project staff and associated experts, as well as by locally based Department of Agricultural Extension workers.

In the case of forest trees, the project produced its own seedlings and conducted research and field trials to determine which tree species were most appropriate for the local environment and needs of the people for fuel, lumber, soil conservation and improvement, etc. All species promoted in the project area were fast-growing, the main ones being *Eucalyptus camaldulensis* and *Leucaena leucocephala*, which were also used for forest rehabilitation. In beekeeping areas, *Calliandra calothyrsus* and *Eucalyptus deglupta* were also promoted. The project supplied approximately 167,950 seedlings to farmers and to schools. Instruction in planting and maintenance was provided. Schools were a focus of this activity, and teachers were involved. Some project staff taught agroforestry, and the school children planted trees on school grounds. Several agroforestry trials were established as experimental and demonstration plots, e.g., Leucaena hedges on contours of steep hillsides intercropped with maize, and maize intercropping in Eucalyptus plantations.

Response to these forestry extension efforts was lukewarm at first. As there was no critical shortage of woodfuel in the area, there was little perceived need to plant trees for this purpose. Even if they planted trees, the people felt they would not benefit because, in their experience, it was illegal to fell trees in a reserved forest area. The very concept of agroforestry was alien to these maize farmers, who felt that to plant trees in their fields would interfere with tractor plowing. Gradually, however, they became interested. One contributing factor was the expanding practice of beekeeping, which provided a motive. According to a sample survey conducted in 1985, 56% of the sampled farmers had at least started to establish hedgerows around their homelots and 51% had started to plant forest trees. Only 3%, however, had participated in establishing village woodlots.

Although the practice of planting trees had become fairly well established by 1985, silviculture as such had not yet become a source of income. It was said that some farmers did indeed market some of their trees, but the income generated was so insignificant (mainly because the trees sold were small) that they lost all motivation to continue. Clearly, the economics of silviculture had not been calculated. To help remedy this situation, the project hired a marketing expert to study the supply and demand for wood products of fast-growing trees in Northeast Thailand. This was completed in December, 1985.

The study confirmed that tree farmers in Nakhon Ratchasima Province currently faced many problems. They were currently selling predominantly to only two buyers, the Phoenix Pulp and Paper Company in Khon Kaen Province (Northeast Region) and the Thai Plywood Industry Co. in Samut Prakan Province (Central Region). Prices received were low in relation to costs, including middlemen brokerage fees and transportation (one quarter of the sale price). Farm gate prices, which excluded the cost of felling, preparation, loading and unloading, and transportation, were 450 baht/ton ($US 17.30) and factory gate prices were 600 baht/ton ($US 23.08). The economic analysis concluded that in spite of these problems, the net profit per unit of land used for tree plantation would be high compared to its use for other crops, in the long term. Assuming a 2 x 2 m spacing of trees on a three-year rotation, and a minimum price of 450 baht/ton, the internal rate of return would be 20-25%.

This optimistic view of potential economic opportunities from silviculture was bolstered by the study's projections of greatly increased demand over the next 15 years for trees for housing and furniture, pulp and paper production, fuelwood, and charcoal. In order for farmers to benefit from this increased demand, however, other conditions needed to be met. Farmers needed low-interest loans to tide them over until trees reached maturity, and more wood-consuming industries in the Northeast (e.g., pulp and paper mills and perhaps wood-fired thermal energy electricity generating plants) needed to be built. Other industries consuming large quantities of woodfuel investigated by another project study include earthenware factories, lime kilns, and tobacco-curing factories, all located in or close to the Northeastern Region. Marketing assistance would certainly be required before unsophisticated farmers could tap such outlets. From a policy perspective, the marketing study concluded that private sector (farmer) tree plantation should be promoted to meet the increasing demand because the only possible alternatives, increasing wood imports or drawing on existing forest resources, were unacceptable.

Charcoal Production

While the idea of planting trees for sale to wood-based industries might have been viewed with skepticism by the farmers, the alternative of transforming the trees into charcoal for their own use or for sale would appear to be more immediately attractive. Farmers have a considerable understanding of the economic value of these products and are familiar with the process of making charcoal. The prevailing market price of charcoal was 45 to 60 baht ($US1.73 to $2.31) per 50 kg bag, with the price differential apparently due to transportation costs. According to a survey of 244 households conducted in early 1986, there was widespread use of charcoal for home cooking in the project area. Only 12.7% of these households used only firewood for cooking. All the others used charcoal either exclusively (60.25%) or partly, with firewood also being used. The average charcoal consumption per household per year was about 14 bags (700 kg). Charcoal and fuelwood were equally available, but charcoal was the preferred fuel as it was cheap and more convenient to use. Over 28% of the households interviewed purchased their charcoal. Most of the charcoal consumed was produced from local forest wood by the people using very simple and inefficient earth-mound kilns. The pattern of production appears to conform to what one generally finds in rural areas of Thailand, and reveals an element of specialization. Some rural people prefer to buy charcoal rather than produce it themselves. Others produce it as an occupation or sideline.

While this situation might appear to be innocuous, it entails problems of considerable consequence for the forest resources of Thailand. The plundering of forests for firewood and charcoal production has been a contributing factor in deforestation. The activity is illegal, and sizable movements of charcoal are presumed to be related to the illegal felling of trees. However, charcoal is an essential commodity, especially in rural areas, where alternative sources of energy such as electricity or liquefied gas are unavailable or too expensive. There is therefore a crying need to normalize charcoal production. The social forestry project provided an ideal setting to work out a solution consistent with both the forestry and developmental objectives of the project and which could be replicated in other parts of the country. Although this element was included in the original project document, its implementation was delayed to the last few months of Phase II in 1986 because of difficulties in recruiting a local dendro-energy specialist. There was not enough time in this brief interval to see the potential of this activity brought to full fruition.

The approach selected was to get charcoal producers to establish and use plantation tree species and to introduce more efficient but inexpensive technologies. A minimum objective was to meet local domestic demand for charcoal. Eventually this could be expanded into a substantial source of income by catering to the considerable demand for charcoal by households, restaurants, and industries throughout the country.

The most popular types of kilns in Thailand are the brick beehive kiln, the mud beehive kiln, and the earthmound/rice husk mound kilns, which, according to RFD figures, account for 17.4, 36.5 and 35.6%, respectively, of all charcoal production in the country. The brick kiln is the most efficient, with a 35% rate of recovery of raw material, but it is less popular as it involves an initial investment of 3,000 to 5,000 baht ($115-192) depending on the size of the kiln (from about 2 m^3 to 8 m^3). The earthmound kiln used in the project area is preferred by many, especially in forest land areas, because it is cheap, requires only family labor, and is not very conspicuous, an important consideration in an illegal activity. It is very

inefficient, however, with only a 12.5% rate of recovery. According to RFD figures, this type of kiln consumes 52% of the annual total wood raw material for charcoal production to supply only 35.6% of what is produced. The third and most popular type of kiln, the mud beehive kiln, is twice as efficient as the earthmound kiln with a 25% rate of recovery. This is lower than that of the brick kiln, but family labor is the only investment required to build it. If this kiln were used instead of the earthmound kiln nationally, it would save nearly 5 million tons of wood raw materials per annum.

Due to the advantages mentioned, the mud beehive kiln was selected for the project area. Two sizes, 2 and 3.7 m^3, were proposed. Kilns of this size were considered adequate to produce enough charcoal for household consumption or for sale as a supplementary source of income. Larger sizes would have been unacceptable to the RFD, presumably because not enough trees were grown privately to supply them and they could not be operated without drawing on the natural forest or on reforestation plantations. A total of 7 kilns of both models were built with at least one of each model located in each of the three sectors of the project area for training and demonstration. Demonstrations were conducted using locally grown eucalyptus, and samples of the charcoal produced were given to users in the area.

Fourteen farmers were trained in mud beehive kiln construction and operation in the first half of 1986 with the expectation that they would train others. In addition, over 300 persons from the project area and from outside came to observe the operation of the kilns. A simple illustrated manual on the construction and operation of the kilns was prepared and distributed to farmers.

A small survey of 68 farmers was conducted to determine their willingness to build the kiln. The response was less than enthusiastic. Reasons given for not wanting to build the kiln were lack of money (34), lack of available space (10), and fear of getting in trouble with the RFD (22), all of which lacked basis in fact. The kiln required only family labor to build, occupied very little space, and was approved by the RFD for wood from the farmers' private tree plantations. If the responses were indeed candid, they indicate that more time and more extension efforts are needed. There could be deeper reasons for the resistance, however. As mentioned, charcoal is used extensively in the project area, and there is sufficient illegal production to meet the demand. This illegal production provides a good source of income of about 20-30 thousand baht ($770-1,154) a year. It is, moreover, almost impossible to control, and old habits tend to be persistent. In the absence of any perceived sense of urgency to change, the real issue is how to persuade the people to abandon their forest-destructive practices and make charcoal in more efficient mud beehive kilns with wood from their own plantations. This requires more than the usual rhetoric about forest conservation. The provision of economic incentives based on demonstrated opportunities offers the best promise of success.

Silvopastoral Activity

The reforestation program of the project provided considerable scope for the promotion of silvo-pastoral activity in tree plantations. Surveys conducted at the inception of the project reported little cattle raising. But when the project area was visited in mid-1986, cattle raising had become a more important activity, especially in the southern sectors. One farmer in Khok Samran/Khao So village had more than 200 cattle. This was exceptional, but herds of 40-50 were not unusual. Some farmers even raised water buffalo for sale to farmers in lowland areas. The project supported this activity by allowing the farmers to graze their cattle in tree plantations under controlled conditions. This was mutually beneficial as it contributed to weed control. Moreover, about 16 ha of forest plantation were planted in guinea grass and other forage plants to improve grazing.

Fruit-Tree Planting

Besides promoting forest tree plantations by the people, the social forestry project also supported the planting of fruit trees and the establishment of orchards. This was encouraged, especially in relation to the new agroforest villages and other consolidated villages, as a means to foster permanent settlements. As in most rural areas in Thailand, local people were already growing some fruit trees and did not need to be convinced of the usefulness of planting them. As one of the project progress reports pointed out, fruit trees cater to the interests of foresters and the people, and lead to better mutual understanding and more positive attitudes about each other.

Fruit tree seedlings were distributed to the farmers at the onset of the project as a good-will gesture. Twenty-eight farmers and two project staff were trained in plant propagation in May, 1982. After the training, the farmers were supplied with grafting material from good varieties of mangoes to do their own propagation. The project also supplied four grafted mangoes of popular varieties to most households to grow and use as a source of grafting material. Many farmers subsequently produced their own grafted trees, especially mangoes, instead of paying 50 to 100 baht/seedling for them. As farmers became self-supporting in the development of fruit orchards, the project lowered its priority in the supply of seedlings, even though this aspect had been rated high in the project document.

As of April, 1986, approximately 60,000 seedlings had been distributed to the farmers and the area in fruit orchards was estimated to be 590 ha. Many different kinds of fruit and other tree crops, such as coconut, cashew, and bamboo (for bamboo shoots) were being grown. The most popular fruits were mangoes, jackfruits, custard apples and sweet tamarinds, but others were also produced, including papayas, bananas and limes. An agribusiness firm tried to get the farmers in one village (Pong Wua Daeng) to establish a 32-ha cashew plantation, but the seedlings proved too delicate to handle.

By 1986, several farmers who had planted fruit trees at the beginning of the project period were earning money. One farmer visited in mid-1986 had sold 40,000 baht ($1,540) worth of mangoes and was able to purchase his own pick-up truck for delivering them. Such visible benefits are probably responsible for the considerable enthusiasm for orchard plantation, and many farmers procured fruit tree seeds and seedlings from other than project sources. One farmer said he was planting all of his land in fruit trees. He said he could not grow maize indefinitely without exhausting the soil and that there was more security in fruit trees. He saw this as the crop of the future for the area. This had not yet happened, of course, and the acreage in fruit orchards was still relatively small (11%) compared to that in field crops. It had grown significantly, however, and several successful fruit farmers demonstrated that it could provide a good source of income.

Beekeeping

An apiculturist with the Faculty of Agriculture of Khon Kaen University assumed responsibility for promoting bee culture in the project area. In March, 1982, eight farmers were sent to Khon Kaen University for a one-week training. Each farmer was loaned three bee colonies purchased by the project, and trial bee keeping with the European bee, *Apis mellifera*, commenced early in April. Three project staff members were trained in beekeeping to assist in extension work. By the end of the year, 30 hives had been purchased for use in the project area. Success in beekeeping was highest in the south-western sector, apparently because fewer crops requiring heavy applications of insecticide were grown there. It was decided not to increase the number of beekeepers for the time being until the problem of providing alternative sources of feeding during periods when pollen supply was inadequate could be overcome. In the meantime, 3,570 seedlings of Calliandra calothyrsus and 500 seedlings of sunflower were produced for beekeepers interested in providing bee forage during the dearth period from May to September.

The project apiculturist visited the beekeepers regularly. Second-year results showed that the honey flow period lasted one month longer than that of the first year because of the increasing number of nectar-bearing plants grown. Also, the honey yield per colony was higher as a result of more experience and better honey bee colony management. The take-off year for apiculture was 1984. Ten additional farmers were trained in beekeeping in June, but the growth of the industry became a self-sustaining process. Other farmers were trained by those already trained. New bee colonies were formed from existing colonies. Even beekeeping boxes and frames were produced by local farmers for sale to fellow farmers. The October, 1984-March, 1985 Project Progress Report describes this development as follows: "There were 11 beekeepers in July, 1984; by December the number of beekeepers had risen to 18; between January and March 1985, the number had further increased to 28. Within a period of nine months (July 1984-March 1985) the number of beekeepers had increased from 11 to 28; the number of bee colonies increased from Mr. Nong's 10 colonies (in 1983-84) to 211 colonies." By March 1986, the number of beekeepers had reached 41, and the number of colonies 350. It is interesting to note that although the training in beekeeping had been given to men, it came to be practiced more by women, who presumably acquired the skill from their menfolk.

Table 1 illustrates the economics of beekeeping in the project area in early 1985. The data are from a well-organized beekeeper, but they demonstrate without doubt that beekeeping as a sideline can perhaps be more lucrative than farming itself. Gross income from 20 months of beekeeping was 75,100 baht ($US 2,888). The net income (if the shadow cost of labor is discounted) was 65,420 baht ($2,516). To generate such levels of income, beekeeping had to be more than a hobby. No data are available on actual earnings of all beekeepers, but there was probably considerable discrepancy among them, depending on the intensity and diligence with which it was practiced.

Honey production in 5 of the 6 villages from October 1985 to March 1986 was 2,522 kg. At this level of production, marketing problems occurred. The price of honey varied from 70 baht/kg ($2.70) at the peak of the production period in January - February to 100 baht ($3.85) after the end of March. During this production period, the project office assisted in the sale of 595 kg of honey through informal channels, obtaining top prices, and was exploring the possibility of finding more regular and permanent marketing arrangements after termination of Phase II of the project. Honey producers met in mid-1986 and expressed concern about falling prices, but they were still optimistic. Even at 70 baht/kg, the margin of profit was still attractive enough to motivate them to expand.

Conclusions

One of the major challenges of the project was to assist farmers having full access to only 15 rai of land to achieve an economically viable situation. A number of agroforestry-related enterprises introduced to the area show promise in this respect. Some were accepted and demonstrated as profitable by farmers. Although not yet affecting the population as a whole due to the relatively small number of acceptors, their potential for enhancing the income of the whole population is considerable. The successful demonstration of an innovation by the people themselves is the surest way of assuring the diffusion and acceptance of that innovation throughout the village population. Although the data are not available to fully document it,

Table 1. One farmer's accounts over a twenty-month period.

Inputs	Baht	
Timber for production of 9 colonies	270	
9 boxes of 9 frames each w/metal cover	900	
250 wax foundation sheets at 25 Baht each	6,250	
Sugar for supplementary feeding when pollen is scarce (May-July)	50	
Paint for boxes	900	
310 bottles (approx. 750 ml.) at 1 Baht each	310	
Labor cost for 8 days/mo. for 22 mos. at 40 Baht/day (shadow cost)	7,040	
Miscellaneous	1,000	
Total	16,720	($643)

Returns	Baht	
Nov. 1983-April 1984: 250 bottles honey at 100 Baht from 9 colonies	25,000	
Nov. 1980-March 1985: 200 bottles honey at 100 Baht (colonies increased to 17)	20,000	
Sale:		
15 colonies (1 queen bee + 5 full frames) at 1,500 Baht each	22,500	
2 colonies (full) at 3,000 each	6,000	
wax produced: 4 kg at 150 Baht each	600	
50 queen bees at 200 Baht each	1,000	
Total	75,100	($2,888)

Net Income	Baht	
Including shadow labor cost	58,380	($2,245)
Discounting shadow labor cost	65,420	($2,516)

Source: THA/81/004 Progress Report, Oct. 1984-March 1985.

beekeeping, fruit tree plantation, and forest grazing of cattle seem to belong to this category.

One hopes that other proposed forestry-related innovations, such as private forest tree plantation for sale to wood-consuming industries, for use as construction material or fuelwood and charcoal production, or intercropping cash crops with trees will prove important in the future. Although some enthusiasm for forest tree plantations began to appear in the final years of the project, it appeared to be waning for lack of demonstrated profitability. Clearly, more time and more forestry extension efforts are needed. As mentioned, such agroforestry approaches are quite alien to these maize farmers and the transition to a "tree farmer" mentality does not come naturally.

The long-term objective of the social forestry project is to bring all farmers in the project area (about 1,500 households in 1986) under the STK program, with each 15-rai holding becoming financially viable for the occupant through the agroforestry-related enterprises described. Any land left over would constitute a pool of communal land to be rented out to the people for orchards, village woodlots, etc. While theoretically sound, the practical difficulties of implementing the plan are formidable. Although the agroforestry component of the project as a means of generating income shows real promise of eventually gaining widespread acceptance among area farmers, it does not follow that these farmers will willingly submit to the landholding size limitation of the land-licensing program. The history of their landholding and land-use practices since 1981 offers little hope that this will happen.

Let us summarize some of the findings of the economic studies conducted in the project area in 1981 and 1985. From 1981 to 1985, the mean size of landholdings increased from 3.98 ha to 5.02 ha. In 1985, about 50% of a sample of 300 farms were larger than 3.2 ha; 16.7% were roughly in the STK program permissible size range--1.6 to 3 ha; the remaining 33.3% were marginal farmers. In the same interval, mean annual income from farm enterprise almost quadrupled, going from US $326 to $1,130. Maize cultivation continued to be the basic enterprise throughout this period. Maize farmers increased their incomes primarily by expanding the area under cultivation, illegally in the majority of the cases. There is a clear correlation between income and size of holding. Most of the

farmers occupying holdings of the size permitted by the STK program were in the "break-even" category, with annual household financial balances, either negative or positive, not exceeding $192. This suggests that the main beneficiaries of the agroforestry innovations described were not the targeted smallholders but rather the more substantial farmers.

Despite agroforestry's potential to generate enough income to make a smallholding financially viable, the STK program's policies on the permissible size of holdings and stringent limitations on the rights of the holder to dispose of this land have never been popular in the project area. As of 1986, 68% of the agricultural land earmarked for implementation remained unallocated because of local resistance. To expect that the farmers controlling this area will voluntarily cede their claim to it without legal and police action would be clearly unrealistic. More flexible alternatives are obviously needed.

REFERENCES

Amyot, J. 1987. *Forestland for the people: A social forestry project in northeast Thailand*. Rome: FAO.

Boonruang, Prem. 1985. Supply and demand for wood products in northeast Thailand. Unpublished project document.

Feder, G., Tongroj Onchan, Yongyuth Chalamwong, and Chira Hongladrom. 1986. *Land ownership security, farm productivity, and land policies in rural Thailand*. Bangkok: World Bank, Kasetsart University, and Thammasat University.

Hoamuangkaew, Wuthipol. 1986. The comparative study between the economic conditions of the villagers living in the area covered by the Diversified Forest Rehabilitation Project, N.E. Thailand, in 1982 and 1985. Unpublished project document.

Kaenmanee, Sumeth, L.K. Danso, and G. Kuchelmeister. 1982. Economic survey of the area covered by the Diversified Forest Rehabilitation Project, N.E. Thailand. Unpublished project document.

Project THA/81/004. 1980-1986. Quarterly progress reports, September 1980-March 1986. Unpublished project documents.

Research (Consulting Firm). 1986. Farm and fuelwood in Nakhon Ratchasima Province, Thailand: Analysis of costs, benefits, problems and prospects of Diversified Rehabilitation in Northeast Thailand. Consultancy report THA/81/004.

Songboonkaow, Vera. 1986. A case study of the present level of fuelwood and charcoal production and consumption in Thailand substantiated by data from action research in Nakhon Ratchasima and field visits. Unpublished project document.

UNDP. 1981. Project THA/81/004 Document: Development of Diversified Forest Rehabilitation Northeast Thailand. Unpublished project document.

Growing Multipurpose Fruit Trees in Bangladesh: Farmers' Perceptions of Opportunities and Obstacles

Kibriaul Khaleque

University of Dhaka, Department of Sociology
Dhaka, Bangladesh

Bangladesh homestead forests are being overcut to meet increasing demands for fuelwood and timber. A survey was made of the conditions of these forests, farmers' knowledge and experience of tree growing, their preferences and perceptions, and obstacles to planting in seven districts that form a cross-section of Bangladesh's agro-ecological zones. The survey revealed that almost every homestead contains a combination of different tree species, a bamboo grove, and woody shrubs. Farmers generally prefer to grow fruit trees, because they also can provide fuel, fodder, and timber. Multipurpose trees are thus more important to them. Bangladesh farmers are aware of the value of trees and want to plant more. A lack of quality seedlings, fencing materials, financial support, and extension services are major constraints preventing more tree planting. A homestead forestry program should therefore be designed in Bangladesh to strengthen existing extension services and provide desired seedlings and other support to farmers.

Natural forests in Bangladesh occur in (a) the eastern hilly region, which covers parts of Chittagong Hill Tracts, Chittagong, and Sylhet Districts; (b) the delta region in Khulna District, and (c) the small patches of *Shorea robusta* (sal) forests in the central region (Dhaka, Mymensingh, Tangail, and Jamalpur Districts) and in the northern region (Rangpur, Dinajpur and Rajshahi Districts). More than 80% of Bangladesh's total population live in villages far away from these natural forests. Facilities for transporting forest products to these villages remain inadequate, and the majority of rural dwellers cannot procure products from these natural forests. Furthermore, the natural forests, which are being managed for commercial products, are an inadequate fuelwood source even to nearby residents.

Bangladesh's farmers have always had to depend on the trees in their homesteads. Byron (1984) estimated that homestead forests produce 65%-70% of the sawlogs and about 90% of the total fuelwood and bamboo consumed in Bangladesh. To meet increasing demands for fuelwood and timber, overcutting of trees from homesteads is now rampant. Hammermaster (1981) estimated that 8.9% of village and homestead forest volume was harvested in 1981, a percentage well in excess of growth.

This study assessed the present condition of homestead forests and farmers' knowledge about growing trees. The study also focused on farmers' preferences and perceptions of both the opportunities for and obstacles to planting trees. A sample survey was taken in seven villages in Barisal, Chittagong, Dhaka, Mymensingh, Pabna, Rangpur and Sylhet districts, providing a cross-section of Bangladesh agro-ecological zones. Fifty household heads were interviewed in each of the seven villages. Thus, 350 households out of the total 2,254 households in the 7 villages were included in the study.

Tree-Growing Practices among Farmers

Farmers generally plant trees around their homesteads or "baris." A bari is the area around the houses where members of extended family units live. Generally, it is built on a mound raised above the surrounding farmland, in the hope of avoiding flood waters. The extra earth required to raise the ground is usually obtained by digging ponds. The mound area is surrounded by fences and/or trees, bamboos, and shrubs. The typical homestead thus contains the houses of different pairs of the extended family, their vegetable gardens, threshing grounds, cowsheds, ponds, trees, bamboos, and shrubs. Trees are planted in the backyards, on the pond-sides, and around the cowshed area. Small areas are generally planted with trees, specifically bamboo, to meet household fruit, fuel, timber, and fodder requirements.

The size of a homestead varies from 0.12 to 3 ha, depending on the number of sub-families. The survey revealed that an individual household's share in homestead land ranged between 0.012 and 1 ha. The average was 0.10 ha of homestead land per household. Except for 11 households (3.1%), all others owned their own homestead (Table 1). Even some of the households that did not have their own farms owned their homestead land, although their homestead area was smaller.

Various tree and bamboo species are present around the homesteads (Table 2). The particular species mix is determined by tradition and area. Farmers generally plant trees and bamboos in the same location that their ancestors planted the same species. However, some farmers have discovered that other species can be grown more successfully in locations other than the traditional locations.

The survey showed that about 70% of farmers had planted an average of 10 trees within the previous 5 years. One-third of all households had planted 11 or more trees (Table 3). About 68% of the respondents said they had space in their homesteads for planting more trees, and 44% of all farmers have room for 11 or more trees. Both males and females planted trees in about 70% of the households interviewed. Tree planting was limited to males in 22% of the households and to females in 8%. Thus, women play a significant role in tree planting. Males usually plant and tend trees farther away from the residence, while females plant trees closer to dwellings. Tree care is generally the responsibility of all family members.

Farmers' Perceptions of Opportunities and Preferences

The predominance of mango (*Mangifera indica*), jackfruit (*Artocarpus heterophyllus*), and other fruit trees in nearly all homesteads throughout the country (Table 2) indicates a preference for fruit trees. Most farmers, however, also plant a few timber and bamboo species around their homesteads. Farmers do not generally plant tree species specifically for fuel or fodder (Table 4). The increasing scarcity of cooking fuel was mentioned by 80% of the respondents, but only 8.5% of farmers assigned top priority to fuelwood species. Farmers indicated that fruit trees were their first preference (76.5%), followed by timber trees (9.1%) and bamboo (5.9%). None of the farmers specifically mentioned fodder tree species, but

difficulties in securing cattle feed were mentioned by 71% of the respondents.

Because no female investigators were involved in the study, it was not possible to interview the female members of the households. Local custom discourages male strangers from talking to female household members. Interviews with the male respondents indicated that females would agree with the preferences for tree species given by their husbands. Collecting fuelwood and fodder is the responsibility of women, particularly in the poorer households, and they thus face the drudgery more than their husbands do. Women may therefore prefer fuelwood and fodder species, but they may not if they get fuel from other trees (i.e., fruit and timber species) or from other sources.

There is clearly a contradiction between farmers' needs and preferences. This contradiction may be explained because a combination of cow dung, rice straw, paddy husks, bamboo residues, tree leaves and twigs, jute residues, and other minor fuels traditionally were used as cooking fuels by rural people in Bangladesh. Tree fuels are often mixed with other fuels in the wet season when other materials are not readily available, combustibles preserved for the wet season are exhausted, and when farmers do not have other fuels. The last situation is particularly true in the case of households with smaller farms and homesteads. In the past, these poorer families were allowed to collect rice straw and cow dung from the land of richer farmers. Now that these combustible materials are becoming scarce, this gathering is no longer possible. Poor people thus suffer the worst in the fuel crisis and have no alternative but to purchase tree fuels. The poorer small farmers therefore usually give top preference to fuelwood species. Some other farmers also prefer growing fuelwood because increasing market value has made it more profitable.

Bangladeshi farmers do not plant single-purpose species even amid acute shortages of fuelwood. They can obtain tree fuel by cutting the branches of fruit and timber trees, or by felling unproductive fruit trees. Fuelwood also can be procured from shrubs and tree species that naturally grow around the homesteads and are usually regarded by farmers as baje gachs or akatha (trees not suitable for timber or that do not bear edible fruits). These tree species were classified in Table 2 as fuelwood species. Farmers generally grow these species on

Table 1. Amount of homestead land owned.

Household's share (acres)	No. of households	% of total
0	11	3.1
less than 0.10	45	12.9
0.10-0.20	121	34.6
0.20-0.50	153	43.7
0.50-1.00	15	4.3
1.00-1.50	3	0.9
1.50-2.00	2	0.5

Source: Survey conducted by author, 1986-87.

Table 2. Species common to homesteads.

Scientific name	Local name	No. homesteads w/species	% of total
Fruit			
Mangifera indica	Aam (mango)	324	92.6
Artocarpus heterophyllus	Kanthal (jackfruit)	285	81.4
Areca catechu	Shupari (betel nut)	223	63.7
Cocos nucifera	Narikel (coconut)	177	50.6
Phoenix dactylifera	Khejur (date palm)	112	32.0
Eugenia spp.	Jaam	109	31.1
Zizyphus jujuba	Boroi	107	30.6
Psidium guajava	Peara (guava)	97	27.7
Palmyra spp.	Taal	72	20.6
Aegle marmelos	Bel	58	16.6
Annona spp.	Ata	42	12.0
Emblica officinalis	Amra	33	9.4
Timber			
Albizia spp.	Koroi	124	35.4
Samanea saman	Rendi	120	34.3
Azadirachta indica	Neem	95	27.1
Fuel			
Erythrina spp.	Mandar	157	44.9
Lannea coromandelica	Jhiga/Kafula	146	41.7
Salmalia malabarica	Shimul	135	38.6
Tamarindus indica	Tetul	102	29.1
Streblus spp.	Sheora	75	21.4
Lagerstroemia speciosa	Jarul	72	20.5
Acacia nilotica	Babla	58	16.6
Bamboo			
Bambusa vulgaris	Barak	209	59.7
Bambusa nutans	Talla/Makla	162	46.3
Melocanna bambusoidea	Muli/Nali	120	34.3

Table 3. Number of trees planted during five years prior to survey and percent of space for new trees.

No. of trees	% households that planted trees during last five years	% households w/space for new trees
0	30.3	32.2
1-10	36.3	23.7
11-20	17.1	22.3
21-30	10.0	10.3
31-50	4.6	6.6
51-100	1.7	3.4
more than 100	0	1.4

Table 4. Multiple use score of various tree species.

Species	Use*				
	Fruit	Fuel	Timber	Other	Score
Mangifera indica	100	86	72	0	258
Artocarpus heterophyllus	100	53	100	0	253
Eugenia spp.	100	81	70	0	230
Phoenix dactylifera	74	80	0	73	227
Areca catechu	100	100	0	24	224
Cocos nucifera	100	100	0	24	224
Palmyra spp.	100	80	0	40	220
Zizyphus jujuba	100	100	0	0	200
Psidium guajava	100	100	0	0	200
Erythrina spp.	0	100	0	100	200
Lannea coromandelica	0	100	0	100	200
Salmalia malabarica	0	100	0	100	200
Azadirachta indica	0	78	100	0	178
Samanea saman	0	75	100	0	175
Albizia spp.	0	7	100	0	175

* % of respondents using species.

Note: Fodder was not considered as a specific use or preference in the questionnaire (see text).

sites where no other valuable timber or fruit trees can be grown. They do not take as much care of these trees as they do of trees planted for other uses. Most baje gachh and shrubs do not need any care.

Grass, rice-straw, rice-bran, other crop residues, and oil cakes are the most common fodders. Cattle are fed on bamboo and tree leaves only when the above are unavailable. Ordinarily, tree leaves are not regarded as good fodder except for feeding goats. Since leaves for fodder can be obtained from other multipurpose trees, farmers do not plant single-purpose fodder species, which might explain the lack of stated preferences for fodder species.

Timber can be obtained from many fruit trees, but certain timber species are preferred. Higher market values and increased income potentials are the main reasons for preferring timber species. Although it takes longer for a timber tree to mature, farmers view timber trees as long-term investments. Before they reach maturity, the branches of timber trees are often cut and used as fuelwood. Timber trees thus constitute a form of family insurance, with the occasional bonus of fuelwood. Although every farmer wants to plant timber species, this is only possible for those with large homestead areas who can wait for long-term returns.

In summary, if a farmer has space in his homestead for planting one tree, he would plant a fruit tree. If he has space for more than one tree, he would plant a timber tree or bamboo. A farmer would not plant a single-purpose fuelwood or a fodder species unless it was a cash crop. Tree species that provide fruits, e.g., mango (*Mangifera indica*) and jackfruit (*Artocarpus heterophyllus*), are perceived as multipurpose trees because they provide fruits for eating and for the market, their green leaves can be used as fodder for goats, their dried leaves and twigs can be used as fuel, their branches can be cut and used as fuel, and the tree itself can later be used as timber. Jackfruit is considered a good timber for furniture, doors, and windows, as is mango to a lesser extent.

Among baje gachh, certain species are regarded as multipurpose. Mandar (*Erythrina* spp.) is often planted as live fences on the edges of vegetable gardens because the thorns of these trees prevent cattle from entering. Mandar is a fast-growing species that needs little care; if a branch from a larger tree is planted, it grows naturally. The branches can be cut every year and the original tree meant for fencing remains. Both the branches and the whole tree can be used as fuel. Shimul (*Salmalia malabarica*) is also considered a multipurpose tree because it provides cotton wool for making pillows and mattresses. Shimul trees can also be used as fuelwood, although it does not provide good-quality fuel. At present, there is an increasing demand for shimul wood in match factories.

Farmers' Knowledge of Growing Trees

All farmers are familiar with tree cultivation. Knowledge of tree growing is transmitted from one generation to the next, and new ideas and skills are added in every generation. Farmers always learn from their own experiences. They know that leaving a tree to grow naturally after planting is not desirable because it needs care and protection. The most commonly mentioned measures for tree growing were selecting good-quality seedlings and suitable locations, preparing sites for planting, watering, putting manure (usually cow dung or compost) on seedlings, weeding, fencing, and protecting seedlings from cattle. Most farmers said they tried to adopt these required measures for tree growing, but that tree cultivation often failed for reasons discussed in the next section.

Obstacles

About 70% of farmers reported that growing trees is becoming more difficult. Although the remaining 30% did not consider tree-growing to be difficult, they alluded to constraints to tree cultivation. Most farmers mentioned several constraints to tree cultivation, and they were asked to rank these.

Lack of Seedlings

Lack of seedlings was the prime constraint perceived by 47% of farmers. Farmers generally try to raise seedlings around their homesteads, but frequent destruction by livestock makes it necessary to collect seedlings from relatives or neighbors or purchase them from the local market and/or government nurseries.

Most farmers who purchase seedlings mentioned the difficulties of bringing them home, particularly where sources are distant and transportation is poor. The costs involved prevent many farmers from procuring seedlings.

Lack of Land for Planting

This obstacle is faced mainly by farmers who have no homestead land or own only small areas. Small homesteads are mainly occupied by the houses and other structures of the different sub-families and there is no room for trees. Landless, smaller farmers cannot expand their homesteads. This obstacle was assigned the highest rank by 30.3% of farmers.

Lack of Labor

Lack of adequate labor for tree-growing was ranked first by 12.6% of respondents. Because the tree-planting and rice-growing seasons coincide, it is difficult for farmers to do both. This is particularly true of farmers with a limited labor supply or who work on other farms. A considerable amount of labor is required in clearing sites for seedlings and weeding and tending them. Trees are generally planted in wet weather so that watering is not essential. If there is a prolonged drought during the monsoon or if trees are planted in the dry season, more laborers are needed for watering.

Seedling Destruction

Destruction of seedlings, both before and after planting, was mentioned by 10% of the farmers as a major difficulty. Lack of adequate fencing to keep out livestock is the main reason.

Tree twigs and bamboo are used as fencing materials, but are themselves becoming scarce. Most farmers cannot afford fencing materials. Those who do have trees or bamboos have to burn their twigs for household cooking and sell their bamboo to meet more pressing needs.

Seedling destruction is also caused by people. For example, neighbors might kill a potentially large-crowned seedling if they fear it will evenutally shade their land.

Disease is another cause of seedling destruction. Most farmers do not know how to save diseased seedlings. It mattered less in the past when seedlings could easily be replaced, but this is no longer possible for reasons listed above.

Conclusions and Recommendations

Tree-growing is traditional among Bangladesh farmers. They prefer fruit trees with multiple functions, but also plant other trees and bamboos. They usually do not plant fuelwood trees as such. The best way to enhance fuelwood supply would be to encourage planting of multipurpose fruit trees. Farmers would even grow exotic trees if they were multipurpose. The basic knowledge and awareness of opportunities are present; farmers need little motivation or training to plant trees, except for non-traditional species.

Farmers also are eager to improve their homestead tree growth and should be provided with good-quality seedlings of the most desired species either free or at a reasonable rate. Seedlings should be delivered to their villages or within walking distance. Fast-growing species are best because farmers have little land to spare and cannot wait for long-term returns. Subsidies and loans should be provided for planting, fencing,and other costs.

The sustainability of homestead forest resources in Bangladesh is threatened by drastic overcutting without proper replacement. This calls for immediate action. The existing government program for homestead forest improvement should be strengthened and new programs established. All should be designed to help farmers increase their income-generating opportunities.

REFERENCES

Byron, N. 1984. People's forestry: A novel perspective of forestry in Bangladesh. *ADAB News*: 11(20):31-37.

Hammermaster, E.T. 1981. *Village forestry inventory of Bangladesh: Inventory results*. Field Documentation No. 5, UNDP/FAO Project BGD/78/0/20. Dhaka: Food and Agriculture Organization of the United Nations.

Introducing Multipurpose Trees on Small Farms in Nepal

Pradeepmani Dixit

TU/IDRC Farm Forestry Project, Tribhuvan University
Kathmandu, Nepal

Farmers in the hills and plains of Nepal have different attitudes about tree planting. Hill farmers are more supportive and accept new species more readily. Land for tree planting is available on almost all farms in one form or another, but farmers have not given serious consideration to the management of the lands not used for crop production. These lands need to be exploited with trees as the main component. Labor requirements for rice and tree planting fall at the same time of the year, however, and tree planting must be managed so that it does not compete with rice. Planting of species not eaten by animals is recommended at sites frequented by unattended animals. Readily available plant residues like cornstalks have been used successfully to protect individual plants. Simple training programs and site visits to successful farm forestry models generate tree-planting interest among undecided farmers. The inclusion of a limited number of exotic species with traditional species allows farmers to compare and select between the two groups. The involvement of the farmer at each step of the operation is important to achieve success. Otherwise he starts to differentiate between the trees as "ours" and "theirs."

The Farmers

Hill farmers grow many different tree species to meet their daily needs of fodder, fuelwood and composting material. Animal raising and tree growing are very important parts of their farming system. Tree fodder is an important animal feed source. Terrace bunds and areas around homesteads are used for tree planting. These farmers accept new species more readily and it is easier to convince this group about the advantages of tree planting. Common species found on hill farms and their uses are listed in Table 1.

Terai (Plain) farmers grow only a few specific species of trees because most available land is already used for fruit trees (e.g., mango, litchi, jackfruit) or *Dalbergia sissoo*. Since animals are not fed tree leaf fodder, few farmers plant trees to produce fodder alone. Bamboo, which is found everywhere, is used for fodder, however. Terai farmers are generally reluctant to accept unfamiliar species. A few trees are planted in small blocks just outside the village. Planting on field bunds or around homesteads is not very common. Animals are fed with crop residues, such as straw mixed with grass scraped from the ground. Common fruit trees and other multipurpose trees found on small farms in the plains are presented in Table 2.

Available Land for Tree Planting

Many farmers in Nepal own less than one hectare of land. However, 3-15% of the area of most farms could be used for tree planting. The type of land found on the farm can be divided into four main categories. "Khet" are lowland areas where rainwater can be collected to plant crops like rice. "Pakho" are upland areas (usually sloping) where water cannot be collected and crops like corn, soybeans, millets are grown. "Bari" are flat uplands where corn, soybean, mustard are the main crops. "Kharbari" (marginal land) is not under any productive use and is covered by small bushes and grass. This area is used by the farmers to collect grass for cattle and thatch roofs. Development of intercropping systems between trees and crops in "Pakho" areas and silvipastoral systems for "kharbari" should be given serious consideration. The landholding and land-type distribution in a village that cooperates with the project are given in Table 3.

Decision Makers on Small Farms

It is necessary to understand the decision-making process and to identify the decision maker in each farm house to achieve success with the introduction of multipurpose trees. The decision maker in the family may be anyone from the grandfather to the grandson, depending upon the domestic situation. Usually, the decision maker is the chief income earner or the owner of the

Table 1. Uses of multipurpose tree species common to the hill farms of Nepal.

Species	Fodder	Fuel	Timber	Fruit	Oil
Aesandra dutyraceae	x	x	x	x	x
Artocarpus lakoocha	x	x	x	x	
Bauhinia purpurea	x	x	x	x	
B. variegata	x	x	x	x	x
Ficus lacor	x	x			
F. semicordata	x	x			
Garuga pinnata	x	x			
Litsea monopetala	x	x			

Table 2. Uses of multipurpose tree species common to the Terai.

Species	Fodder	Fuel	Timber	Fruit	Oil
Acacia catechu			x		
A. nilotica	x	x	x		
Anthocephalus chinensis	x	x	x		
Artocarpus integra	x	x	x	x	
Dalbergia sissoo	x	x	x		
Emplica officinalis		x		x	
Mangifera indica	x	x	x	x	
Tamarindus indica		x	x	x	
Zizyphus mauritiana	x	x		x	

property. Other members take over if the main member is either sick or incapable of making decisions. The general pattern observed, in order of importance, was grandfather, grandmother, son, first wife, second wife, grandson or granddaughter.

The decision to plant trees is generally unanimous (the women are almost always in favor), although there is usually disagreement about which species to plant.

Popular Species

There is a wide variety of popular species growing in the hill farms, and the majority supply good quality fodder. The trees produce fresh leaves in large quantities during the dry, hot months when animal feed is scarce. Important multipurpose trees, with their lopping season and the methods of propagation in the hills, are presented in Table 4. In the farms of the plains, the number of popular species is limited. Beside a few popular fruit trees like mango, litchi and jackfruit, multipurpose trees like *Dalbergia sissoo* and *Acacia nilotica* are popular. A few trees of *Anthocephalus chinensis, Artocarpus lakoocha*, and various *Ficus* spp. may found in some farms. *Azadirachta indica* is also found here and there. The most popular multipurpose tree in the plain is *D. sissoo*.

Reasons for Tree Planting

Hill and Terai farmers were asked why they planted trees. Most hill farmers replied that they grew trees for fodder and fuelwood (Table 5), as most animals are stall-fed with fodder from trees. Almost 50% of the animal feed requirement in some farms is met from privately owned trees, 33% is from agricultural by-products, and 17% is grass from field bunds.

Tree planting in the Terai is mostly for commercial production of timber and fruit (Table 5). A few selected trees are separated for domestic use, and the rest are sold. Branch loppings from *Dalbergia sissoo* and broken branches from fruit trees are used for fuelwood. Fruits from a few selected trees are consumed at home, the rest are sold.

Reasons Not To Plant Trees

A questionnaire circulated among hill and Terai farmers in the project area indicated that there were several constraints to tree planting (Table 6).

Lack of Land

One of the simplest answers given by farmers for not planting trees was they do not have land to plant trees. This may be true in some cases, but it is not the main reason. Some farmers who own only lowlands or rice paddy fields do have difficulties, but those who own uplands and marginal land should not have given this reason. Every farm has some land for tree planting. Marginal lands and land around roads, homes and field and sand bunds in each farm need to be exploited.

Lack of Time and Labor

Tree-planting and rice-planting season coincide. Rice is the staple food and gets top priority. Tree planting is therefore not done or neglected. Labor is not only scarce at this time, but expensive (up to double the normal rate) so farmers do not use hired labor to plant trees. It is necessary to take this into consideration when a calendar is made for tree planting.

Lack of Know-How

Some farmers say they do not have any idea what should be planted and where it should be planted. Such farmers are usually involved in specific, fairly successful enterprises, such as growing vegetables or dairy farming.

Lack of Decision Makers

This situation is common on farms cultivated by farmers who have rented the land from landowners, a fairly common practice in the Terai. The farmers who cultivate the land are not the owners (the owners live in cities far away), so they cannot make a decision for the landlord as to whether trees should be planted.

Lack of Inputs

Many small farmers lack the necessary inputs to take up tree-planting. Seedlings of the species they would like to plant are unavailable. Sometimes

Table 3. Total land holdings and distribution in a project village in the plains.

Total land holdings (ha)	Khet[1]	Pakho[2]	Irrigated Khet	Kharbari[3]
0.74	0.67	--	--	0.07
1.42	0.67	0.67	--	0.08
0.50	--	0.30	0.20	--
0.33	--	0.33	--	--
0.68	0.34	0.34	--	--
1.01	0.17	0.84	--	--
1.69	0.33	--	0.68	0.68
1.18	--	--	--	1.18
2.20	1.35	1.35	--	--
0.98	0.02	0.54	--	0.42

[1] lowland where rainwater is collected to plant crops such as rice.
[2] upland (usually sloping) where water cannot be collected. Corn, millet, and soybean are grown.
[3] marginal land under no current productive use.

Table 4. Lopping season and propagation methods of important multipurpose tree species.

Scientific name	Common name	Lopping season	Propagation method*
Artocarpus lakoocha	Badhar	Nov.-Feb.	A,S
Bauhinia variegata	Koiralo	Nov.-March	A,S,L
Ficus lacor	Kavro	June-Aug.	S,C,L
F. semicordata	Khanyo	Jan.-March	S,C,L
F. subincisa	Barulo	Nov.-March	S,C
Garuga pinnata	Dabdaba	Jan.-May	S,C
Grewia subinaequalis	Fosro	Dec.-March	S
Litsea monopetala	Kutmero	Nov.-March	A,S
Premna spp.	Ginderi	Oct.-Feb.	A,S,C

* A = coppicing, C = cutting, L = layering, S = seed.

seedlings are available but are small and sickly, and the farmers are reluctant to plant them. Although some banks have started to provide loans for tree-planting, farmers have not shown much interest because they have to wait a long time before the trees mature and start giving returns.

Repeated Failure

Besides unavailability of seedlings, protection of seedlings was the second most common problem identified by farmers because animals are allowed to wander freely. After repeated failures farmers have given up tree planting.

Introducing Multipurpose Trees on Small Farms

This section deals with the approach the Farm Forestry Project has adopted in introducing multipurpose trees on small farms. The whole approach has been planned with the farmers' direct participation so they are familiar with what is done. This also allows farmers to select what is best for their land and needs. The main reasons given by the farmers for not planting trees have served as guidelines, and the project has tried to remove the constraints one by one.

Farmer's Position in the Community

A careful investigation was done to determine the position of each farmer in the community. Three points were taken into consideration: (1) farm size, (2) relation with fellow farmers, and (3) the amount and type of non-crop area held.

The first point classified the farmer as big, medium, or small-scale. It also enabled us to judge the farmer's ability to take risks while trying new species or in investing money to plant and protect trees in poor, degraded areas. Bigger- or medium-sized farmers were more prepared to invest time and money. The second point enabled us to find out whether the farmer was honest and respected or distrusted by the community. If he fell into the first category, we enlisted him as a cooperating farmer and planted demonstration plots on his land. We avoided the second group. The honest farmer became our extension worker. The third point was needed to develop specific farm-forestry models using appropriate species and designs.

The Land Factor

Every farm has some land for trees. Non-agricultural land has not received due attention as a useful resource either from agriculturists or foresters. Field-bund planting in lowlands, agroforestry practices in the "bari" or sloping uplands and silvipasture in the "kharbari" (unused bush areas) provide a wide range of possibilities. In degraded agricultural lands and river-damaged areas, successful block plantations using native species like *Dalbergia sissoo* and *Acacia catechu* have been established. The empty area around homesteads, pond bunds, and along roads and paths can accommodate trees. Tree planting on these lands has been encouraged. Farmers need to be made aware of their unused resources. Species that require supervision and protection (like many fodder species) are planted near the home. Common multipurpose species planted close to the home are *Artocarpus lakoocha, Ficus semicordata, Bauhinia variegata,* and *Bauhinia purpurea.* Trees grown for timber or fuelwood are planted further away from the home. The species selected are hardy, fast-growing, non-browsed or able to recover quickly after being browsed, such as *Eucalyptus* spp., *Melia azedarach,* and *Azadirachta indica.*

Time and Labor

Time and labor problems encountered by the farmers for tree planting during rice-planting season are overcome by supplying the seedlings at the beginning of the pre-monsoon showers. One-year-old seedlings were made available. The farmers are glad to receive these well-developed seedlings when the labor demand for crop production is less intense. Seedlings planted at the beginning of the monsoon have a high survival rate and grow vigorously. School children who have summer holidays can supplement the labor requirement for tree planting while their parents are busy with agriculture. This has been encouraged.

Traditional and New Species

Farmers stick to their traditional species if they do not know about new species. Many farmers who have taken up tree planting continue to use one or two traditional species. Results are not usually satisfactory. The project has made a point of meeting 80% of the demand of each farmer with traditional species and the remaining 20% with

Table 5. Survey of farmers' reasons for planting trees by percent.

Farmer group	Fuel/ fodder	Monetary (timber/ fruit)	Environment protection
Plain	15*	73	12
Hill	70	15	15

* fuelwood.

Table 6. Tree planting constraints identified by heads of households.

Constraint	Hill Farmers		Plain Farmers	
	No.	%	No.	%
Unavailability of seedlings	16	64	13	52
Protection of seedlings	4	16	10	40
Land shortage	3	12	1	4
Lack of time/labor*	2	8	1	4
Total	25	100	25	100

* Labor problems are encountered mostly by large farmers.

Table 7. Average growth of exotic and local species on five farms thirty months after planting.

Species	Height (m)	Dbh (cm)
Dalbergia sissoo	6.8	8.2
Eucalyptus camaldulensis	7.4	5.2
Leucaena leucocephala	6.1	5.4
Tristania conferta	5.4	5.1
Acacia catechu	3.9	3.9
Cassia siamea	2.7	2.7
Ceiba pentandra	3.1	4.6

promising new species. This has allowed the farmers to compare the performance of new species with the familiar. They can then judge which species they want for future plantings. Many farmers have included these new species in subsequent plantings. The new species occupy only a small portion of their total requirement because farmers are still aware of the risk factor. Farmers in Karmaiya have included *Leucaena leucocephala* with *Dalbergia sissoo*; In Parwanipur they have included *Melia azedarach* with *D. sissoo*.

This idea of letting farmers test new species in their land and compare it with the most popular traditional species was adopted to save time in waiting until results are available from research stations. This method also provides a general idea of species performance in particular land types and farms, allows farmers to make their own selection based on personal observation, and discourages mono-specific planting even on a small scale. The performance of some exotic species as compared to local species is presented in Table 7.

Training and Visits

A simple practical training followed by visits to successful farm-forestry models has proved beneficial to farmers who claimed lack of know-how as the reason for not planting trees. After the training and visits, many farmers who were ignorant about planting trees have started to plant them. The free discussion that took place between the cooperating farmers and the visitors was an important factor in convincing the visitors about the advantages of trees. The farmers listened to each other.

Protection of Planted Trees

One of the following methods was employed to protect trees, depending on the number, area, and species:

o Protection of individual trees. Locally available materials like corn stalks and mature stalks of Cassia and Eupatorium were used successfully to barricade each plant.

o Protection at areas regularly frequented by stray animals. Species not eaten by animals, such as *Cassia siamea, Eucalyptus* spp., *Acacia auriculiformis, Melia azedarach*, and *Azadirachta indica*, were planted along roads,

paths, and fields adjoining public grazing grounds.

o Using live hedges. Large areas that need protection can easily be fenced using quick-growing hedge plants. Some species that have been used successfully are *Caesalpinia floribunda, Dodonaea viscosa*, and *Duranta repens*. These species grow vigorously and are not eaten by animals, and most also have strong thorns. They are more effective than barbed wire.

o Positioning a guard/workman. This has become the most effective protection at those sites that can be affected by floods and siltation. The guard/workman not only protects the planted seedlings but also weeds, cultivates, and cares for individual seedlings when the need arises. If quick-growing species are planted, this also becomes very cost-effective when compared to barbed wire.

Decision Making Factor

The government has reduced the land tax from about $US2 per 0.66 ha to about 2 cents for the same area if trees are planted. Such incentives have influenced many farmers to plant trees. Good agricultural lands that have been abandoned as unproductive after siltation or erosion have been planted with trees. This has not only decreased the land tax, but it also has given hopes to farmers for later returns. Farmers need a lot of convincing that the trees they plant belong to them and not to the Forest Department. When they are convinced of this, their decision is generally in favor of trees.

Session VI: Research Strategies for Filling Information Gaps

Chairman: C. Devendra

Discussants: Cherla Sastry
 Session Chairmen

Session VI Summary

C. Devendra

International Development Research Centre, Singapore

Keeping in mind that this workshop is the first of its kind, it is perhaps understandable that the sessions are general. On the positive side, this provides the unique opportunity to appreciate multipurpose tree species (MPTS) across programs and disciplines, and therefore underline a definition of the scope for future efforts. An important central issue in the overall objective of the use of MPTS is the search for directions for small-farm use.

Focusing now on the small farms, and in the context of multipurpose trees, the following questions can be asked:

o Does research start with the perceived needs of the farmer? or

o should there be a critical assessment of the perception of farmers and their aims, objectives, and rationale of choice of operations?

How sure are research and development specialists on the choice of technologies and entry points, such as MPTS, animals or irrigation, in enhancing farm productivity and income generation? Based on the perception of farmers, how do we determine the allocation of technologies and mechanisms that can ensure success and generate increased productivity? What rationale have we used in identifying the choice of technological priorities? How sure are we that our choice of these technologies is going to be advantageous and not disastrous? In particular, do we have an adequate research base to justify the extension of appropriate technology? I very much hope that these and other issues will be raised in the discussions and set the stage for future efforts.

Regarding animal production, I would like to stake a claim that the use of MPTS by animals is possibly the most important function of these trees. Some of the main advantages are described below:

o The use of tree forages is part of traditional systems of feeding animals because of the ready availability of these feeds on small farms.

o A variety of important forages are identifiable, supplying critical nutrients (N in particular), energy, minerals and vitamins, often during periods of feed scarcity and enabling feeding strategies that can be sustained throughout the year.

o The use of these forages represents a most unique way to reduce the cost of feeding and increases the margins of profit from ruminants (buffalo, cattle, goats, sheep) and possibly other herbivores, especially in feeding systems based on roughages like cereal straws.

o Use in fence lines including harnessing of animals, and

o Provision of shade, especially in the tropics, where high temperature can be a major limitation to animal performance.

Using the example of animals, we see there is inadequate information on inventories of MPTS, roles within patterns of farming systems, agronomic requirements, and quantitative and qualitative characteristics and priorities in the use of selected cultivars. With respect to animals, fodder production potential, high digestibility, minimum deleterious effects, and economic relevance are important advantages.

Given these background comments and the discussions that have been held, it is important to outline a set of suggestions and recommendations to enhance the future role and contributions of MPTS, specifically in the Asia region.

These steps need to emphasize the specific priorities and directions for research and development of small-farm systems. This will enable a more critical discussion of specific topics at future workshops, build on the results that are presently available, and strengthen levels of communication that have already been established here. I personally hope that our efforts can help realize results for small farms that are demonstrably superior to existing ones and that are targeted to uplifting the rural poor.

Methodological Issues in Determining Growth, Yield, and Value of MPTS

Patrick J. Robinson and I.S. Thompson

Forestry Research Project
Kathmandu, Nepal

Tree growth is related to the nutrient, moisture, light, and temperature regime of a site. Site management aspects also influence growth. Yield is a function of growth and a range of harvesting variables. Growth and yield comparisons between multipurpose trees need to consider all these factors and their interactions. The problems faced by on-farm research are highlighted in that context. Lack of time, resources, and information restrict the applicability/feasibility of this approach. Controlled trials on research stations suffer from the inability to study the interactions of the important variables on growth and yield. The value to farmers of the various products of MPTS is a function of a further range of factors that are location-specific. A multifaceted pragmatic approach is therefore required to provide the minimum data set necessary to interpret those factors conditioning yield and its value. The approach may include a combination of farmer surveys, formal trial sites, on-farm monitoring, and on-farm trials. Methodological problems of these approaches are discussed and the need for caution in the application of results in extension and development are stressed.

Research on multipurpose tree species (MPTS) can be divided into two broad categories. The first aims to provide extension agents and farmers recommendations on the potential growth, yield, and value of indigenous and exotic MPTS for particular environments.

To be valuable, research results should closely predict the growth, yield, and value of tree products on sites of varying ecological, management, and economic conditions. Hence, research results should be comparable between species, sites, and management regimes. For results to be comparable under different site and management conditions, they need to be interpretable under different conditions. In areas where farmers already use indigenous MPTS, one needs to be able to rank potential new species and subsequently monitor and compare their impact against those already existing in the farming system.

Improvements in the availability of MPTS products can be achieved without extension services. For millennia, farmers have been introducing new MPTS on their farms and perfecting their management (Robinson 1985a). More recently, extension agents have been introducing exotics and encouraging the multiplication of indigenous species and the use of improved management techniques. So far, however, the choice of species and management recommendations have largely followed subjective rather than objective selection criteria.

More objective ventures have been initiated recently. Some examples are the exploration and seed collection of MPTS in the dry zone of Central America (Hughes and Styles 1984) with a view to worldwide testing in different environments, and range-wide seed collections of *Gliricidia sepium* for international provenance trials (Hughes 1987). In Asia, the International Union of Forest Research Organizations (IUFRO) has identified the need for more objective and comparable evaluation of MPTS (Burley and Stewart 1985). This has led to the establishment of the Forestry/Fuelwood Research and Development (F/FRED) Project, the report of the MPTS Network Trials Planning Meeting (F/FRED 1986), a guide for MPTS F/FRED research cooperators (Lantican and Yantasath 1987), and this workshop. MacDicken (1986) has summarized the stages necessary for the planning and implementing of MPTS field trials related to Asian regional networks. Huxley (1986) and Burley and Wood (1976) are essential background documents to the exploration and evaluation of MPTS.

It is sobering that even for the star among MPTS, *Leucaena leucocephala*, recommendations concerning the best cutting height for maximum foliage production are contradictory for both Hawaiian and Peruvian types (Robinson 1985a). To honestly justify their work, researchers must be able to provide more useful, concrete, interpretable, and predictive results than farmers and innovative extension agents.

The second category aims to understand the processes underlying growth and yield responses to different management regimes. Only by understanding these biological processes can one interpret the different responses of species to varied environments and managements.

The aim of a scientific approach to the evaluation of MPTS performance must be the provision of a minimum data set for the various species tested on different sites. Lantican and Yantasath's guide (1987) is taken as a basis for discussion.

Fig. 1 schematically describes the three steps for the evaluation of growth, yield, and value of MPTS and the various quantitative and qualitative factors that influence them. This paper discusses issues related to the methodologies required for their assessment and the relative merits of research station trials, farm trials, and surveys, given the range of species to be tested and the ecological and socioeconomic situations to be included.

Species Selection

Because F/FRED has recommended that research should concentrate on three to four species for each of two environmental zones in the Asia region (humid/semi-humid tropics and arid/semi-arid tropics), a brief discussion of species selection is necessary. On a national basis, species selection for exploration and evaluation should start from the eco-zone coverage of the country and from the identification of key constraints to land-use systems that MPTS can likely address (Robinson, Burley, and Wood 1985). These criteria are regionally and nationally variable. Certain species still receive high ranking on MPTS priority lists for research because more is known about them. In some countries of Asia, such as Nepal, this has resulted in a concentrated research effort on exotics, sometimes those that do not produce the products most needed. In Nepal, less than 25% of the tree species on which research was done between 1970 and 1979 were indigenous (Hudson 1987). However, between 1980 and 1986, close to 50% of the species investigated were

indigenous. This is partly due to the slow realization that many indigenous species have considerable potential, and some perform as well, if not better, than recommended exotics. In Nepal, farmers consider several indigenous MPTS that taxonomists recognize as one species as comprised of two or more distinct types with different ecological ranges and/or value in terms of fodder (Upton, in press).

Inventories of locally used species, with farmers' evaluations concerning their relative value for different products and growth, still have to be made in many areas. Upton (in press), found that 70 tree species were viewed by various farmers as being one of the best 5 species for fodder in a small area of Nepal (two sides of a hill, 1,000 m altitudinal range).

From a small farmer's point of view, there are two important reasons why a range of species should be used on their farms. First, there is a need to spread the risk, so it is inappropriate to narrow the genetic base of tree resources. Second, for some products, such as flowers (for honey) or fodder, a range of species is required to provide a good-quality product during various seasons. Scientists should not therefore confine themselves to research on a few species.

Concurrently, promising indigenous species should be evaluated against better known exotics and indigenous ones in replicated species elimination and testing trials in different environments.

Site Assessment and Management

The purpose of site assessment is to correlate growth and yield parameters to site factors. This allows comparison of growth and yield of different species between similar sites, and growth and yield of the same species between different sites. The information gained should allow prediction of growth and yield under different site conditions along a cline of the various environmental factors that have been measured. Growth and yield are related to temperature, moisture, light, and nutrient regimes of the site.

Discussions and recommendations concerning environmental-factor evaluation of research sites have been made for tropical forestry by Greaves and Hughes (1976) and adapted for MPTS by

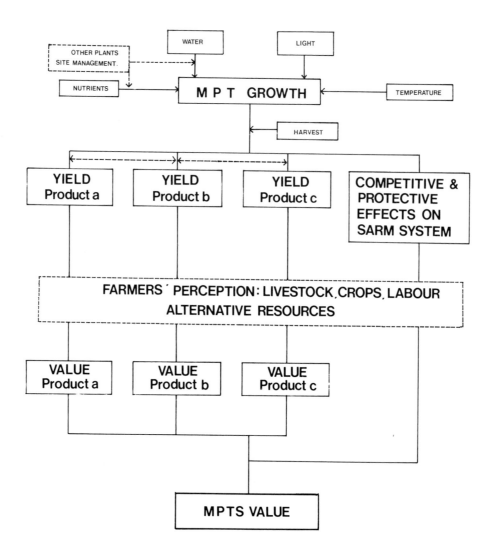

Fig. 1. Procedure for obtaining local minimum data set for MPTS.

Robinson (1985b) and Young (1983, 1985). In relation to the F/FRED Regional Cooperative Research, recommendations also have been made (Lantican and Yantasath 1987, F/FRED 1986). The following comments therefore refer to the difficulties encountered in applying such recommendations to MPTS research in a country like Nepal, which has a highly varied environment, largely caused by wide altitudinal variations. However, such variations also occur in some areas of most countries of the Asia Region.

Climate

It is not possible to have climatological stations close to the many research sites to cover the range of existing climatic environments. Where enough climatological stations exist to cover the range of climates (caused by differences in latitude, continental effects, altitude, aspect, location in relation to mountain ranges causing orographic precipitation and rain shadow effects), it is possible to extrapolate the likely climate of a site. However, significant variations can exist within short distances. It is more valuable to have climatological data from the year since yearly fluctuations in climate can be considerable in many areas of the region. If it has not been disturbed, existing vegetation may be a good guide to local climate if the relationship between vegetation and climate is understood. Considerable differences can exist on a micro-scale. As an example, air and soil temperature can differ greatly between trees planted on the top and bottom of a terrace bank (these are common in Nepal) on north-facing slopes. Such differences could significantly affect growth, and unless a very careful description of the planting pattern is made, growth between such sites is not likely to be comparable.

Soils

Soils in mountainous regions vary widely in nutrients and moisture, depending on land use, altitude, aspect, slope, soil parent material, depth, and stoniness. Again, big differences exist within short distances. Malla (1980) found that the top half of terrace banks produces 5.7 times more fodder from an exotic herbaceous species than the bottom half of the same bank due to different micro-site conditions. While the roots of most MPTS are likely to grow deeper that those of herbaceous plants, such micro-site differences are still likely to affect growth and yield, particularly in

the early years. The age of terrace banks also influences soil fertility and texture. Recently created terrace banks are less fertile and have not developed the soil texture and organic matter found in the top soil layer of older banks. Terrace soil fertility usually varies with distance from the farm house because more manure is applied to fields close to the house.

In relation to soil nutrients and/or moisture conditions (retention capacity, availability) it is now believed in Nepal, particularly in the hills, that physical rather than chemical properties (slope aspect and gradient, soil depth, texture, and degree of stoniness) are the key parameters to assess (Howell, in press). On flat sites, the depth of the water table is an important factor. Elsewhere too, Greaves and Hughes (1976) have suggested that attempts to relate tree growth to chemical content of soil as derived from soil analysis have not been very rewarding, while more easily assessed soil features, such as soil texture, have given better results (even though soil depth and effective rooting depth are difficult to assess). Nevertheless, where chemical analysis of soil is feasible, it is recommended for reasonably homogenous sites, such as are found at more traditional research sites. Presently in Nepal, chemical information is largely academic. In the future, it could help in selecting species. Unless scientists measure nutrient levels at every logical and feasible opportunity, they will never know how site preferences are related to nutrient status.

However, when trials are located on farmers' land, site variability can be much greater even within short distances. Terrace bank differences have already been discussed. Furthermore, farmers apply different quantities of manure on their fields, and the timing of the application within the yearly cropping cycle may vary, as do crop rotations, crop types, cropping intensity, and irrigation applications. The proportion of crop residues and weeds that are plowed in or harvested for fodder also vary. All these factors make meaningful soil assessment strategies complex, and the resources required for such site assessment are likely beyond the capability of research projects in similar environments in the region.

Site Management

A number of site management factors can considerably influence survival, as well as initial and subsequent growth and yield.

Cultivation, Pitting, and Weeding

It is not possible to recommend a uniform treatment for all research sites since practical labor and economic considerations vary both in large-scale plantings and on small-scale plantings on farms. Research site conditions and management must reflect what is practical in any area, or research results may suggest a different ranking of MPTS from that which would occur under realistic management.

In Nepal (monsoon climate), complete site cultivation is not feasible in the hills either on degraded grassland/shrubland or on terrace banks. However, in the Terai (subtropical) zone, complete cultivation is highly desirable for most species, especially where imperata grass is particularly detrimental to tree growth.

In the Bhabar Terai, results have shown that for a number of MPTS, height growth at 18 months was less than half for trees planted in pits 0.5 m wide and weeded regularly for 18 months compared to those in plots that were regularly and fully cultivated over the same period (Forest Survey and Research Office 1986). In the flat Terai zone, complete mechanical or manual cultivation is economically feasible on plantation sites, particularly if cultivation is followed by a sequence of crops in the inter-rows for a few years. Where intercropping is recommended following cultivation, the area that must remain unplanted around the trees must, to some extent, depend on the crop type (e.g., maize or mustard), and different crops will have different effects on tree growth. However, even in the Terai, the vegetation of the planting site will determine whether pitting and weeding are sufficient.

Pit size, where complete cultivation is not practical or necessary (such as on hill sites), influences survival and growth. However, it is necessary to monitor growth for a number of years, particularly on poorer and drier sites, since differences due to pit size may only become apparent after a few years.

Watering and Fertilizers

Watering planted seedlings during the dry season is not an option with general application. Nevertheless, in the case of farmers who concentrate on planting a few seedlings each year, planting prior to the monsoon with regular watering can make a considerable difference to growth in the first year. Labor constraints make it difficult for farmers to plant trees in the early part of the monsoon, however, and a good proportion of private planting is often done one-third to halfway through the monsoon. This limits the growth potential in the first year and restricts the survival capability the following dry season.

Fertilizer application at planting time is not generally practiced. However, some farmers see a considerable advantage in applying farmyard manure to the planting pit. Trees grow considerably faster, are out of browsing reach sooner, and start yielding desirable products earlier. On non-cropland and on sites suited to more traditional afforestation, the application of small quantities of fertilizer in the pit at planting has been found to make a difference to growth of some species, creating the possibility of more desirable species than the pines presently planted. Careful supervision is required in experiments using natural fertilizer to ensure that the same quantity is applied to each pit. Even for quite small trials, the amount of manure required cannot be bought from one farmer. Since the quality of the manure can vary considerably between farmers, nutrient availability due to manure application may vary even for a given weight or volume applied.

Growth, Yield, and Research Approaches

Growth

Although recurrent yield of useful products is more important than growth itself, initial growth and survival are important factors to be considered in species elimination trials and species and provenance testing trials.

Survival, height, basal diameter, diameter at 30 cm, and dbh should be assessed at regular intervals. Initial height of seedlings at planting also should be recorded. The frequency and timing of measurements should be dictated in part by the physiological status of the species. In areas where there is a yearly pattern of growth, measurements should be taken during the most dormant period of the year rather than following rigid calendar periodicity. Timing of survival assessment should be related to periods of stress due to different factors (e.g., winter cold and dry-season moisture stress). This has important effects on manpower requirements for experiments.

In Nepal, a number of valuable MPTS (e.g., *Morus alba, Cassia siamea, Ficus semicordata, Ficus auriculata*, and *Prosopis juliflora*) tend not to grow as single erect stems, particularly in early years. Since the extent of stem multiplicity can be influenced by whether the tree originates from seed or rooted cuttings, the nature of the planting stock should be recorded. Some species (e.g., *Ficus semicordata*) tend to prostrate growth in early years, in which case length of stems rather than vertical height should be measured. For some species, multiple stems can be singled out at the nursery stage or early after planting. For others, the multiplicity of stems must be considered since it may be a valuable trait (with central or inner stem protected by outer stems against browsing). It is important to find simple measurements that singly or in combination can be correlated with stem, branch, foliar, and total biomass. Basal diameter (D0) or diameter at 30 cm (D30) can in some cases overcome the problem of multiplicity of stems, but not in all cases. Correlations with total biomass or biomass of important products have been found for various tree and shrub species from diameter at ground level, at the root crown, immediately above the basal swelling, and at 15, 30, 50, and 60 cm height. It is recommended that diameter immediately above the ground swelling and D30 should be recorded. The best way to treat these measurements has not yet been determined. The recommendation for dealing with multiple stems below dbh in traditional forestry is the square root of the sum of the squared diameters of the stems. This is a simplification that, while sufficient for the infrequent need in traditional single-stem forestry practice, will greatly distort comparisons between species that vary in multiplicity of stems. Pre-harvest results must be interpreted in the context of tree habits, and firm conclusions must await actual harvest results.

Experimental trees that have been browsed should be excluded or treated separately in the calculations. It is essential not to make decisions concerning differences in growth between species,

sites, and treatments until the trees are well established and have reached a reasonable size. Differences in pit size may take a few years to show an effect. Conversely, differences in fertilizer application may clearly show early differences, which narrow in later years as the additional nutrients are insufficient for the developing tree biomass. The benefit may be removed in an initial first harvest.

Yield

Harvesting at ground level is only one of many management regimes that farmers apply to MPTS, and then often only after they have harvested some products from the tree for many years. *Alnus nepalensis* stem branches are lopped periodically in Nepal for mulch, animal bedding, and fuelwood until the tree has reached a suitable size or there is a demand for its use as timber.

For any growing tree or shrub, source-path-sink relationships exist between the various components of the tree (Cannel 1985). Harvesting any part or product of the tree influences these source-sink relationships and subsequent dry matter accumulation and growth of component parts. Yield is dependent on growth and harvesting intensity.

A large number of harvesting variables influence yield and subsequent growth and yields. Taking foliage as a product for mulch or fodder, the following variables are considered important:

o time of first harvest,

o season of harvest in relation to the seasonal developmental stage of the tree and in relation to the climate, which may vary significantly between years,

o frequency of harvest,

o quantity of harvest,

o residual foliage after harvesting,

o number of buds left after harvesting and the competition between these growing points for regrowth, and

o manner of harvest in relation to tools used.

The effect of a number of these variables is certain to be interactive. For *Leucaena leucocephala*, it is already known that there is an interactive effect between cutting height and frequency (Horne, Catchpoole, and Ella 1986). Their review also refers to an interactive effect with planting density. It is also likely that interactive effects between harvesting regimes will be found with a number of inherent site conditions and site management regimes, since shoot or root biomass relationships are widely known to be affected by soil nutrients, moisture, and light (Cannel 1985). Furthermore, the nature and degree of these interactions are likely to be species- and perhaps even provenance-specific.

It must be emphasized that such interactive effects of harvesting variables and site conditions are relevant to any tree or shrub product. Further, because of set source-sink relationships between the various component parts of trees (resulting in dry matter partitioning patterns that are species- and provenance-specific), the harvesting of any one product influences the subsequent production of others. The multipurpose concept in terms of products and even service functions (Torres 1983) must acknowledge the trade-offs between availability of products. For instance, there is a trade-off between fruit and flowers, foliage, branch wood, stem wood, and exudates (Cannel 1985, Ford 1985).

There also are difficulties in standardizing harvesting procedures to study foliage yields between treatments. How does one standardize the harvesting of branches and foliage to leave a set proportion of branches and leaves on a lopped tree (residual branches and foliage)? Even taking the easier situation where pollarded trees are only allowed (by repeated pruning intervention) to have three branches of regrowth and one removes two branches at each pollarding sequence, the residual branch wood and foliage biomass will not be exactly 33% of the pre-pollard branch and foliage biomass. Furthermore, the percentage will vary between trees, thereby increasing the variability in subsequent growth and yield between trees within the treatment.

Various studies have studied the effect of lopping frequency on branch and foliage yield of *Leucaena leucocephala*. However, the majority have used set time periods as a benchmark of frequency. Since climatic conditions usually vary somewhat throughout the year, harvesting at regular intervals will result in different regrowth biomass between time periods, which causes variability between periods in the effect of harvesting on subsequent growth and yield. For increased standardization to improve the likelihood of detecting patterns in yield caused by different lopping treatments involving frequency of harvest, cutting frequency based on predetermined stages of regrowth makes more sense (Horne, Catchpoole, and Ella 1986).

The need for standardization is even more crucial in regional research networks where results are to be compared between different sites and investigations involving various research institutions. In relation to harvesting frequency, while one site may dictate two harvests a year, a harvest five times a year from a better site may be comparable in relation to the speed of regrowth of the plant, thereby maintaining more comparability in the partitioning of dry matter to various plant components.

If comparisons are to be made on the effect of the same harvesting variables and schedules between species, the problems of standardization are further confounded. Different species develop in different ways (e.g., yearly growth patterns, branching structure, regrowth patterns).

Research Approaches

For a national research strategy on MPTS involving many species, a wide range of ecological and socioeconomic conditions, as well as a number of treatment variables, surveys, and both station and on-farm research need to be considered.

Research Station Trials

For species elimination trials and for assessments of survival and early growth, small plots are sufficient. However, as soon as one enters the realm of assessing MPTS yield, including various treatment combinations, experiment size becomes daunting. Table 1 gives the number of trees of *Leucaena leucocephala* required per plot (treatment combination) to give reasonable certainty of distinguishing height and foliage biomass differences between treatments for two levels of expected percentage error of the mean.

Table 1. Number of *Leucaena leucocephala* trees required per plot.

	Height	Foliage biomass
Coefficient of variation	15[a]	50[b]
No. of trees, 10% error	11	117
No. of trees, 25% error	2	19

[a] Hawkins, 1985 (Nijgad variety, CV of height at 22 mos.).
[b] Hawkins, personal communication.

Table 2. Determination of lopping trial size for *Leucaena leucocephala*.

Plot size	19 tree line plot + 2 border trees = 21 trees (derived from Table 1)
Treatments	A: 3 branching structures x 3 lopping frequencies = 9 treatment combinations. B: 3 branching structures x 2 lopping frequencies x 2 residual foliage = 12 treatment combinations.
Replication	3 required for both options
Spacing	2 m within line, 5 m between lines
Clear surrounds	5 m

A: 2 treatment variables, 9 treatment combinations
- 8 x 5 m interline distances + 10 m = 50 m
- 21 trees x 3 replications = 63 trees
- 62 between tree distances + 10 m = 134 m
- 50 m x 134 m = 0.67 ha

B: 3 treatment variables, 12 treatment combinations
- 11 x 5 m interline distances + 10 m = 65 m
- 21 trees x 3 replications = 63 trees
- 62 between tree distances + 10 m = 134 m
- 65 m x 134 m = 0.87 ha

The high inherent variability in leaf biomass necessitates large plots.

A lopping trial to consider the effect of only 2 or 3 harvesting variables in one species (9 or 12 treatment combinations) would require a (factorial) experiment size of 0.67 or 0.87 ha, respectively, for an expected level of error of 25% to have a reasonable chance of detecting statistical differences between treatment combinations (Table 2). The CV 50% found by Hawkins (personal communication) for foliage was for whole tree harvests of *Leucaena leucocephala*. As indicated in a previous section, yield studies looking at harvesting variables would increase the plot variability, requiring a plot size larger than 19 trees. Hawkins' site was a homogenous Terai site; hence, any such harvesting experiments on more varied sites, such as throughout the hills of Nepal or on farmer's fields, would require bigger plot sizes and more replications, further increasing the area of the trial.

Clearly, the number of trials that can be initiated are limited, given the lack of resources, the shortage of land available for research purposes, the high labor demand for biomass studies, and the high level of supervision required at harvest to meet the requirements for standardization necessary to minimize within-treatment variability. The choice of species on which such trials should be initiated should therefore reflect the relative national importance of the species available, as well as the range of types of species in terms of characteristics including branching and growth and light or heavy shade. The harvesting variables to be investigated should be carefully chosen to include those likely to differ most in subsequent growth and yield and most feasible in terms of farmer options. The only way to reduce inherent variability, and therefore plot and experiment size, would be to use planting material propagated vegetatively. More efficient experimental designs, such as confounded factorials, will reduce the areas required to some extent but are more complex and will still require large areas.

In the hills of Nepal, many of the most valuable MPTS do not grow on the only sites that would be available for the sole purpose of research and that are of poorer quality. All suitable sites for such species are cultivated.

On-Farm Trials

On-farm trials of the type described in the previous section are not appropriate for many reasons. Hill farms are small (average less than 1 ha), not all farmland is available for tree planting, and site variability within and between farms is high. However, more important reasons are related to farmers' needs. A farmer generally requires or prefers a range of species to provide the needed products, to take account of the time of their availability, and to reduce risks. In relation to investigations on harvesting, a number of difficulties also arise. Regarding fodder trees, a farmer's demand is for a regular but small amount of tree fodder to supplement poorer fodder from other sources (dry grass from the forest and agricultural residues). Hence, a farmer cannot harvest all the foliage of a large tree in one day, or harvest all the foliage from a number of trees within a short period to satisfy a researcher's requirement for timely harvesting to minimize between-tree variability in treatment. The shading of crops by fodder trees may require the farmer to manipulate the tree canopy in a way that does not fit in with the researchers' harvesting schedules and which varies from year to year on any field with the varied crop rotations.

Trees that already exist on farms are valuable for yield studies based on the way trees are actually managed by farmers. The need for frequent weighing, especially if the products are harvested daily, will require a high level of research input. The use of such trees for yield studies to look at treatment variables and interactions between treatments is not feasible, however. Trees are of different ages and have had different harvesting histories.

The types of trials that are feasible and desirable on farmers' land are therefore different from those suitable for research station conditions.

Surveys

Given that little is known about the many valuable MPTS that cover a wide range of ecological conditions, surveys play an important role in quickly providing information about a number of characteristics. In relation to the various products, such as fodder, farmers can rank species for quality, yield, and growth rate. However, their perception of quality has been found to be influenced by season (Upton, in press) since different species are suitable or available during different seasons. While their own perception of yield and growth rate may be clear, it is much harder to interpret the information on these two factors, since yield is obviously related to size and growth rate is related to the intensity with which they harvest the trees. The intensity will vary between farmers, depending on what alternative sources of fodder they have. Farmers may not take site into account when describing relative growth rates, and different farmers may know the same species from different sites and hence estimates its value differently. Surveys thus have to be carefully designed and pretested, and the interviewers well-trained and motivated.

Value to Farmers

The value of MPTS products to farmers is a function of the inherent characteristic of the tree products (such as yield and quality) and other location-specific factors.

To be comparable between different locations, the assessment of value requires objective comparisons where possible based on absolute values. Yield is one factor that has already been discussed. However, in relation to fodder, one needs to know not just the foliage yield, but also the proportion of twig that is consumed by different animal types (Oli 1987) and the proportion of the feed wasted by livestock. These vary significantly with location according to a number of factors (alternative feed available, feeding system, etc.). The evaluation of quality may be easier for some products but complicated for others. For instance, the evaluation of feed quality of tree fodder is fraught with difficulties. The chemical composition of leaves varies with season, location within the tree canopy, and with harvesting and shading. Even assuming one can get a reasonable average leaf (and edible portion of twig) sample from which to analyze constituents, the chemical methods of feed-value evaluation available do not give meaningful results in terms of true feed value to animals (Robinson 1986). The use of *in vitro* measurements is constrained by large, systematic errors when applied to data sets other than those used to derive them (Devendra 1987). Work has been done on very few fodder tree species relating *in vivo* studies to *in vitro* measurements. Furthermore, different ruminant types (sheep, buffalo, goats, cattle, camel) show considerable

differences in their ability to digest and utilize the nutrients from foliage of the same fodder tree species (Robinson 1986).

At the farmer level, any attempt to valuate a range of MPTS products must relate to farmer needs and perceptions, which varies with location, individual, time, and circumstances. For quantity and quality fodder, leaves for bedding and composting, and for fuelwood, the biggest factor in a farmer's assessment of the value of a particular MPTS is access to alternative sources of these products (on or off his farm). This relates to socioeconomic issues, such as labor availability, yearly land management and feeding problems, restrictions on access to forest, or type of livestock owned. Farmers state that certain species of fodder trees provide good fodder for one type of livestock but not for another. Many inconsistencies among areas arise from surveys on the perceived fodder quality of various MPTS. Panday (1982) states that *Bauhinia variegata* decreases milk yield, yet in some areas farmers say it is a good fodder for buffalo to produce milk.

Timber species in the hills of Nepal are in infrequent demand, but they play a vital role as an accumulation of a cash source and for house-building. Medicinally valuable trees are found but not necessarily on every farm (e.g., *Litsea cubeba*). The value to the individual farmer of this community benefit is unknown, but could be high in terms of social standing.

Even where no markets are accessible, a few fruit trees are highly valued and other trees have a number of special and crucial uses (leaf plates, handles, plows). A number of ecological or service roles (terrace bank stabilization, providing a climbing frame for garden vegetables or for storing dry fodder) are also perceived by many farmers.

Experience in Nepal suggests that the best way to assess farmers' valuation of the range of products provided by MPTS is by talking to them and getting relative rankings, both for different species providing the same products and between product types. Because of the great variability in response due to different socioeconomic conditions, ecological zones, MPTS composition and frequency, and product demand, large-scale surveys have to be conducted regularly at different seasons. The interviewer has to have a good understanding of the farming system so that he can probe for missing items in responses. To get some degree of comparability between areas and interviewers, the same checklist of topics to be covered must be used. For a number of topics, the question has to be asked in exactly the same way. Such surveys are time-consuming. Farmers are usually suspicious of outsiders probing in their affairs and often have little spare time to sit and talk. Again, compared to a brief questionnaire interview, a prolonged informal visit is more likely to glean reliable information, but it will cost the farmer and the researcher more time. Surveys have shown that women in a household give different answers than men concerning the number of fuel and fodder loads collected. Women's ranking of the value of products are also known to differ from men's, yet access to their valuation is often difficult.

Conclusions

The difficulties of applying statistically valid experimental design to trees as they occur in small farms in the hills of Nepal and elsewhere concentrate the researcher's mind on an essential point: the difference between statistical and practical significance and how this relates to his role of providing information that can be applied for practical gain. There is a great dearth of knowledge on MPTS and how they interact in farming systems. There is great variation in the trees and their ecological and social effects. We have seen the complexity in investigating and even assessing the growth, yield, and value of trees due to this variation and the interactions of the factors involved. The research sector is faced with an immense task in which the priority is applied research. Despite the problems involved, the three main types of research (on-station, on-farm, and surveys) must each be followed, but it remains to allocate priorities to these. We suggest that, under these circumstances, the detection of practical differences (essentially those perceived by farmers) must be given high priority so that research impact is maximized. This can be achieved by farmer surveys and small plots of relevant treatments applied on many farms. The recording of these experiences on database systems where the researcher attaches a rank (1-5) of importance to each factor of a comprehensive checklist will allow the distribution of this experience. It is critical that the checklist of factors, which may influence the growth, yield, or value, is comprehensive. We see this as the "informational database" of Rose and Cady (1986) and recommend that it is given a

much higher priority. Factors crucial in Nepal may have no significance in other parts of the region, but a regional checklist should be drawn up as a priority.

It is especially important that the researcher is fully aware of the limitations of his or her work and the validity and applicability of findings. The thrust of research toward simple experiments on many small farms allows the local farmers to be their own extension agents.

REFERENCES

Burley, J. and J. L. Stewart, eds. 1985. *Increasing productivity of multipurpose species*. Vienna: IUFRO.

Burley, J. and P.J. Wood. 1976. *A manual on species and provenance research with particular reference to the tropics*. Tropical Forestry Paper No. 10. Oxford: Oxford Forestry Institute.

Cannel, M.G.R. 1985. Dry matter partitioning in tree crops. In *Attributes of trees as crop plants*, eds. M.G.R. Cannel and J. E. Jackson, pp. 160-193. Huntingdon, UK: Institute of Terrestrial Ecology.

Devendra, C. 1987. Strategies for the Effective Utilization of Small Ruminants and Feed Resources in the Highlands of Asia. Expert Meeting on Himalayan Pasture and Fodder Development. Mimeograph. Kathmandu: ICIMOD.

F/FRED. 1986. Report of the pilot program participants. MPTS Network Trials Planning Meeting, Bangkok.

Ford, E.D. 1985. Branching, crown structure and the control of timber production. In *Attributes of trees as crop plants*, eds. M.G.R. Cannel and J. E. Jackson, pp. 228-252. Huntindon, UK: Institute of Terrestrial Ecology.

Forest Survey and Research Office. 1986. Adabhar Trial Site Results 1980-1986. Mimeograph. Kathmandu: Forest Survey and Research Office.

Greaves, A. and J.F. Hughes. 1976. Site assessment in species and provenance research. In *A manual on species and provenance research with particular reference to the tropics*. Tropical Forestry Paper No. 10. Oxford: Oxford Forestry Institute.

Hawkins, T. 1985. Leucaena variety trial in Nepal. *Leucaena Res. Rept.* 6:66-67.

Horne, P.M., D.W. Catchpoole, and A. Ella. 1986. Cutting management of tree and shrub legumes. In *Forages in Southeast Asia and South Pacific agriculture*, eds. G.J. Blair et al., pp. 164-169. ACIAR Proceedings No. 12. Canberra, Australia: ACIAR.

Howell, J. (In press). Choice of species for afforestation in the mountains of Nepal. *Banko Janakari* Vol. 1(3).

Hudson, J. 1987. Forest research in Nepal up to 1986. *Banko Janakari* Vol. 1(2):3-14.

Hughes, C.E. 1987. Biological considerations in designing a seed collection strategy for *Gliricidia sepium* (Jacq.) Walp. (*Leguminosae*). *Commonwealth Forestry Review* 66:31-48.

Hughes, C.E. and B.T. Styles. 1984. Exploration and seed collection of multipurpose dry zone trees in Central America. *Int. Tree Crops J.* 3:1-31.

Huxley, P.A., ed. 1986. *Methodology for the exploration and assessment of multipurpose trees.* Nairobi: ICRAF.

Lantican, C.B. and K. Yantasath. 1987. A Guide to Multipurpose Tree Species Research Cooperators. Mimeograph. Bangkok: F/FRED Project.

MacDicken, K.G. 1986. Planning and implementing MPTS field trials. In *Forestry Networks: Procs. first workshop of the Forestry/Fuelwood Research and Development Project (F/FRED)*, eds. N. Adams and R.K. Dixon, pp. 25-31. Arlington, Virginia: Winrock International.

Malla, Y.B. 1980. *Studies on the production of exotic grasses grown on terrace banks at Pakhribas Agricultural Center.* PAC Technical Paper No. 48. Pakhribas, Dhankuta, Nepal.

Oli, K.P. 1987. Work of the livestock section, Pakhribas Agricultural Center. In *Proceedings of the first fodder tree working group meeting*, ed. P.J. Robinson, p. 4. Kathmandu: Forest Research and Information Center.

Panday, K.K. 1982. *Fodder trees and tree fodder in Nepal.* Berne: Swiss Development Corporation.

Robinson, P.J. 1985a. Trees as fodder crops. In *Attributes of trees as crops plants*, eds. M.G.R. Cannel and J.E. Jackson, pp. 281-300. Huntingdon, UK: Institute of Terrestrial Ecology.

_____. 1985b. The assessment and choice of experimental sites. In *A methodology for the exploration and assessment of multipurpose trees*, ed. P.A. Huxley, Nairobi: ICRAF.

_____. 1986. Fodder tree research in Nepal: A brief review of past work and future needs. *Proceedings of first farming systems workshop*, pp. 171-185. Kathmandu: FSRDD-ARP-DOA.

Robinson, P.J., J. Burley, and P.J. Wood. 1985. Methodology for the exploration and collection of MPTS. In *Methodology for the exploration and assessment of multipurpose trees*, ed. P.A. Huxley. Nairobi: ICRAF.

Rose, D.W. and F.B. Cady. 1986. Overview of DBMS for MPTS. In *Procs. first workshop of the Forestry/Fuelwood Research and Development Project (F/FRED)*, eds. N. Adams and R.K. Dixon, pp. 63-70. Arlington, Virginia: Winrock International.

Torres, F. 1983. Role of woody perennials in animal agroforestry. *Agroforestry Systems* 1:131-163.

Upton, P.L.J. (In press). A survey of fodder tree use: Changes with altitude, aspect and season in South Lalitpur, Nepal. Kathmandu: Forest Research and Information Center.

Young, A. 1983. *An environmental database for agroforestry.* Working Paper No. 5. Nairobi: ICRAF.

_____. 1985. Various sections. In A methodology for the exploration and assessment of multipurpose trees, ed. P.A. Huxley, pp. 90-104. Nairobi: ICRAF.

Poster Session Presentations

Palms as Multipurpose Cash and Subsistence Tree Crops

Dennis Johnson

U.S. Department of Agriculture Forest Service/
U.S. Agency for International Development Forestry Support Program
Washington, D.C., USA

Tropical forestry development has tended to exclude palms despite their role on small farms as significant sources of edible fruit, oil, green vegetable, fiber, thatch, construction wood, fuelwood, and numerous other useful products. Palm species furnish various products of subsistence value and provide raw materials for cottage industries, large-scale commercialization, and export. Worldwide more than 50 domesticated, managed, or wild palms are known to have potential for integration into forestry and agricultural development projects.

In selecting palms suitable for small farmers in agroforestry systems, it is important to consider the amount of knowledge available about a candidate species. A good indicator is the degree to which a palm has been domesticated. Generally, the more advanced the stage of domestication, the easier the species can be cultivated successfully.

Tables 1-3 classify 55 palms in three levels of domestication: improved cultivated palms; unimproved cultivated or managed palms; and semi-wild or wild palms.[1] Palms cultivated exclusively for ornamental purposes are excluded.

Table 1 (improved cultivated palms) lists five palms that have been the object of botanical and agronomic research. Cultivars have been named, species improved, and agronomic potential well-developed. These palms, to varying degrees, are included in agroforestry systems, and more advances are expected as strategies for improving palm productivity are integrated with complementary cropping and grazing systems.

Although commonly omitted from agroforestry species lists, the date palm is included here because of its potential in tropical semi-deserts and deserts. These improved, cultivated palms are multipurpose although their minor products are not yet well-developed. Palms in this group may serve as models in programs designed to develop other species.

Table 2 (unimproved cultivated or managed palms) lists 20 species either cultivated or occurring in wild stands under some degree of management, which may be described as "incipiently domesticated." These have received considerably less attention, and in many cases taxonomy is uncertain. Cultivars are not recognized although some rudimentary crop improvement may have been achieved. All are multipurpose, encompassing the full range of tropical habitats. Most of their products are consumed locally, and little effort has focused on increasing their productivity or quality. Generally, this group of palms has not been evaluated with regard to multiple-cropping potential. Thus, an excellent opportunity exists to combine plant improvement and agroforestry objectives in development programs that include palms.

Table 3 (semi-wild or wild palms) lists 30 species exploited for a variety of products. Information on these palms is incomplete and their taxonomy uncertain. Further research is needed to evaluate their potential for improvement. Despite their underdeveloped status, the piassava (*Attalea funifera*) and raffia (*Raphia* spp.) palms support export fiber industries, and the rattans (*Calamus* spp.) supply commercial quantities of cane. Various palms show promise as reforestation plants, such as the mazari palm (*Nannorrhops ritchiana*) in mountainous areas and the thatch palms (*Thrinax* spp.) on dry, limestone soils.

---

[1] These tables first appeared in "Multipurpose palms in agroforestry: A classification and assessment," in *The international tree crops journal* 2 (1983): 217-244. An edited version appeared as an appendix to "Multipurpose palm germplasm," in *Multipurpose tree germplasm*, eds. Burley and von Carlowitz (Nairobi: ICRAF, 1984): 249-278.

Table 1. Improved, cultivated palms.

Scientific name	Common name	Origin & native habitat	Products & yields	Status/ comments	Sel. ref.
Areca catechu L.	areca palm, betel palm	S.&S.E. Asia: inferred trop. rain forest, to 900 m; unknown wild.	Seed as masticatory (1,037 kg/ha/yr); edible heart; leaves for thatch; leaf sheaths for hats, containers; trunk for wood; seed in veterinary medicine; dye source.	Cultivated alone or w/annual or perennial crops; seed propagation; cultivars based on seed quality; limited research on selection, breeding; ornamental; solitary feather palms.	12, 19, 20, 50, 91, 94
Bactris gasipaes H.B.K.	peach palm, pejibaye	Central & S. Amer.: inferred trop. rain forest, to 1,200 m; uknown wild.	Edible fruit (25-30 t/ha/yr), oil & flour source, animal ration; edible heart, trunk for wood.	Widely cultivated Trop. Amer., propagation by seed or suckers; fruit local product; hearts exported; suckering feather palms.	4, 22, 25-27, 39, 80-81, 83, 91
Cocos nucifera L.	coconut palm	S.E. Asia: inferred trop. rain forest, coastal sites, to 300 m; unknown wild.	Edible oil, fruit, drink (2,500-7,500 nuts/ha/yr, copra yields to 1,200 kg/ha/yr); edible heart; leaves for thatch, weaving; trunk for wood; minor products.	Most widely cultivated palm, alone or w/annual, perennial crops; seed propagation but progress w/vegetative propagation; breeding objective to increase oil yield; numerous cultivars; solitary feather palms.	20, 24, 41, 43, 51, 78-79, 82, 84, 88, 91, 102
Elaeis guineensis Jacq.	oil palm, African oil palm	W. Africa: trop. rain forest, esp. open, wet sites, to 800 m; semiwild only.	Edible oil (5 t/ha/yr); sap for wine; edible heart; leaves for thatch, weaving; petioles for fencing, construction.	Cultivated in pure stands on large estates; smallholders intercrop w/annual crops; seed propagation, promising research on tissue culture; excellent improved cultivars; solitary feather palms.	28, 44, 52, 62, 75, 80, 91-92

Table 1. Continued.

Scientific name	Common name	Origin & native habitat	Products & yields	Status/ comments	Sel. Ref.
Phoenix dactylifera L.	date palm	Middle East: sub-trop. semi-desert, desert; unknown wild.	Edible fruit (20-100 kg/tree/yr); sap for wine; leaves for thatch, weaving; trunk for wood; minor products.	Widely cultivated in arid regions, alone & w/peren-nial crops; vege-tative propagation w/suckers; many named cultivars based on fruit quality; suckering feather palms.	69, 77, 83, 89, 91

Table 2. Unimproved cultivated or managed palms.

Scientific name	Common name	Origin & native habitat	Products & yields	Status/ comments	Sel. Ref.
Arenga pinnata (Wurmb.) Merr.	sugar palm, gomuti palm	S.&S.E. Asia: inferred trop. rain forest into dry forest, to 1,200 m; putatively wild in Assam & Burma.	Sap for sugar, wine (3-6 l/tree/day); starch from trunk (75 kg/tree); fiber from leaf sheath; edible heart.	Widely cultivated India, S.E. Asia; sometimes planted after shifting cultivation; fiber exported; sugar, wine, starch local products; inedible fruit; solitary, terminal flowering feather palms.	19-20, 35, 39, 72, 81, 91, 105
Borassus flabellifer L.	palmyra palm	S.Asia: trop. dry forest into savanna to 750 m; naturalized in S.E. Asia & widely cultivated.	Sap for sugar, wine (150 l/tree/yr); fiber from leaf stalk; edible fruit; leaves for thatch, weaving; trunk for wood, fuel; minor products.	Often planted India; fiber exported; sugar, wine local products; solitary fan palms.	15, 20, 29, 32, 48, 65, 91
Borassus sundaicus Becc.	lontar palm	Indonesia: trop. dry forest.	Sap for sugar, wine (150 l/tree/yr); fiber from leaf stalk; edible fruit, leaves for thatch, weaving; trunk for wood, fuel; minor products.	Cultivated in Indonesia; local products only; solitary fan palms.	48
Calamus spp.	rattan	Old World: trop. rain forest to 1,000 m; common in secondary forest.	Rattan canes (to 6 t/ha); edible fruit some spp.; fruit for medicine.	Cultivated small-scale S.E. Asia; about 16 economic spp. Malay Peninsula; canes and finished products exported; climbing feather palms.	20, 36-38, 58, 91, 109

Table 2. Continued.

Scientific name	Common name	Origin & native habitat	Products & yields	Status/ comments	Sel. Ref.
Carvota urens L.	fishtail palm, toddy palm	S.&S.E. Asia: trop. rain forest, esp. primary forest, to 1,500 m.	Sap for wine, sugar (20-27 l/tree/day); starch from trunk (100-150 kg/tree); fiber from leaf sheath; edible heart	Casual cultivation, underdeveloped; fiber exported; wine, sugar, starch local products; solitary terminal flowering feather palms.	20, 29, 33, 81, 91, 97
Copernicia prunifera (Mill.) Moore	carnauba palm	S. Amer.: trop. dry forest into savanna, esp. floodplains & coastal plains.	Wax from leaves (100 g/tree/yr); leaves for weaving; trunk for wood.	Incipient plantation cultivation; wax exported; woven goods local products; solitary fan palms.	13, 59, 91
Corypha umbraculifera L.	talipot palm	S.&S.E. Asia: inferred trop. rain forest, to 600 m; unknown wild.	Sap for sugar, wine; starch from trunk (90 kg/tree); leaves for matting, paper.	Widely cultivated; sugar, wine, starch local products; solitary, terminal flowering fan palms.	15, 20, 29, 91
Elaeis oleifera (H.B.K.) Cortes	American oil palm	Central& S. Amer.: trop. rain forest lowlands.	Edible & industrial oil.	Limited cultivation in native area; excellent germplasm for hybrids w/*E. guineensis*.	28, 52, 87, 91
Eugeissona utilis Becc.	bertan palm	S.E. Asia: trop. rain forest, esp. disturbed sites, to 1,000 m.	Starch from trunk; edible fruit; leaves for thatch.	Rudimentary cultivation in Borneo; starch staple among some groups; suckering feather palms.	20, 35, 66

Table 2. Continued.

Scientific name	Common name	Origin & native habitat	Products & yields	Status/ comments	Sel. Ref.
Euterpe edulis Mart.	palmito branco, jucara	S.Amer.: trop. rain forest into subtropics, to 1,000 m.	Edible heart (1 kg/tree); cellulose.	Limited culti-vation; heart exported; solitary feather palms.	68, 81
Euterpe oleracea Mart.	acai, palmito da Amazonia	S.Amer.: trop. rain forest sites subject to flooding.	Edible heart (1 kg/tree); edible fruit.	Limited culti-vation; heart exported; fruit local product; attractive orna-mental; suck-ering feather palms.	2, 21, 56, 81
Hyphaene thebaica (L.) Mart.	doum or dum palm, ginger-bread palm	Africa: semi-deserts, desert, to 600 m.	Edible fruit and heart; sap for wine; fruit for medicinal use; leaves for weaving.	Cultivated since ancient times; local products only; promising desert palm; branched fan palms.	29, 42, 91
Metroxylon sagu Rottb.	sago palm	S.E. Asia: trop. rain forest swamps; from S. Pacific Islands through Melanesia into Indonesia, Malaysia, & Thailand	Starch from trunk (300 kg/tree); leaves for thatch.	Cultivated & managed stands; starch exported; suckering, terminal flower-ing feather palms.	20, 45, 80, 91, 97, 100-101
Nypa fruticans Wurmb.	nipa palm	S.E. Asia: west to Sri Lanka, Bay of Bengal; including N. Australia: trop. rain forest, banks of brackish & tidal rivers.	Edible fruit; sap for sugar (0.6-1.8 l sap/palm/day); leaves for thatch, weaving.	Sometimes planted; local products only; suckering feather palms.	19-20, 29, 46, 67, 81, 86, 91

Table 2. Continued.

Scientific name	Common name	Origin & native habitat	Products & yields	Status/ comments	Sel. Ref.
Orbignya phalerata Mart.	babassu palm	S.Amer.: trop. rain forest, up-land sites, managed stands.	Edible oil (10-35 kg/fruit/tree/yr); leaves for thatch; misc. other products.	Cultivated & managed stands; oil exported; other products local; soli-tary feather palms.	5, 9, 39, 49, 70, 80-81, 91
Phoenix sylvetris (L.) Roxb.	wild date palm, silver date palm	India: trop. rain forest to 1,500 m.	Edible fruit; sap for wine, sugar (40 kg sugar/tree/ yr); leaves for weaving.	Limited culti-vation; local products only; solitary feather palms.	29, 31
Roystonea spp.	royal palm	Caribbean, Central Amer.: trop. rain forest into drier forest & savanna.	Fruit for live-stock; edible heart; starch from trunk; leaves for thatch, weaving; trunk for wood.	Cultivated mainly as ornamental; local products only; solitary feather palms.	3, 81, 91
Salacca zalacca (Gaertn.) Voss	salak palm	S.E. Asia: trop. rain forest, in dense shade, to 300 m.	Edible fruit; leaves for thatch, mats.	Widely culti-vated; fruit eaten fresh or pickled; local products only; suckering feather palms.	20, 73-74, 83, 91
Syagrus coronata (Mart.) Becc.	ouricuri palm, licuri palm	S.Amer.: trop. dry forest to savanna.	Edible oil; wax from leaves.	Limited culti-vation; oil, wax exported; wax substitute for carnauba wax; solitary feather palms.	17, 57, 63
Trachycarpus fortunei (Hook.) Wendl.	windmill palm, chusan palm	S.Asia: mts. of S. Central China	Fruit for wax, drug (25-50 kg/ tree/yr); leaf-base fiber; trunk for wood; many other products.	Widely culti-vated in China for products & as ornamental; local products only; popular subtropical ornamental; solitary fan palms.	40

Table 3. Semi-wild or wild palms.

Scientific name	Common name	Origin & native habitat	Products or potential use	Comments	Sel. Ref.
Acrocomia sclerocarpa Mart.	macauba, mucaja coco de catorro	S.Amer.: trop. dry savanna to trop. rain forest.	Edible oil, fruit, heart; leaves for thatch.	Sp. name illegitimate, needs clarification; local products only; solitary feather palms.	5, 23, 60, 81
Astrocaryum jauari Mart.	jauari, awarra	S.Amer.: trop. rain forest, sites subject to floods.	Edible oil, leaves for thatch.	Local products only; solitary feather palms.	5, 87, 99
Astrocaryum murumuru Mart.	murumuru	S.Amer.: trop. rain forest, sites subject to floods.	Edible oil, fruit for livestock, leaves for thatch.	Local products only; solitary feather palms.	5, 91, 99, 104
Astrocaryum vulgare Mart.	tucuma	S.Amer.: trop. rain forest, up-land sites.	Edible oil, fruit, heart; fiber from rachis; leaves for thatch.	Local products only; solitary feather palms.	23, 81, 99, 104
Attalea funifera Mart.	piassava, Bahia piassava	S.Amer.: trop. rain forest, coastal sites.	Leaf base fiber; leaves for thatch; edible oil.	Fiber exported; solitary feather palms.	63-64, 91, 96, 103
Borassus aethiopum Mart.	African fan palm	Africa: trop. wet savanna.	Edible immature fruit; sap for wine; leaves for thatch, weaving.	Local products only; solitary fan palms.	47, 91
Caryota mitis Lour.	fishtail palm	S.E.Asia: trop. secondary forest.	Starch from trunk (small amounts); edible heart, seed; leaf base fiber for stuffing & tinder.	Local products only; solitary feather palms.	20, 105
Ceroxylon alpinum Bonpl. ex DC.	S.Amer. wax palm	S.Amer.: trop. mts.	Wax from trunk; mt. reforestation.	Wax substitute for carnauba wax; solitary feather palms.	76, 91

Table 3. Continued.

Scientific name	Common name	Origin & native habitat	Products or potential use	Comments	Sel. Ref.
Copernicia spp.	yarey palms	Caribbean: trop. dry savanna.	Fruit for livestock; leaves for weaving; trunks for pilings; compatible w/grazing systems.	Local products only; fan palms.	30
Cyrtostachys lakka Becc.	sealing wax palm	S.E.Asia: trop. peat swamp forests.	Wood for pillars, flooring; swamp stabilization.	Common name misnomer, no wax produced; can be vegetatively propagated; outstanding ornamental; suckering feather palms.	20, 34, 91, 105
Daemonorops spp.	rattan	S.E.Asia: trop. rain forest.	Rattan canes; some spp. have edible fruit; fruit scales yield dragon's blood, former dye & Chinese medicine.	Canes exported; about 5 economic spp. Malay Peninsula; climbing, solitary or suckering feather palms.	20, 36-37, 58, 91, 109
Iriartea deltoidea Ruiz & Pavon	pambil, barrigona, huacrapona	Central&S.Amer.: trop. rain forest, swampy sites.	Trunk wood for parquetry, construction wood, rafts; edible heart; leaves for thatch.	Local products only; solitary stiltrooted feather palms.	11, 71
Jessenia spp.	seje, pataua, milpesos	S.Amer.: trop. rain forest, upland sites.	Edible oil; fruit made into beverage; leaves for weaving; could be grown w/ *Astrocaryum*.	Local products only; solitary feather palms.	6-8, 10, 23, 81

Table 3. Continued.

Scientific name	Common name	Origin & native habitat	Products or potential use	Comments	Sel. Ref.
Leopoldinia piassaba Wall.	piassava, Para piassava	S.Amer.: trop. rain forest, sites subject to floods.	Edible fruit; leaf base fiber; leaves for thatch.	Fiber exported; other products local only; solitary feather palms.	18, 63, 91, 93, 96, 104
Licuala spp.	licuala	S.E.Asia, east to Vanuatu: trop. rain forest undergrowth.	Edible heart; walking sticks from trunk; leaves for weaving & wrapping food.	Local products only; some spp. possible ecological indicators; ornamental; suckering fan palms.	20, 34, 105
Livistona spp.	serdang	S.E.Asia, east to Niggela; trop. rain forest, coastal and mt. sites.	Edible fruit, heart; fruit for livestock; leaves for fans; trunk for wood.	Local products only; *L. saribus* (Lour.) Chev. grows in poor soils; *L. speciosa* Kurz, good ornamental; solitary fan palms.	19-20, 105
Manicaria saccifera Gaert.	temiche, guagara	S.Amer.: trop. rain forest, swampy sites.	Edible fruit; starch from trunk (3 kg/tree); leaves for thatch, sails; spathe for cloth.	Local products only; starch emergency food; solitary feather palms.	106-108
Mauritia flexuosa L.	moriche, buriti, muriti, aguaje	S.Amer.: trop. rain forest, sites subject to floods.	Edible oil, fruit, heart; starch from trunk (60 kg/tree); leaf fiber for rope; petiole for cork; trunk for wood.	Local products only; indicator plant of trop. rain forest swamp; solitary fan palms.	5, 23, 54, 63, 80-81, 97, 99
Maximiliana martiana Karst	inaja	S.Amer.: trop. rain forest, dry, sandy, upland sites.	Edible fruit, heart, oil; leaves for thatch, weaving.	Local products only; solitary feather palms.	23, 104

Table 3. Continued.

Scientific name	Common name	Origin & native habitat	Products or potential use	Comments	Sel. Ref.
Nannorrhops ritchiana Griff.	mazari palm	S.Asia: subtrop. mts. to 1,500 m.	Edible fruit, heart, young inflorescence; leaves for thatch, weaving; leaves, petioles, trunk for fuel; erosion control on mt. slopes.	Local products only; shrub-like suckering fan palms.	15, 53
Oenocarpus spp.	bacaba	S.Amer.: trop. rain forest second growth, upland sites.	Edible oil, fruit; trunk for wood; potential leaf-base fiber source.	Local products only; solitary feather palms.	5, 7-8, 23, 99, 104
Oncosperma horridum (Griff.) Scheff.	bayas	S.E.Asia: trop. rain forest, in-land to 500-1,000 m only.	Edible heart; trunk for wood.	Local products only; suckering feather palms.	20, 34, 105
Oncosperma tigillarium (Jack) Ridley	nibong, nibung	S.E.Asia: trop. rain forest, coastal sites.	Edible heart; trunk for wood resistant to salt water.	Local products only; suckering feather palms.	15, 20, 34, 105
Phytelephas macrocarpa Ruiz & Pavon	ivory nut palm, tagua	S.Amer.: trop. rain forest, esp. sites sub-ject to floods, to 1,500 m.	Edible immature fruit; seeds former source of vegetable ivory.	Local products only; solitary feather palms.	1, 61, 91
Pinanga spp.	pinang	S.E.Asia: trop. rain forest, mts. to 1,200 m; some spp. in sites subject to floods.	Edible heart; leaves for weaving; trunk for walking sticks & wood; erosion control.	Local products only; forms thickets; small stature, sucker-ing feather palms.	20, 34, 105
Pritchardia spp.	Fiji fan palm	Polynesia: trop. mt. slopes, wet to dry.	Edible immature seed; leaves for thatch, weaving; erosion control.	Local products only; grow on dry, rocky, steep sites; popular orna-mentals; soli-tary fan palms.	14, 55

Table 3. Continued.

Scientific name	Common name	Origin & native habitat	Products or potential use	Comments	Sel. Ref.
Raphia spp.	raffia palms, African piassava	W.Africa: trop. rain forest, swamps, sites subject to floods.	Edible fruit, oil; sap for wine; leaves for fiber, thatch; petiole, leaf rachis for building material.	Fiber exported; fruit, oil, wine local products; suckering terminal flowering feather palms.	64, 85, 90-91, 98
Sabal spp.	palmetto	N.&S.Amer.: trop. & subtrop., wet to dry.	Edible fruit, heart; leaves for thatch, weaving, brush fiber; reforestation.	Local products only; at least 25 spp.; solitary fan palms.	16, 18, 29, 81
Scheelea spp.	corozo, coroba	S.Amer.: trop. rain forest, riverbanks, savanna.	Edible oil (3 kg/tree/yr); leaves for thatch; trunk for wood, fuel.	Local products only; about 40 spp.; *S. excelsa* Karst., *S. marco-carpa* Karst., *S. princeps* (Mart.) Karst. most promising for oil; solitary feather palms.	5, 18, 29, 81
Thrinax spp.	thatch palms	Caribbean: trop. & subtrop. dry forest.	Leaves for thatch, weaving; reforestation.	Local products only; some spp. grow on dry limestone sites; solitary fan palms.	95

This study classifies multipurpose palms that have potential for small farmers by considering their agronomic, economic, and related agroforestry potential. The 55 palms represent a variety of tropical products of long-standing value to local people, diverse tropical habitats, and increasing promise in untapped domestic and international export markets. Serious consideration of one or more of these multipurpose palms in agroforestry development schemes can ensure the broadest array of products for small farmers who live in areas where palms grow or can be grown.

Gathering information on palms is somewhat difficult because references are scattered throughout the world's scientific literature.[2] The selected references below are intended as lead-in information sources for research on particular species.

REFERENCES TO TABLES

1. Acosta-Solis, M. 1948. Tagua or vegetable ivory--a forest product of Ecuador. *Econ. Bot.* 2:46-57.

2. Anderson, A.B. 1988. Use and management of native forests dominated by acai palm (*Euterpe oleracea* Mart.) in the Amazon estuary. *Adv. Econ. Bot.* 6:152-162.

3. Bailey, L.H. 1949. *Manual of cultivated plants.* New York: Macmillan.

4. Balick, M.J. 1976. The palm heart as a new commercial crop from Tropical America. *Prin.* 20:24-28.

5. _____. 1979a. Amazonian oil palms of promise: A survey. *Econ. Bot.* 33:11-28.

6. _____. 1979b. Economic botany of the Guahibo: I: Palmae, *Econ. Bot.* 33:361-376.

7. _____. 1981. Une huile comestible de haute qualite en provenance des especes *Jessenia* et *Oenocarpus. Oleag.* 36:319-326.

8. _____. 1986. Systematics and economic botany of the *Oenocarpus-Jessenia* (Palmae) complex. *Adv. Econ. Bot.* 3:1-140.

9. _____. 1988. The use of palms by the Apinaye and Guajajara Indians of northeastern Brazil. *Adv. Econ. Bot.* 6:72-97.

10. Balick, M.J. and S.N. Gershoff. 1981. Nutritional evaluation of the *Jessenia bataua* palms: Source of high quality protein and oil from Tropical America. *Econ. Bot.* 35:261-271.

11. Balslev, H. and A. Barfod. 1987. Ecuadorean palms--an overview. *Opera Bot.* 92:17-35.

12. Bavappa, K.V.A., M.K. Nair, and T.P. Kumar. 1982. *The arecanut palm.* Kasaragod, India: Central Plantation Crops Res. Inst.

13. Bayma, C. 1958. Carnauba. *Produtos rurais.* No. 9. Rio de Janeiro: Ministerio da Agricultura.

14. Beccari, O. and J.F. Rock. 1921. A monographic study of the genus *Pritchardia. Mem. B.P. Bish. Mus.* 8:3-77.

15. Blatter, E. 1926. *Palms of British India and Ceylon.* London: Oxford University Press.

16. Bomhard, M.L. 1950. Palm trees in the United States. *Agr. Inf. Bull.* No. 22. Washington, D.C.: U.S. Dept. Agr.

17. Bondar, G. 1964. Palmeiras do Brasil. *Bol. Inst. de Bot.* No. 2., Sao Paulo.

18. Braun, A. 1968. Cultivated palms of Venezuela. *Prin.* 12:39-103.

19. Brown, W.H. 1951. *Useful plants of the Philippines.* Vol. 1, Tech. Bull. no. 10. Manila: Dept. Agr. and Nat. Res.

20. Burkill, I.H. 1966. *A dictionary of the economic products of the Malay Peninsula.* 2 vols. Kuala Lumpur: Min. Agr. Coop.

21. Calzavara, B.B.G. 1972. As possibilidades do Acaizeiro no estuario Amazonica. *Bol. Fac. Cien. Agr. Para,* No. 5, Belem, Brasil.

22. Camacho, E. and V.J. Soria. 1970. Palmito de pejibaye. *Proc. Trop. Reg. Amer. Soc. Hort. Sci.* 14:122-132.

23. Cavalcante, P.B. 1977. Edible palm fruits of the Brazilian Amazon. *Prin.* 21:91-102.

24. Child, R. 1964. *Coconuts.* London: Longman.

25. Clement, C.R. 1986. The pejibaye palm (*Bactris gasipaes* H.B.K.) as an agroforestry component. *Agrofor. Sys.* 4:205-219.

26. Clement, C.R. and J. Mora Urpi. 1982. The pejibaye (*Bactris gasipaes*) comes of age. *Prin.* 26:150-152.

27. _____. 1987. Pejibaye palm (*Bactris gasipaes*, Arecaceae): multi-use potential for the lowland humid tropics. *Econ. Bot.* 41:302-311.

28. Corley, R.H.V., J.J. Hardon, and B.J. Wood, eds. 1976. *Oil palm research.* Amsterdam: Elsevier.

29. Corner, E.J.H. 1966. *The natural history of palms.* London: Weidenfeld and Nicolson.

30. Dahlgren, B.E. and S.F. Glassman.1963. A revision of the genus *Copernicia*: West Indian species. *Gent. Herb.* 9:43-232.

31. Davis, T.A. 1972. Tapping the wild date. *Prin.* 16:12-15.

---

[2] A major source of technical and general information is *Principes,* quarterly journal of the International Palm Society, P.O. Box 368, Lawrence, Kansas 66044. Two other useful sources are N.W. Uhl and J. Dransfield, *Genera palmarum: A classification of palms based on the work of Harold E. Moore, Jr.* (Lawrence, Kansas: Allen Press, 1987) and M.J. Balick and H.T. Beck, comps., *An applied, annotated bibliography of the useful palms of the world* (New York: Columbia University Press, forthcoming).

32. Davis, T.A. and D.V. Johnson. 1987. Current utilization and further development of the palmyra palm (*Borassus flabellifer* L., Arecaceae) in Tamil Nadu State, India. *Econ. Bot.* 41:247-266.

33. Dissanayake, B.W. 1977. Use of *Caryota urens* in Sri Lanka. In *Sago-76: Papers of the first international Sago symposium*, ed. K.Tan, pp. 84-90. Kuala Lumpur: Kemajuan Kanji.

34. Dransfield, J. 1976. Palms in everyday life of West Indonesia. *Prin.* 20:39-47.

35. _____. 1977. Dryland sago palms. In *Sago-76: Papers of the first international sago symposium*, ed. K. Tan, pp. 77-83. Kuala Lumpur: Kemajuan Kanji.

36. _____. 1979. *A manual of the rattans of the Malay Peninsula*. Kuala Lumpur: For. Dept. Min. Prim. Ind. Malaysia.

37. _____. 1984. *The rattans of Sabah*. Sabah, Malaysia: For. Dept.

38. _____. 1988. Prospects for rattan cultivation. *Adv. Econ. Bot.* 6:198-208.

39. Duke, J.A. 1977. Palms as energy sources: A solicitation. *Prin.* 21:60-62.

40. Essig, F.B. and Y-F. Dong. 1987. The many uses of *Trachycarpus fortunei* (Arecaceae) in China. *Econ. Bot.* 41:411-417.

41. Familton, A.K., A.J. McQuire, J.A. Kininmonth, and A.M.L. Bowles, eds. 1977. *Coconut stem utilization seminar*. Wellington: Min. For. Aff.

42. Fanshaw, D.B. 1966. The dum palm *Hyphaene thebaica* (Del.) Mart. *East Afr. Agr. For. Jol.* 32:108-116.

43. Food and Agriculture Organization. 1985. *Coconut wood: Processing and use*. FAO for. paper, no. 57. Rome: FAO.

44. Ferwerda, J.D. 1977. Oil palm. In *Ecophysiology of tropical crops*, eds. P.T. Alvim and T.T. Kozlowski, pp. 351-382. New York: Academic Press.

45. Flach, M. 1983. *The sago palm*. FAO plant prod. prot. paper, no. 47. Rome: FAO.

46. Fong, F.W. 1987. An unconventional alcohol fuel crop. *Prin.* 31:64- 67.

47. Fosberg, F.R. 1960. Random notes on West African palms. *Prin.* 4:125-132.

48. Fox, J.J. 1977. *Harvest of the palm*. Cambridge, Mass.: Harvard Univ. Press.

49. Gonsalves, A.D. 1955. *O babacu*. Ser. Estud. Ens. Min. Agri, no. 8. Rio de Janeiro.

50. Gowda, M. 1951. The story of pan chewing in India. *Bot. Mus. Leaf. Harv. Univ.* 14:181-214.

51. Grimwood, B.E.1975. *Coconut palm products*. FAO agri. dev. paper, no. 99. Rome: FAO.

52. Hartley, C.W.S. 1977. *The oil palm*. 2nd ed. London: Longman.

53. Hawkes, A.D. 1967. On *Nannorrhops. Prin.* 11:137-138.

54. Heinen, H.D. and K. Ruddle. 1974. Ecology, ritual, and economic organization in the distribution of palm starch among the Warao of the Orinoco Delta. *Jol. Anth. Res.* 30:116-138.

55. Hodel, D. 1980. Notes on *Pritchardia* in Hawaii. *Prin.* 24:65-81.

56. Hodge, W.H. 1965. Palm cabbage. *Prin.* 9:124-131.

57. _____. 1975. Oil-producing palms of the world--a review. *Prin.* 19:119-136.

58. International Development Research Centre. 1980. *Rattan: A report of a workshop held in Singapore 4-6 June 1979*. Ottawa: IDRC.

59. Johnson, D. 1972a. The carnauba wax palm (*Copernicia prunifera*). *Prin.* 16:16-19, 42-48, 111-114, 128-131.

60. _____. 1972b. O comercio de frutas em Fortaleza. *Pesq. Agropec. Nord.* 4:75-81.

61. _____. 1975. Some palm products of the Peruvian Amazon. *Prin.* 19:78-79.

62. _____. 1980. Tree crops and tropical development: The oil palm as a successful example. *Agr. Adm.* 7:107-112.

63. _____. 1982. Commercial palm products of Brazil. *Prin.* 26:141-143.

64. Kitzke, E.D. and D. Johnson. 1975. Commercial palm products other than oils. *Prin.* 19:3-26.

65. Kovoor, A. 1983. *The Palmyrah palm: Potential and perspectives*. FAO plant prod. prot. paper, no. 52. Rome: FAO.

66. LIPI. 1978. *Palmen Indonesia*. Bogor: Lembaga Biologi Nasional.

67. Loomis, H.F. 1949. The nipa palm of the Orient. *Nat. Hort. Mag.* 28:4-10.

68. Macedo. J.H.P., F.O. Rittershofer, and A. Dessewffy. 1975. *A silvicultura e a industria do palmito*. Porto Alegre, Brasil: Instituto de Pesquisas de Recursos Naturais Renovaveis.

69. Makki, Y.M., ed. 1983. *Proceedings of the first symposium on the date palm*. Saudi Arabia: King Faisal Univ.

70. May, P.H., A.B. Anderson, J.M.F. Frazao, and M.J. Balick. 1985. Babassu palm in the agroforestry systems in Brazil's mid-north region. *Agrofor. Sys.* 3:275-295.

71. Mejia C.,K. 1988. Utilization of palms in eleven mestizo villages of the Peruvian Amazon (Ucayali river, department of Loreto). *Adv. Econ. Bot.* 6:137-143.

72. Miller, R.H. 1964. The versatile sugar palm. *Prin.* 8:115-147.

73. Mogea, J.P. 1978. Pollination in *Salacca edulis. Prin.* 22:56-63.

74. _____. 1982. *Salacca zalacca*, the correct name for the salak palm. *Prin.* 26:70-72.

75. Moll, H.A.J. 1987. *The economics of oil palm*. Wageningen, The Netherlands: Pudoc.

76. Moore, H.E., Jr. and A.B. Anderson. 1976. *Ceroxylon alpinum* and *Ceroxylon quindiuense* (Palmae). *Gent. Herb.* 11:168-185.

77. Munier, P. 1973. *Le palmier-dattier*. Paris: Maisonneuve and Larose.

78. Murray, D.B. 1977. Coconut palm. In *Ecophysiology of tropical crops*, eds. P.T. Alvim and T.T. Kozlowski, pp. 383-407. New York: Academic Press.

79. Nair, P.K.R. 1979. *Intensive multiple cropping with coconuts in India*. Berlin: Paul Parey.

80. _____. 1980. *Agroforestry species: A crop sheets manual*. Nairobi: International Council for Research in Agroforestry.

81. National Academy of Sciences. 1975. *Underexploited tropical plants with promising economic value*. Washington, D.C.: NAS.

82. Nayar, N.M., ed. 1983. *Coconut research and development*. New Delhi: Wiley.

83. Ochse, J.J., M.J. Soule, M.J. Dijkman, and C. Wehlburg. 1961. *Tropical and subtropical agriculture*. 2 vols. New York: Macmillan.

84. Ohler, J.G. 1984. *Coconut, tree of life*. FAO plant prod. prot. paper, no. 57. Rome: FAO.

85. Otedoh, M.O. 1974. Raphia oil: Its extraction, properties and utilization. *Jol. Nig. Oil Palm Res.* 5:45-49.

86. Paivoke, A.E.A. 1984. Tapping patterns in the nipa palm (*Nypa fruticans* Wurmb.). *Prin.* 28:132-137.

87. Pesce, C. 1985. *Oil palms and other oilseeds of the Amazon*. Translated by D.V. Johnson. Algonac, Michigan: Reference Publications.

88. Plucknett, D.L. 1979. *Managing pastures and cattle under coconuts*. Boulder, Colorado: Westview Press.

89. Popenoe, W. 1920. *Manual of tropical and subtropical fruits*. New York: Macmillan.

90. Profizi, J-P. 1985. *Raphia hookeri*: A survey of some aspects of growth of a useful swamp Lepidocaryoid palm in Benin (West Africa). *Prin.* 29:108-114.

91. Purseglove, J.W. 1972. *Tropical crops: Monocotyledons*. 2 vols. New York: John Wiley.

92. Pushparajah, E. and Chew Poh Soon, eds. 1982. *The oil palm in agriculture in the eighties*. 2 vols. Kuala Lumpur: Incorporated Society of Planters.

93. Putz, F.E. 1979. Biology and human use of *Leopoldinia piassaba*. *Prin.* 23:149-156.

94. Raghavan, V. and H.K. Barvah. 1958. Arecanut: India's popular masticatory--history, chemistry and utilization. *Econ. Bot.* 12:315-345.

95. Read, R.W. 1975. The genus *Thrinax* (Palmae: Coryphoideae). *Smith. Cont. Bot.* No. 19. Washington, D.C.: Smithsonian Inst. Press.

96. Rizzini, C.T. and W.B. Mors. 1976. *Botanica economica Brasileira*. Sao Paulo: Ed. Univ.

97. Ruddle, K., D. Johnson, P.K. Townsend, and J.D. Rees. 1978. *Palm sago: A tropical starch from marginal lands*. Honolulu: Univ. Press of Hawaii.

98. Russell, T.A. 1964. The *Raphia* palms of West Africa. *Kew Bull.* 19:173-196.

99. Schultes, R.E. 1977. Promising structural fiber palms of the Colombian Amazon. *Prin.* 21:72-82.

100. Stanton, W.R. and M. Flach, eds. 1980. *Sago: The equatorial swamp as a natural resource*. The Hague: Martinus Nijhoff.

101. Tan, K., ed. 1977. *Sago-76*. Papers of the first international sago symposium. Kuala Lumpur: Kemajuan Kanji.

102. Thampan, P.K. 1975. *The coconut palm and its products*. Cochin, India: Green Villa Pub.

103. Voeks, R.A. 1988. The Brazilian fiber belt: Harvest and management of piassava palm (*Attalea funifera* Mart.). *Adv. Econ. Bot.* 6:262-275.

104. Wallace, A.R. 1853. *Palm trees of the Amazon*. London: John Van Voorst.

105. Whitmore, T.C. 1973. *Palms of Malaya*. London: Oxford Univ. Press.

106. Wilbert, J. 1976. *Manicaria saccifera* and its cultural significance among the Warao Indians of Venezuela. *Bot. Mus. Leaf. Harv. Univ.* 24:275-335.

107. Wilbert, J. 1980a. The palm-leaf sail of the Warao Indians. *Prin.* 24:162-169.

108. _____. 1980b. The temiche cap. *Prin.* 24:105-109.

109. Wong, K.M. and N. Manokaran, eds. 1985. *Proceedings of the rattan seminar*. Kepong: Forest Research Institute of Malaysia.

How To Run a Small Forest Farm with Multipurpose Trees

Fu Maoyi and Lan Linfu

Subtropical Forestry Research Institute
The Chinese Academy of Forestry

In China, there are large areas of barren land due to a long history of repeated cuttings and unsustainable land-use practices. By 1949, forests covered only 8.6% of the total land area. Since then, the forest area has increased to 12.7% through the efforts of people and the government.

Still, the current forest obviously cannot fulfill the needs of a large country of over one billion people, and more work remains. From 1978 to mid-1984, 20 million hectares of barren, mountainous land was allocated to 50 million farm families for private cultivation. Another 50.67 million hectares of such land was contracted to farmers for tree planting, and 175,000 cooperative tree farms were established to manage 16.67 million hectares. More than 4,000 county-owned tree farms contain 46 million managed hectares.

Of the more than 8,000 tree species found in China, approximately 800 have been used in forestry. How to use this rich resource effectively to plant barren lands is this paper's subject. A small forest farm owned by the government in the northern part of Zhejiang Province served as the study area.

General Conditions of the Forest Farm

The Linfengshi Forest Farm in Anji County is located in the northeast of Zhejiang Province (30°23'N, 119°53'E). It contains a management area of 3,166 hectares and a staff of 196 people in 5 organizational divisions (planning and finance, production, security, purchasing and selling, and the general farm office). There are 9 production units on the farm, which are divided by environmental conditions and management objectives.

At present, the forest standing volume is 89,004 m^3 and the standing volume of bamboo is 1.99 million culms. The annual production value is 1.1 million Yuan, with a profit of 300,000 Yuan/year

($US1 = 3.7 Yuan). It is a small tree farm at about the middle level of management intensity in China. Details about land use are described in Table 1.

Management Analysis

The bamboo forest produces the highest income and profit (Table 2). Intensive management measures, such as organic and inorganic fertilization, pest and disease control, annual weeding, and silvicultural management, have been adopted. Using these methods, the average number of culms/ha is 3,270 and the fresh weight production is 31.5 t/ha/2 years, among the highest production rates in the country.

To prevent snow and typhoon damage to bamboo, the tops of new culms are usually trimmed every year. At least 14 pairs of branches must be kept, however, to avoid depression of photosynthesis. The main pests and diseases of bamboo in this region are *Ceratosphaeria phyllostachydis* Zhang sp. nov., *Algedonia coclesalis, Pantana sinica, Besaia goddrica, Norraca retrofusca* de Joannis, and *Hippota dorsalis.*

To meet an increased market demand for bamboo shoots, a portion of the bamboo forests have been converted. The density of these stands has been reduced to 3,000 culms/ha, the soil is cultivated to a depth of 20 cm and fertilized annually with 7 tons of buried weeds and 750 kg/ha of a NPK compound. These methods increase the production of bamboo shoots by as much as 5 times.

The timber tree forest can be cut and thinned now, but it would produce only 1,000 m^3 because the forests are in the middle years of rotation. The timber forest still employs as many workers as the bamboo forest. However, an intensive management plan that will be implemented soon is

Table 1. Land-use at the Linfengshi forest farm.

Use	Area (ha)	% total area
Forestry		
timber	863.6	27.3
reserved	1,020.2	32.2
bamboo	678.5	21.4
windbreak	60.8	1.9
economic (tea, etc.)	50.8	1.6
Unforested	202.1	6.4

Table 2. Annual production values and profit of various farm crops.

Management objective	Area (ha)	No. workers	Annual Prod. Value (1,000 yuan)	Annual Prod. % total	Profit Value (1,000 yuan)	Profit % total
Bamboo forest	606.9	80	567	51.5	185	61.7
Timber forest	863.6	80	380	34.5	109	36.3
Rice	13.3	15	18	1.6	-10	-3.3
Economic forest (including tea)	50.8	*	135	12.3	16	5.3

* Contracted by farmers.

expected to greatly increase timber production and profits.

To strengthen and broaden its economy, the farm also has started several business ventures, including forest products, a store, hotel, and beverage factory.

The farm's experience shows that, with proper management, small tree farms are practical in barren, mountainous lands. But the following factors must be considered:

o Multipurpose tree species best suited for the location should be selected, and elite varieties and clones should be used.

o At least one of the selected tree species should be managed on a short rotation to provide some early income.

o Intensive management should be adopted if possible.

o Although the main business of a tree farm is growing trees, other business activities will strengthen its economy.

Cost-Benefit Analysis of Forest Plantations in a Watershed in Northeastern Thailand

Arjen Sterk and Pieter van Ginneken

FAO Integrated Development of the Phu Wiang Watershed Project
Khon Kaen, Thailand

The Government of Thailand, assisted by the United Nations Development Programme and the Food and Agriculture Organization of the United Nations, is carrying out an integrated development plan in the Phu Wiang watershed located 70 kilometers northwest of Khon Kaen in northeastern Thailand. It covers an area of 300 km^2 and consists of a central valley of 100 km^2 surrounded by mountains that form part of a forest reserve. Total population of the area is about 17,000.

Farmers in the central valley of the watershed have been encroaching on the forest reserve to grow cassava (*Manihot esculenta*). The government is establishing plantations of fast-growing tree species, including *Eucalyptus camaldulensis*, *Leucaena leucocephala* and *Acacia auriculiformis*, to reforest the area and protect the watershed.

Generally, conditions in the watershed are similar to those of the northeastern region of Thailand. Agriculture is the main occupation, soils are sandy and susceptible to erosion, rainfall occurs from May to November, and unemployment is high during the dry season.

Forestry and agroforestry are considered ecologically more stable than traditional cropping systems, particularly on sloping lands. However, a complete appraisal of the financial, economic, and social feasibility of land-use changes would require considerable reliable information, which is not always easily available. Nevertheless, to assess the potential of the Phu Wiang Forest Reserve and to demonstrate the structure of a cost-benefit study for forest plantations, several simulations have been carried out with *Eucalyptus camaldulensis*, a species about which relatively much information is available in Thailand.

In an economic analysis conducted from a national point of view, various indirect costs and benefits, additional to the costs and benefits already included in the financial analysis, should be considered. Quantification and valuation of intangible benefits (and also costs) are outside the scope of this study. However, an attempt has been made to identify indirect--and often intangible--costs and benefits.

For an economic analysis of the Phu Wiang Forest Reserve, the following five situations should be considered:

o pure agriculture, cassava cultivated by farmers

o plantation forestry, trees planted by the Royal Forest Department (RFD)

o agroforestry, trees planted by the RFD

o agroforestry, trees planted by farmers

o management and exploitation of the natural forest

Table 1 presents an overview of identified indirect costs and benefits in relation to the situations listed above. Most of this paper, however, deals with a financial analysis of Eucalyptus plantations, upland cassava cultivation, and intercropping of Eucalyptus with cassava.

Financial Analysis

The financial analysis covers forestry, agricultural, and agroforestry land uses with regard to charcoal, pulpwood, and timber production of Eucalyptus and dry chips produced from cassava.

Eucalyptus Plantations

Basic data required for a financial analysis of tree plantations can be broadly classified into the following groups:

Table 1. Indirect costs and benefits of land-use systems in the Phu Wiang forest reserve.

Land-use system	Indirect costs and benefits*
Pure agriculture, cassava cultivated by farmers	-- destruction of capital--natural resources-- in the form of standing forest; -- other costs are related to soil erosion and water quality; -- contribution to foreign exchange earnings, since all cassava is exported.
Plantation forestry, trees planted by RFD	-- reduction of erosion; -- soil improvement; -- improved water quality and supply; -- job creation; -- fodder provided for buffalo; -- costs are associated with loss of other productive land-use alternatives, such as replacement of cassava by plantations.
Agroforestry, trees planted by RFD	-- income for local farmers from intercropping cassava; -- reduced encroachment of natural forest because of partially satisfied cash needs; -- better forest protection because farmers have planted the trees whose products they benefit from; -- increased replanting of area over same period of time and thus improved control of soil erosion, water and soil quality, more jobs and income opportunities; reduced encroachment of natural forest and thus positive effects on genetic diversity and availability of minor forest products.
Management and exploitation of natural forest	-- timber of valuable hardwood species; -- fuelwood and charcoal; -- fruits, vegetables, medicinal herbs, ornamental plants, wild animals, etc.; -- creation of jobs related to above benefits; -- ecological effects on soil and water; -- maintenance of genetic diversity; -- recreational opportunities.

* Because quantification and valuation are impossible at this stage, costs and benefits have not been listed separately.

o costs of plantation establishment and management

o tree growth and yield

o prices of tree products

Data from the Phu Wiang plantations indicated that over 95% of costs involved in plantation establishment are labor related.

Papers on the growth of different species are in preparation. The average diameter and height were calculated from a sample of four-year-old trees planted at a 4 x 4 m spacing (100 trees per rai) in sub-unit 1 of the project. (1 rai = 0.16 ha, 1 ha = 6.25 rai.) Volume was 4.2 m^3 per rai, and the Mean Annual Increment (MAI) per tree was 0.011 m^3. The young age of the plantations in Phu Wiang and the lack of a local volume table made it necessary to use an estimated average MAI of 0.027 m^3 per tree. This value is less than those obtained from other research in the northeast but higher than those suggested by preliminary data of young plantations in Phu Wiang or by plantations on bad sites. A MAI of 0.027 m^3 would result in an annual growth of 16.7 m^3 per ha, a reasonable estimate for Eucalyptus.

Prices of Wood Products

Charcoal. A survey in the Phu Wiang region, including Khon Kaen, indicated that charcoal prices averaged about 1.25 baht per kg. Prices varied from 0.90 to 2.00 baht, depending on season, quality, and unit-quantity of selling. Khon Kaen prices averaged 2.5 baht and ranged from 2 to 3.5 baht.

Pulpwood. The Phoenix Pulp and Paper Mill in Khon Kaen Province was working at 80% capacity in March 1987. Its management expressed a clear preference for bamboo and kenaf as raw materials. The price for pulpwood of exotic trees, such as Eucalyptus and Casuarina, was 500 baht per ton green weight.

However, there are no reported intentions of using exotic trees as a raw material. Sometimes these trees can be used in a mix with bamboo or kenaf (10% exotic trees, 90% bamboo or kenaf), but such mixes are not generally preferred.

Timber. Information on the quantity and prices of exotic timber species for both Khon Kaen Province and Thailand as a whole is scarce. Most plantations with exotic trees are still too young, and the trees are often grown for uses other than timber. The average timber price at the Phu Wiang sawmill is 1,100 baht per m^3, with 800 baht for softwood and 1,500 baht for hardwood. Sawmills in the Phu Wiang region are expecting an increase in timber prices this year.

Data on Cassava

To compare forestry, agroforestry and agricultural land-use systems, cassava monocropping and intercropping with trees have to be considered.

Cassava yields are high for the years immediately following encroachment of the natural forest, but decrease rapidly in subsequent years of continuous cultivation. In the first year, yields of fresh cassava are estimated at 25 tons per ha, but drop to no more than 7 tons per ha after 6 years. Most farmers in the Phu Wiang area use family labor to process the cassava roots into chips. The price for dry cassava chips was 1.9 baht per kg in April 1987.

It has been observed that cassava can be intercropped with Eucalyptus in the first three years after establishment. Based on a cassava survey in the Phu Wiang watershed and on other information, yields of 4.4, 3.8, and 3.4 tons per hectare of cassava chips have been assumed for the first, second, and third years, respectively.

MULBUD Simulations

MULBUD (Multi-enterprise, multi-period budgeting) is an interactive computer package designed to assist the economic appraisal of perennial crops and agroforestry land-use systems where many tree species, crops, or livestock are grown together. MULBUD acknowledges the time dimension, is user friendly, and is suitable for sensitivity analysis.

Table 2 shows the major results of the MULBUD simulations with respect to forestry, agroforestry, and agricultural land uses. The highest Sum of Net Present Values (SNPV) and amortized values (AV) per rai per year were achieved with timber. The average labor requirements per rai and per year are related to the length of the rotation. From the column "SNPV/days," the Return to Labor (RTL) can be deduced by considering the incorporated wage rates of 35 baht per day in forestry and agroforestry combinations and 20 baht per day for cassava monocropping. The Eucalyptus-timber simulation return was over 62 baht per day (The SNPV/day of the Eucalyptus-timber enterprise is 27 baht; the wage rate is 35 baht). The other timber simulation (multi-enterprise) and the simulations with pulpwood total 50 baht or more per day, while the simulations with charcoal are just over 40 baht per day.

The RTL of cassava cultivation was above 35 baht per day in the first four years, but decreased to 24 and 21 baht per day in years 5 and 6, respectively, as yields declined.

With regard to the benefit/cost ratio, timber and pure cassava simulations for the first four years after encroachment gave the best results (between 2.3 and 1.8). The benefit/cost ratios for simulations of charcoal and pure cassava in years 5 and 6 were 1.1 and 1.2, respectively, and were the lowest obtained. Apart from the cassava monocropping simulations, the Eucalyptus (pulp)-cassava simulation gave the highest (25%) Internal Rate of Return (IRR) because of the relatively short period of analysis (8 years) compared to the simulation with timber as the main product. All others had an IRR of 20-23%, except the eucalyptus-charcoal simulation, which had an IRR of 16%

Consequences of the Results

The results of Table 2 can further compare the natural forest (dry evergreen), upland cropping (cassava), and Eucalyptus plantation land-use systems. Theoretically, land resources have a current market value equal to the present value of their expected future land rents (Barlowe 1972). Using this concept, land rent may be defined simply as the economic return that accrues or should accrue on production land. The rate of discount used is 10%.

Based on a growing stock of 55.6 m^3 per hectare and a profit margin on concessions of 2,000 baht per m^3 of wood, the value of dry evergreen forest has been estimated at 112,000 baht per hectare. The period of regrowth to obtain the same volume of the stand is assumed as 30 years. Year 0 is defined as the time when selective

logging by concessionaire takes place. The land value for year 0 is 112,000 plus 112,000 discounted over 30 years plus 112,000 discounted over 60 years, etc. The value for year 1 is 112,000 discounted over 29, 59, 89, etc. years. The same can be done for years 2-29, inclusive. The value for year 30 will be the same as for year 0. In year 30, a concession is again assumed to take place (as in years 60, 90, etc.), so the cycle can start again.

Given a decline in yields, cassava cropping is assumed to last continuously for 8 years, followed by 4 fallow years, and 2 years of cassava production in a continuous cycle. The yields between the fallow periods are assumed equal to years 4 and 5 of the MULBUD pure cassava simulations, or 17.2 and 6.3 fresh tons per hectare. All future net benefits are discounted back to the year of reference.

The third major land-use system within the forest reserve consists of forest plantations intercropped with cassava during the first three years. Again, all future net benefits are discounted back to the year of reference.

Fig. 1 shows the results in a graph form. Year 1 is the initial year for all compared land-use types, selective logging by the concessionaire, cassava growing, and establishment of forest plantations. At first, the cassava results are favorable because of relatively high yields in the first years after encroachment, but after 8 years the results are nearly equal to Eucalyptus-charcoal plantations. The timber plantations give the best results in terms of net present value. The curve representing pulpwood plantations (not displayed for reasons of clarity) lies in between charcoal and timber plantations. It is important to stress that the natural forest produces benefits other than timber that are not considered here.

Employment Generation

Generation of employment in rural areas is an important issue for the Royal Thai Government, particularly in the Northeast. Over a 30-year time frame, the labor demand for the charcoal alternative is the highest, with 5,340 days per hectare, an average of 178 days per year. This compares to 142, 125, and 109 days annually for pulpwood, cassava, and timber plantations, respectively. Compared to the forest plantations,

Table 2. Overview of main results of MULBUD simulations.[1]

Land use	Years	SNPV[2] baht	AV[3]	Average labor days/yr	SNPV/ days baht	RTL/ day[4]	B/C ratio[5]	IRR %[6]
Agriculture								
Pure cassava								
year 1	1	1,810	1,990	114	16	36	1.8	--
year 2	1	1,937	2,131	96	20	40	2.0	--
year 3	1	1,681	1,849	87	19	39	2.0	--
year 4	1	1,383	1,522	77	18	38	1.9	--
year 5	1	181	199	43	4	24	1.2	--
year 6	1	43	47	40	1	21	1.1	--
Forestry								
Eucalyptus-charcoal	6	656	151	15	7	42	1.2	16
Eucalyptus-pulpwood	8	1,826	342	13	18	53	1.6	20
Eucalyptus-timber	15	4,922	647	12	27	62	2.3	22
Agroforestry								
E. camaldulensis (charcoal)-cassava	6	1,030	236	29	6	41	1.2	20
E. camaldulensis (pulp)-cassava	8	2,500	469	22	15	50	1.5	25
E. camaldulensis (timber)-cassava	15	5,273	693	18	20	55	1.8	23

[1] Results are based on an area unit of 1 rai (1 rai = 0.16 ha; 1 ha = 6.25 rai).

[2] SNPV = sum of net present values, based on a 10% rate of discount.

[3] AV = amortized value (baht/yr).

[4] RTL = return to labor (baht/yr).

[5] derived by dividing discounted benefit stream by discounted cost stream.

[6] IRR = internal rate of return.

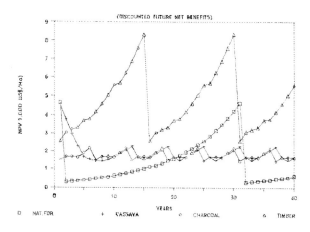

Fig. 1. Net present values of various land uses of the forest reserve.

244

Table 3. Forest land use in Phu Wiang.

Year	Dry evergreen forest	Dry dipterocarp forest[1]	Forest planta-tions[2]	Cassava[3]
1976[4]	16,540	6,480	--	100
1982[5]	14,469	6,100	--	1,463
1984[6]	13,529	5,322	152	3,009
1987	10,974	5,038	904	5,100
1990	8,544	4,768	1,500	7,200
1995	5,034	3,778	2,500	10,700
2000	2,604	1,708	3,500	14,200
2005	--	--	4,500	17,512

[1] Including decrease of dry dipterocarp forest outside the forest reserve.

[2] Figs. for 1984 and 1987 are actual for established plantations; projections after 1987 are based on a yearly increase of 200 ha.

[3] Figs. only of cassava inside forest reserve. Figs. for 1976, 1982, and 1984 are based on aerial photographs and a ground encroachment of the natural forest for growing 700 ha/yr cassava, slightly below the average annual increase during the 1982-84 period.

[4] Based on aerial photographs of 1976.

[5] Based on aerial photographs of 1976, followed by a ground check.

[6] Based on new aerial photographs of 1984.

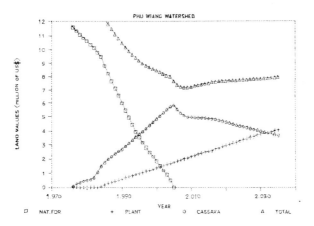

Fig. 2. Total land values of the forest reserve for various purposes.

only the timber alternative has a lower labor demand.

Aggregated Land-Use Values

Thus far, all figures have been expressed per unit area (hectare). To understand their implications for the Phu Wiang watershed as a whole, one needs information on the total land use within the forest reserve.

Table 3 gives the projected forest land use in Phu Wiang over a period of 30 years. From 1976 to 1984, the total area of natural forest in the watershed, inside as well as outside the forest reserve, decreased from 23,020 to 18,851 hectares. This implies a decrease of natural forest from 76.5 to 62.7 percent of the total watershed area. From 1984 to 1987, forest plantation figures coincided with what actually was planted. After 1987, an annual increase of 200 hectares is assumed. The encroachment by farmers in the forest reserve is assumed to continue unconstrained at a rate of 700 hectares per year, slightly below the average increase during 1982-1984. After the disappearance of all the natural forest, it is assumed that the establishment of new forest plantations continues at the same rate of 200 hectares per year at the expense of areas used for cassava growing.

Land Value Assumptions

Table 3 shows existing data and the assumed scenario of further encroachment of the natural forest and of expansion of the forest plantations. Many different scenarios might be chosen. In this case, the main purpose is to show the implications of present trends.

As mentioned above, a value of 112,000 baht per hectare was assumed for dry evergreen forest just before logging by a concessionaire.

No separate distinction was made with reference to the main product of the forest plantations. Instead, an average value of 62,500 baht per hectare was assumed. This approximates the result for pulpwood.

For cassava, 12 different values were used, depending on the number of years after encroachment. From year 12 on, a continuous cycle was assumed with a constant value of 50,000 baht per hectare.

Aggregated Land Values

The results of this aggregation exercise of land values are displayed in Fig. 2 ($1 = 26 Baht). Natural forests have disappeared by the year 2005. In the same year, the value for cassava has peaked at about $5.8 million.

In 2005, cassava cultivation is assumed to take place on an area of 17,500 hectares or almost 80% of the forest reserve, the remainder being forest plantations. The curve representing the value of all lands within the forest reserve declines steeply because of the disappearance of dry evergreen forest. After 2005, cassava lands gradually will be reverted back to forest lands. Given the higher value of the latter, the total value curve then rises slowly again.

Discussion

Based on the assumed data, forest plantations seem able to compete with cassava growing in terms of net present value, RTL, and employment generation. In reality, the assumed scenarios could become much different, depending on government policy and other factors. In view of an expected increase in forestry-related employment opportunities in the near future, pressing questions need to be answered. Is the RFD going to manage the plantations by itself, including the harvesting and selling of the wood products, or will it become possible for farmers to participate in the process of tree growing? The last option implies that benefits of the trees would go to farmers as well.

REFERENCES

Barlowe, R. 1972. *Land resource economics: The economics of real property*. 2nd ed. Englewood Cliffs, N.J.: Prentice-Hall, Inc.

Boonruang, Prem. 1986. *Supply of and demand for wood products in Northeast Thailand*. Development of Diversified Forest Rehabilitation Project Northeast Thailand, Working Paper No. 5, Vol. VI.

Etherington, D. and P. Matthews. 1985. *MULBUD user's manual*. 3rd revised DOS ed. The National Australian University.

Gittinger, J. Price. 1984. *Economic analysis of agricultural projects*. 2nd ed. Baltimore and London: The John Hopkins University Press.

Hoamuangkaew, Wuthipol. 1986. *Reforestation by agroforestry system in Thailand*. Bangkok: Kasetsart University, Faculty of Forestry.

Ngamsomsuke, Kamol, Prasat Saenchai, Panomsak Promburom, and Bunthom Suraporn. 1987. *Farmers attitudes toward forest, plantation, and conservation farming in selected villages of the Phu Wiang valley, Khon Kaen, Thailand*. Project THA/84/002 Field Document 3.

Petmak, Pitaya. 1987. *Research evidences of the possibilities of agrosilvicultural system in Northeast Thailand*. Bangkok: Royal Forest Department.

RESEARCH (consultants). 1986. *Farms and fuelwood in Nakorn Ratchasima Province, Thailand: Analysis of costs, benefits, problems, and prospects of diversified forest rehabilitation on Northeast Thailand*. Projects GCP/RAS/111/NET and THA/81/004.

Royal Forest Department. 1984. *Results of analysis about expenditure on forest plantations of RFD*. (In Thai.) Bangkok: RFD.

Ruedenauer, Michael, Nakorn Imboriboon, and Ananchai Khuantham. 1986. Eucalyptus camaldulensis *plantations in forest reserve areas*. Bangkok: Thai-German Land Settlement Promotion Project, Department of Public Welfare.

Thongmee, Uthai. 1982. *Forest inventory and management*. Project THA/70/016, Working Paper No. 10.

Farmers' Attitudes Toward Planting Multipurpose Trees in Northeast Thailand

Narong Srisawas

Department of Sociology and Anthropology, Kasetsart University
Bangkok, Thailand

In this study, rural farm household heads in 5 provinces of Northeast Thailand were surveyed regarding their access to and use of multipurpose tree species. The farmers surveyed noted a shortage of available seedlings and a lack of proper knowledge about multipurpose trees. Pilot self-help nurseries should be established in selected primary schools, and some multipurpose tree species should be promoted.

Study Method

Ten districts in 5 provinces that had previously been identified as having fuelwood scarcities were selected for the study in 1986. A total of 440 heads of farm households were interviewed.

Results

The farmers surveyed listed the multipurpose tree species that they most wanted (Fig. 1), the quantities they desired (Fig. 2), their reasons for growing trees (Fig. 3), and the uses of the species (Fig. 4).

Nurseries

There were 6 Royal Forest Department nursery centers in the provinces studied. Each center could produce about 500,000 seedlings per year. There were also 10 private nurseries of various sizes producing *Eucalyptus camaldulensis* seedlings at 0.50-0.75 baht per seedling ($US1 = 25 baht).

Farmers' Problems

Farmers said there was a shortage of seedlings and a lack of proper knowledge of multipurpose trees.

Recommendations

o Pilot self-help nurseries producing multipurpose tree species should be set up in selected primary schools of the districts.

o Some multipurpose tree species, such as *Azadirachta indica* and *Cassia siamea*, should be promoted.

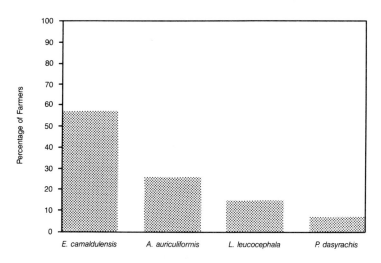

Fig. 1. Multipurpose tree species desired by farmers.

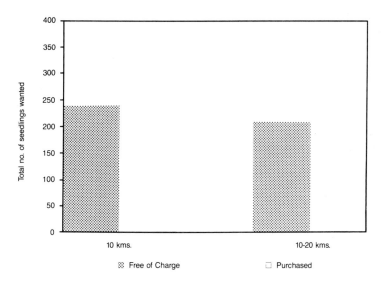

Fig. 2. Average no. seedlings farmers want, by distance.

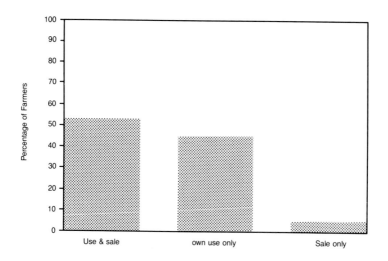

Fig.3. Reasons given by farmers for growing trees.

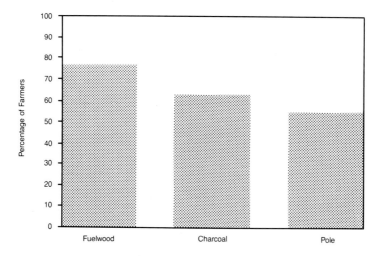

Fig. 4. Uses of multipurpose trees by farmers.

Potential Uses of Nitrogen-Fixing Trees on Small Coconut Plantations in Sri Lanka

L.V.K. Liyanage, H.P.S. Jayasundara, and T.G.L.G. Gunasekara

Coconut Research Institute
Sri Lanka

Establishment of nitrogen-fixing trees (NFT) on coconut plantations has received increasing attention recently, although no quantitative data are available on their growth and yield or specific uses under various management regimes. It is now evident that *Gliricidia sepium, Leucaena leucocephala*, and several other NFT crops are undoubtedly suitable for planting in coconut plantations as multipurpose tree species (MPTS) for small-scale farms. The results from a series of field experiments designed to investigate the growth and yield performance and the multiple uses of NFT species in coconut plantations are reported here.

Gliricidia and Leucaena Trials

Gliricidia and Leucaena were grown in a cover crop evaluation trial for 3 years in 4 locations of 3 different soil types and climates of coconut-growing areas of Sri Lanka. The 4 locations were wet zone, wet intermediate zone, dry intermediate zone, and dry zone. The soil types were ultisol (lateritic) and 2 entisols (sandy and sandy loam). A randomized block design with 4 replications was used. Crops were planted 2.0 x 0.9 m apart in double rows in the coconut avenues, leaving more than 2 m from the rows to the coconut boles. One year after planting, harvesting was started by lopping the plants at 1.0 m height above the ground. Green matter (leaves and tender stems) was separated from fresh firewood.

Lopping was done at 3-month intervals, giving 4 harvests per year. Gliricidia and Leucaena produced 7-10 t/ha and 12-16 t/ha green matter and 8-15 t/ha and 14-20 t/ha fresh firewood during the first and second years, respectively. Gliricidia demonstrated a wide adaptability to all the climates and soil types while Leucaena performed well in dry areas and preferred light soils.

Use of Gliricidia and Leucaena as sources of green manure was tested in 2 experiments. These

were set up in low-yielding mature coconut plantations on ultisols (lateritic) of the intermediate dry zone and intermediate wet zone. The purpose was to study ways to induce new root development and expand the active root zone around the base of the palms. Treatments included opening up quarter-, semi-, and full-circle trenches of 30 cm width and 30 m depth 30 cm away from the bole of the palm. The quarter-, semi-, and full-circle trenches were filled with 31, 62.5 and 125 kg Gliricidia leaves, respectively. A randomized block design with 3 replications was used. There were 9 effective palms per plot. Nut and copra production was measured at 2-month intervals for 2 years.

Incorporation of Gliricidia leaves in different size trenches around the palms produced copra weights significantly higher than those of the control without trenches or added material. The quarter-circle trenches (Fig. 1) produced the highest contents of copra and copra weights compared to controls.

In another experiment, Gliricidia and Leucaena were planted in double rows at a spacing of 2 x 0.9 m under 40-year-old coconuts in the wet intermediate zone on a ultisol (lateritic). A randomized block design with 3 replications and 9 palms per plant were taken to record nut production. Gliricidia and Leucaena were lopped 1 year after planting at 1.0 m height and every 6 months thereafter, and fresh material of 24 kg/palm/year was incorporated to a depth of 22 cm in a 2 m radius around the palm during the southwest monsoons (Fig. 2). Soil physical and chemical measurements were taken at 6-month intervals and coconut/copra production was recorded every 2 months. Although no improvement in nut production was observed, a reduction of the soil bulk density was evident after 6 months.

Use of Gliricidia and Leucaena in integrated farming systems of coconut, pasture, and cattle

Fig. 1. Quarter circle trench cut around a coconut palm to bury Gliricidia leaves.

Fig. 2. Gliricidia mulch applied as a green manure to coconuts.

was studied for 2 years. Two rows of Leucaena were planted 2 m apart in the coconut avenues while Gliricidia and Leucaena were planted 1.5 m apart alternately in a living fence along the boundary of a 1-hectare block. A mixture of *Brachiaria miliiformis/Pueraria phaseoloides* was grown in the avenues between the coconuts. Jersey and Local heifers were reared at 4 animals/ha and fed the forage. When the living fence was 2 years old, it yielded more than 2 t/ha/yr fresh green matter, with harvests every 6 months. Significant increases of live weight gains of over 450 g/head/day were produced when Gliricidia and Leucaena were mixed in the forage mixture during the wet season. More than 300 g/head/day of live weight gains were produced during the dry season. General improvement in the coconut plantation also occurred, and it became evident that inorganic fertilizers were not necessary in this integrated farming system.

Gliricidia stakes of 1.5 m height, planted 25 cm away from the pepper plants grown under coconut and lopped regularly, have also served as successful live pepper supports for over 10 years.

Twenty introduced NFT species were tested for local adaptability in the low country intermediate zone of Sri Lanka for 2.5 years. The experiment consisted of a non-replicated block design, where tree species were planted in plots of 5 x 20 m with 1 x 1 m spacing. *Acacia auriculiformis, A. mangium, Leucaena leucocephala* var. K636, *Gliricidia sepium, Enterolobium cyclocarpum,* and *Calliandra calothyrsus* were identified as promising, yielding more than 2.5 kg/plant green matter and 4.0 kg/plant firewood after 2 years.

Superiority Indices of Some Multipurpose Trees
from The Central Himalaya

A. R. Nautiyal and A. N. Purohit

High Altitude Plant Physiology Research Centre, Garhwal University
Srinagar, India

Although the forest flora of Kumaon and Garhwal, India are reported to include about 200 angiosperm tree species (Osmaston 1926), only a few are being included in social forestry programs due to the lack of information on their growth rate, photosynthetic potential, water use efficiency, transpiration, calorific value, wood density, and ash content. These are some of the important parameters to consider when selecting suitable species for fuelwood and fodder. In the absence of this data on local species, exotics that have been useful elsewhere are being introduced without consideration of the possible inherent risks. This was the case in South Africa, where 8 out of 19 Acacia species introduced from Australia regenerated so vigorously after fire that they suppressed indigenous vegetation (Berg 1977).

The U.S. National Academy of Sciences (1980) has warned that invasive plants should be avoided as much as possible and that local species should be given first priority in trial plantings for fuelwood plantations. Thus, there is clearly a need for a broad database on socioeconomic, ecological, and biological aspects of local species. Desirable characteristics cannot be considered in isolation, however. If one considers only fast growth, the species might not yield higher calorific value per unit area, as wood density is usually negatively correlated with the rate of growth (Burley 1980). Similarly, a multipurpose species might not be a fast-growing species. Keeping in view some of the disparities in combining characteristics, we reviewed the major and minor uses of 84 tree species from the Central Himalaya and selected 20 to study rate of growth, water and light use efficiencies, and properties associated with fuelwood characteristics. Their relative superiority index was computed to identify the species with the optimum combination of desirable characteristics.

Materials and Methods

A utility index of 84 tree species, based on their major and minor uses, was worked out. Values for 20 species are given in Table 1.

Seeds of these species were sown in styrofoam trays containing a 1:2 mixture of garden soil and farmyard manure. The trays were kept in net houses and watered regularly. At the 3-4 leaf stage, seedlings were transplanted to earthen pots (1 seedling per pot) containing a 1:2:3 mixture of river sand, farmyard manure and garden soil. Pots were kept in a net house at Srinagar after planting. Observations on 10 plants were recorded every 2 weeks for height and leaf number for 2 years.

The CO_2 exchange rate (CER) and stomatal conductance were measured simultaneously with the LI-COR LI-6000 Portable Photosynthesis System, which calculates the leaf internal CO_2 concentration as well as transpiration. The light quanta falling on leaf surfaces were measured with the help of a quantum sensor (LI-COR model 190s-1), and after working out the percentage absorptivity of 3 different leaves with the help of an LI-1800 portable spectroradiometer, the total light absorbed by the leaves was determined. Light use efficiency (LUE) was calculated as mole CO_2 fixed per einstein of light absorbed. Based on transpiration and CER values, the water use efficiency (WUE) was worked out as mg CO2 fixed per mg of water lost by transpiration. All these observations were recorded the 15th day of each month during the period when plant foliage was present. Mean values of each parameter represent the average of values recorded during the annual growth period of each species. Calorific value, density, ash content, moisture, and nitrogen content of wood were measured (Purohit and Nautiyal 1987). The Pearson correlation matrix was used to analyze the mean values of the data. Light transmissivity of leaf at 660 and 730 nm was

Table 1. Utility index of 20 tree species based on prevailing major and minor uses in the hills.[1]

Species	Uses[2]									
	E	FD	FL	OL	CT	FB	FR	M	G/R	ST
Toona ciliata		1.0	1.0	0.5	1.0			0.5		
Grewia optiva		1.0	0.5		0.5	1.0			0.5	
Ougeinia dalbergioides		1.0	1.0		1.0	0.5	1.0	0.5	0.5	
Bauhinia retusa		1.0	1.0				0.5	0.5	3.0	
Dalbergia sissoo		1.0	1.0		1.0		1.0	0.5	0.5	0.5
Sapindus mukorossi			0.5						1.0	0.5
Terminala tomentosa		1.0	0.5		1.0					0.5
Ficus bengalensis	0.5	1.0	0.5					0.5	1.0	
Cordia myxa	0.5	1.0	0.5							
Holoptelea integrifolia		1.0	0.5	0.5						0.5
Bombax ceiba	0.5			0.5	0.5	1.0				1.0
Eugenia jambolana	1.0		0.5		0.5				0.5	
Cornus capitata	1.0	0.5	0.5							
Quercus glauca		1.0	1.0		0.5				0.5	0.5
Q. incana	0.5	1.0	1.0		1.0				0.5	1.0
Fraxinus micrantha		0.5	0.5		1.0					0.5
Celtis australis		1.0	1.0	0.5				0.5		1.0
Aesculus indica	0.5	0.5		0.5	1.0					
Prunus cerasoides	0.5	0.5			0.5				0.5	
Alnus nepalensis		0.5	0.5		0.5		1.0		1.0	0.5

[1] Scores 1 and 0.5 represent major and minor uses, respectively.
[2] E = edible, FD = fodder, FL = fuel, OL = oil, CT = commercial timber, FB = fiber, FR = fertilizer, M = medicine, G/R = gums/resins, ST = small timber.

measured with the help of an LI-1800 spectroradiometer.

Results and Discussion

Based on an annual average (Table 2), the highest rate of growth (REG) was recorded in *Toona ciliata* and the lowest in *Aesculus indica*. Photosynthesis (Ph) and light use efficiency (LUE) were maximum in *Dalbergia sissoo* and minimum in *Sapindus mukorossi*. Similarly, transpiration was highest in *D. sissoo* and lowest in *Quercus incana*. The water use efficiency was found highest in *Ougeinia dalbergioides* and lowest in *Sapindus mukorossi*.

For superior firewood, wood should contain high calorific value, high density and low ash, moisture, and nitrogen. These values for the tested species are shown in Table 3. *Toona ciliata* had the highest calorific value, but its wood density was comparatively low. On the basis of wood density, *Quercus incana* ranked first, but it had a considerably higher ash content. *Dalbergia sissoo* had the lowest moisture content and *Prunus cerasoides* contained the least amount of nitrogen in its wood.

From the standpoint of productivity, a species should grow fast with high production potential in terms of photosynthesis and water and light use efficiencies. However, none of the species seems to combine maximum values of desirable characteristics and minimum values of all non-desirable characteristics. Some desirable characteristics, such as REG and wood density, are inversely correlated, and some non-desirable characteristics, such as transpiration, are positively correlated with rates of growth and photosynthesis. Therefore, selecting species with suitable combinations of desirable and non-desirable characteristics is essential.

Based on actual measurements for the first 2 years of growth, the 20 species were given index values of 1-20 for each characteristic measured. An index value of 1 was assigned to the species with the highest value of a desirable characteristic and the lowest value of an undesirable one. Species with the lowest value of a desirable characteristic or the highest value of a undesirable characteristic received an index value of 20.

On the basis of index values of each characteristic, the final index of the species was computed. The results are shown in Table 4. On the basis of the final index value, *Dalbergia sissoo* is the best species although it has the highest transpiration rate. *Prunus cerasoides* is the next best species, followed by *Ougeinia dalbergioides*, which, like *D. sissoo*, has the additional advantage of nitrogen fixation. These species have high coppicing ability as well as good seed germination. They usually grow successfully on eroded soils.

The quality of light transmitted through the canopy of a tree species has a substantial photomorphological effect on the understory vegetation. Tree species with a high ratio of 660/730 nm (R/FR ratio) transmitted light will have better chances for germination and growth of associated vegetation than those with a low ratio. Thus, if a tree species is to be used in agroforestry, information on the ratio of R/FR light passing through its canopy is important. From this viewpoint, *Dalbergia sissoo* stands almost in the middle of the other species, whereas *Ougeinia dalbergioides* has a higher R/FR ratio of the transmitted light through its canopy and allows better under-canopy growth.

From a socioeconomic viewpoint, multipurpose species with higher utility index values will be more acceptable to farmers. *Dalbergia sissoo* and *Ougeinia dalbergioides* have the highest utility index values (Table 1). Therefore, from socioeconomic, ecological, and biological perspectives, these will be the best species from valleys to middle altitudes of about 6,000 feet (1,828 m) in the Central Himalaya.

Table 2. Average values for 20 tree species during active growth period.[1]

Species	Value[2]				
	REG (cm/day)	Ph (mg CO_2/m^2/s)	LUE (mol CO_2/E)	Tr (mg/m^2/s)	WUE (mg CO_2/mg H_2O)
Toona ciliata	0.0214	0.332	0.0059	143.8	0.0021
Grewia optiva	0.0163	0.532	0.0084	168.1	0.0030
Ougeinia dalbergioides	0.0193	0.514	0.0114	125.3	0.0048
Bauhinia retusa	0.0148	0.301	0.0064	104.6	0.0023
Dalbergia sissoo	0.0200	0.826	0.0147	215.9	0.0039
Sapindus mukorossi	0.0144	0.126	0.0015	92.9	0.0012
Terminalia tomentosa	0.0177	0.286	0.0051	106.9	0.0029
Ficus bengalensis	0.0147	0.607	0.0096	151.8	0.0040
Cordia myxa	0.0171	0.573	0.0114	163.7	0.0040
Holoptelea integrifolia	0.0156	0.239	0.0031	112.5	0.0018
Bombax ceiba	0.0185	0.451	0.0095	139.2	0.0035
Eugenia jambolana	0.0150	0.258	0.0051	116.2	0.0022
Cornus capitata	0.0104	0.344	0.0101	144.7	0.0024
Quercus glauca	0.0065	0.298	0.0069	122.8	0.0029
Q. incana	0.0052	0.198	0.0048	77.4	0.0030
Fraxinus micrantha	0.0111	0.147	0.0057	96.6	0.0015
Celtis australis	0.0150	0.334	0.0067	146.8	0.0023
Aesculus indica	0.0013	0.279	0.0061	97.7	0.0030
Prunus cerasoides	0.0191	0.519	0.0111	128.1	0.0040
Alnus nepalensis	0.0212	0.661	0.0098	164.7	0.0043

[1] April-Sept., 550 m elevation.

[2] REG = rate of extension growth, Ph = photosynthesis, LUE = light use efficiency, Tr = transpiration, WUE = water use efficiency.

Table 3. Average values of the characters related to fuelwood energy potential for 20 species.

SI No.*	Species	Calorific value (kJ/g dry wt)	Density (g/cc)	Ash (%)	Moisture (%)	Nitrogen (%)
1	*Toona ciliata*	22.01	0.49	1.4	59.84	0.57
2	*Grewia optiva*	16.87	0.52	0.9	67.22	0.62
	Ougeinia					
3	*dalbergioides*	16.79	0.68	0.6	67.78	0.38
4	*Bauhinia retusa*	16.34	0.75	4.2	50.00	0.36
5	*Dalbergia sissoo*	16.28	0.74	1.3	46.32	0.43
6	*Sapindus mukorossi*	16.12	0.90	1.1	39.18	0.60
	Terminalia					
7	*tomentosa*	15.89	0.50	2.8	52.43	0.27
8	*Ficus bengalensis*	15.71	0.63	3.0	66.67	0.61
9	*Cordia myxa*	15.70	0.40	1.0	72.16	0.87
	Holoptelea					
10	*integrifolia*	15.54	0.54	2.0	67.41	0.57
11	*Bombax ceiba*	15.20	0.61	4.2	59.84	0.47
12	*Eugenia jambolana*	14.39	0.69	2.0	48.98	0.51
13	*Cornus capitata*	22.40	0.74	2.5	52.58	0.16
14	*Quercus glauca*	17.86	0.85	1.6	39.54	0.19
15	*Q. incana*	15.74	1.00	1.6	47.47	0.24
16	*Fraxinus micrantha*	17.07	0.43	0.8	45.04	0.22
17	*Celtis australis*	16.81	0.54	3.4	57.53	0.40
18	*Aesculus indica*	16.22	0.89	2.4	44.67	0.30
19	*Prunus cerasoides*	16.02	0.72	2.7	49.69	0.05
20	*Alnus nepalensis*	15.09	0.43	1.9	61.50	0.43

* SI = superiority index; 1 = highest value of desirable characteristic and lowest value of undesirable characteristic.
Based on data of Purohit and Nautiyal, 1987.

Table 4 .   Final index values of Indian mountain tree species.

Sl No.[1]	Species	REG	Ph	WUE	LUE	CV	Density	Ash	Moisture	N	Tr	Rank
							Characteristics[2]					
1	Toona ciliata	1	11	11	13	2	15	7	13	14	13	10
2	Grewia optiva	9	5	6	8	5	13	3	16	17	19	12
3	Ougeinia dalbergioides	4	7	1	2	7	9	1	18	9	10	3
4	Bauhinia retusa	13	12	9	11	8	5	17	9	8	5	9
5	Dalbergia sissoo	3	1	4	1	9	6	6	5	11	20	1
6	Sapindus mukorossi	15	20	15	18	11	2	5	1	15	2	13
7	Terminalia tomentosa	7	14	7	15	13	14	14	10	6	6	14
8	Ficus bengalensis	14	3	3	5	15	10	15	15	16	16	17
9	Cordia myxa	8	4	3	2	16	16	4	19	18	17	16
10	Holoptelea integrifolia	10	17	12	17	17	12	10	17	14	7	20
11	Bombax ceiba	6	8	5	6	18	11	17	13	12	12	15
12	Eugenia jambolana	11	16	10	15	20	8	10	7	13	8	19
13	Cornus capitata	17	9	8	4	1	6	12	11	2	14	5
14	Quercus glauca	18	13	7	9	3	4	8	2	3	9	4
15	Q. incana	19	18	6	16	14	1	8	6	5	1	7
16	Fraxinus micrantha	16	19	14	14	4	16	2	4	4	3	8
17	Celtis australis	12	10	9	10	6	12	16	12	10	15	18
18	Aesculus indica	20	15	6	12	10	3	11	3	7	4	6
19	Prunus cerasoides	5	6	3	3	12	7	13	8	1	11	2
20	Alnus nepalensis	2	2	2	7	19	16	9	14	11	18	11

[1] SI = superiority index; 1 = highest value of desirable characteristic and lowest value of undesirable characteristic.

[2] REG = rate of extension growth, Ph = photosynthesis, WUE = water use efficiency, value, LUE = light use efficiency, CV = calorific value, N = nitrogen content of wood, Tr = transpiration.

Acknowledgment

Financial assistance from the Department of Non-Conventional Energy Sources, Government of India is gratefully acknowledged.

REFERENCES

Berg, M.A. van den. 1977. Natural enemies of certain Acacias in Australia. In *Proc. of the 2nd national weeds conf. of South Africa*. Stellenbosch.

Burley, J. 1980. Selection of species for fuelwood plantations. *Commonwealth Forestry Review* 59(2):133-147.

NAS. 1980. *Firewood crops: Shrub and tree species for energy production*. Washington, D.C.: National Academy of Sciences.

Osmaston, A.E. 1926. *A forest flora for Kumaon*. Allahabad.

Purohit, A.N. and A.R. Nautiyal. 1987. Fuelwood value index of Indian mountain tree species. *The International Tree Crops Journal* 4:177-182.

Effects of *Gliricidia sepium* Mulch
on Upland Rice Yield and Soil Fertility

Dominador G. Gonzal and Romeo S. Raros

Visayas State College of Agriculture
Baybay, Leyte, Philippines

The effects of Gliricidia sepium *mulch on the grain yield of upland rice (UPLB-Ri-5) and soil fertility were evaluated for two successive croppings at the Agroforestry Research and Demonstration Station, ViSCA, from May 1985 to March 1986. Double-row hedges of Gliricidia were spaced apart 0.5 m in elevation on a 20% average slope. Inter-row spacing of the hedges was 0.5m and within-row spacing was 0.25m. Grain yields of upland rice with Gliricidia mulch for the first and second harvests were 1.72 and 1.60 tons/ha, respectively, compared to 1.27 and 0.90 tons/ha, respectively, for the control plots. Organic matter content of the soil increased consistently in all treatments. However, organic matter accumulation was higher in the mulched plots. Soil phosphorus content was lower in unmulched treatments.*

In upland cultivation, vegetation is usually burned, leaving the area entirely bare. When it rains, run-off rushes downhill and erodes the soil. Because the land becomes unproductive, farming is possible for only two years. Farmers are forced to shift to other areas. The unproductive land is left untilled or fallow for five years. Mulching and contour hedgerow establishment of leguminous species, such as *Gliricidia sepium*, are two common erosion-control methods.

The objectives of this study were as follows:

o Determine the effect of Gliricidia mulch on the growth and yield of upland rice (UPLB-Ri-5) within contour hedgerows of Gliricidia and

o Appraise the ability of Gliricidia mulch and contour hedgerows to rehabilitate impoverished upland soils.

Methodology

The study was conducted in the Agroforestry Research and Demonstration Area of the Visayas State College of Agriculture from May 1985 to March 1986. The site is located on a Gliricidia-based alley cropping farm with three-year-old, double-row Gliricidia hedges. The hedges were planted on contours separated by 0.5m in elevation on a 20% average slope. Spacing between rows was 0.5m and 0.25m within rows. Experimental plots were laid out within the strips (areas between hedgerows) using a randomized complete block design with five replications. Each replication was subdivided into two plots of 4 x 5m. The treatments were upland rice plus Gliricidia mulch and upland rice alone. Rice was dibbled at $0.3m^2$ spacing at a 80 kg/ha seeding rate. The first mulching with Gliricidia was done immediately after planting, the second 45 days later, and the last at booting stage (before panicle initiation). No pesticides, synthetic fertilizer, or any related chemicals were used during the study. The experimental field was weeded by hand.

The crop was harvested after four months with hand sickles. Data gathered were plant height at harvest, herbage yield, number of filled and unfilled grains per panicle, and grain yield. Soil samples, before and after harvesting, were also taken and analyzed.

Results and Discussion

Yield and Yield Components of Upland Rice

Generally, Gliricidia mulch improved the overall growth and yield performance of upland rice. Mulched rice plants exhibited greener, longer, and broader leaves; larger stalks; and better growth than the unmulched plants (Table 1).

Plant Height

Highly significant differences in height between the mulched and unmulched plants were obtained during the first cropping (Table 1). In the second cropping, there was a marked reduction in plant height, although the mulched plants were still significantly taller than the control. The increment in height of the mulched plants, especially during the first cropping, resulted in the overlapping of leaves and higher degree of mutual shading among plants.

Number of Tillers per Plant

Mulching increased the number of tillers produced per plant relative to the control, but the difference was not statistically significant (Table 1). A noticeable increase in the number of tillers produced by the mulched and unmulched plants was observed during the second cropping. This could be attributed to the low germination percentage of the seeds used, which consequently led to less competition for nutrients and space. Hence, more tillers were produced compared to the first cropping.

Panicle Length and Number of Unfilled and Filled Grains per Panicle

Gliricidia mulch did not cause any significant increase in panicle length of the rice plants (Table 1). Mulching reduced the number and percent of unfilled grains per panicle compared to the control during the first and second croppings, but the reduction was not statistically significant. The number of filled grains per panicle of the mulched plants during the first cropping was significantly higher than that of the control. In the second cropping, no significant difference was obtained between treatments. The production of more filled grains per panicle by the mulched plants in both croppings could be due to increased photosynthetic rate of the plants. As mentioned above, mulched plants were taller than unmulched ones and, in most cases, the leaves of the mulched plants were broader and longer than those of the control. Therefore, mulched plants could intercept more sunlight needed for photosynthesis than could the unmulched ones, resulting in production of more carbohydrates, which, in turn, are translocated to the developing grains.

Herbage Yield

Herbage yield of rice plants with Gliricidia mulch was significantly higher than that of the control in both croppings (Table 1). The increase could be due to the better growth of the mulched plants. Mulched plants were taller and more vigorous than

the control and, hence, yield more biomass. Herbage yield during the second cropping, regardless of treatment, was almost 50% lower than that obtained during the first cropping. This could be explained by the relatively lower plant density caused by poor rice germination. The seeds used in the second cropping were from the harvest of the first cropping with an average germination of 65%. Perhaps some seeds were still dormant.

Grain Yield

Relative to the control, the grain yield of mulched rice plants increased by 15.06 and 28%, respectively, during the first and second croppings (Table 1). The grain yield of mulched plants was positively correlated to the production of more filled grains. The relatively lower grain yield in the control plants could be due to the reduction in net assimilation rate of the plants. Moreover, the leaves of unmulched plants were short and narrow, which suggests that the amount of light intercepted by these leaves was lower than for mulched plants. Consequently, a less grain was obtained from unmulched plants.

Soil Fertility

Gliricidia mulch generally increased the soil pH, percent of organic matter, and phosphorus and potassium content (Table 2). Soil pH increased slightly after the first cropping regardless of treatment and remained constant thereafter. Percent of organic matter and potassium content consistently increased with all treatments. However, more organic matter and potassium accumulation were obtained in treatments with Gliricidia mulch. Similarly, phosphorus content of the soil was also higher in mulched treatments than in unmulched ones.

Conclusions and Recommendations

Based on the results obtained, it can be concluded that establishment of Gliricidia contour hedgerows and the use of herbage from the hedgerows as mulch may greatly increase soil fertility and improve growth and yield of upland rice on hilly lands. Hillside farming can therefore become more profitable through alley cropping using nitrogen-fixing trees, such as Gliricidia, as contour hedgerows.

Table 1. Yield and yield components of upland rice (UPLB-Ri-5) using *Gliricidia sepium* mulch.

Treatment	Height (cm)	Tillers/ plant	Panicle length (cm)	Unfilled grains/ panicle	Filled grains/ panicle	Herbage yield (t/ha)
First Crop						
Control	99.70	2.68	23.2	70.30	83.00	16.70
Mulch	111.00[b]	2.80	24.8	61.30	96.50[a]	20.50[a]
CV (%)	4.33	13.00	4.2	33.90	9.98	9.88
Second Crop						
Control	89.70	3.40	20.50	14.10	59.30	6.90
Mulch	108.40[a]	3.80	19.60	12.60	61.90	11.00[a]
CV (%)	7.16	9.73	5.64	11.47	2.47	7.90

[a] = significant
[b] = highly significant

Table 2. Effect of *Gliricidia sepium* mulch on soil characteristics.

Treatment	pH	Organic matter (%)	P (ppm)	K (ppm)
Before Planting	4.55	3.48	0.65	120
First Crop				
Control	4.75	3.62	0.98	150
Mulched	4.60	3.82	1.65	171
Second Crop				
Control	4.60	5.22	0.69	139
Mulched	4.60	5.68	1.57	243

Further research is recommended to substantiate the results obtained. Soil analysis should include calcium, magnesium, and micronutrients. Soil erosion should be monitored to evaluate the effectiveness of Gliricidia hedgerows, as well as mulching, in controlling soil loss. The nodulation of Gliricidia hedgerows should also be monitored to program the planting of upland crops.

An Economic Model for Evaluating Charcoal Production
of Multipurpose Tree Species
Using the Transportable Metal Kiln

W.C. Woon

Forest Research Institute of Malaysia
Kepong, Malaysia

This paper evaluates the use of a micro-computer model for rubberwood (*Hevea brasiliensis*) charcoal production using the transportable metal kiln (TMK) in rubber smallholdings in Peninsular Malaysia. Although the demand for Malaysian rubberwood sawn timber has increased tenfold since 1980, most rubber trees are burned during replanting in rubber smallholdings. As more than 75% of the total area planted with rubber is under smallholdings, a substantial amount of rubberwood is wasted. According to Mohd. (1985), about 50% of the smallholdings are between 3-5 acres and about 45% are under 3 acres. Since the rubber smallholdings are small and transportation costs are high, contractors find it uneconomical to use the rubberwood. In view of this, the Forest Research Institute of Malaysia (FRIM), the Rubber Research Institute of Malaysia (RRIM) and the Rubber Industry Small Holders Development Authority (RISDA) have introduced the TMK for rubberwood charcoal production on rubber smallholdings. This type of production will provide an additional source of income to the smallholder.

The use of the TMK for charcoal production on rubber smallholdings was found technically and economically feasible (Hoi, Low, and Wong 1985; Hoi et al. 1986; and Woon, Hoi, and Low 1987). Two factors were identified as crucial to the success of the TMK: (1) the initial capital investment required and (2) the number of kilns used, given the small acreage of the individual rubber smallholding. The model described in this paper was designed to answer these questions and to determine whether government agencies should buy the kilns and rent them to smallholders.

Overview

The TMK consists of 2 interlocking cylindrical sections, a conical cover, and 8 steel air inlet/outlet channels. Details on the construction of the TMK are given by Whitehead (1980). When assembled,

the kiln is about 2.5 m in diameter and about 2 m high. It has a capacity of about 7 m^3 and can produce about 0.5 tons of charcoal per burn, each burn requiring 3.5 days. Hoi et al. (1986) and Woon, Hoi, and Low (1987) provide more detailed descriptions of the operations involved.

Set Up

This model was developed based on a study by Woon, Hoi, and Low (1987). The scientists investigated the economic feasibility of using the following operating schedules for rubberwood charcoal production using the TMK:

o   1 kiln/2 man

o   2 kilns/3 man

o   3 kilns/3 man

They also considered the options of buying and renting the kilns. The operating schedules are summarized in Table 1.

The following indicators and parameters are presented in graph form in the model:

o   unit cost of production ($ per kg)

o   profitability

o   payback period (months)

o   acreage of rubberwood required to break even

The model is composed of 3 levels with a total of 16 options (Fig. 1). There are 5 options in level 1, 4 in level 2, and 7 in level 3.

Table 1. Method of study.

Schedule	1 kiln/2 man	2 kilns/3 man	3 kilns/3 man
Own kiln(s)	Option Ia	Option IIa	Option IIIa
Rented kiln(s)	Option Ib	Option IIb	Option IIIb

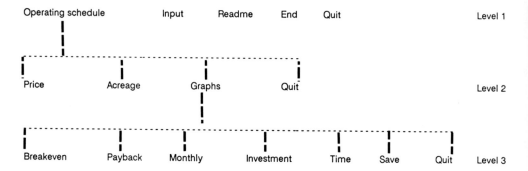

Fig. 1. Layout of the model.

Hardware and Software Requirements

Lotus 1-2-3 (R.2 or later release) system disk is required to use this model. The Printgraph disk is used for printing hard copies of the graphs. The model will run on any IBM-compatible micro-computer with at least 256k of RAM and a 2 floppy-disk system.

How To Use the Model

The user inserts the Lotus system disk in drive A and closes the latch before turning on the computer. When the blank worksheet appears on the monitor, he or she presses the '/ ' key to call up the Lotus 1-2-3 main menu, followed by 'f' for File and 'r' for Retrieve. At the 'what file to retrieve?' prompt, the user types TMKMODEL and presses 'Return.' The following main menu and instructions then appear in the top left-hand corner of the screen:

Operating Schedule

Input Readme End Quit

Operating schedule for 1, 2, and 3 kilns

The user then moves the cursor to the operating schedule option and presses 'Return.' On selection, the level 2 submenu appears and only the instructions on the control panel change. The instructions on the control panel now read as follows:

Price Acreage Graphs Quit

Projected profitability for a given price

The user may change both the selling price and acreage of rubberwood available by selecting the 'Price' and 'Acreage' options, respectively. When he selects the price option and presses 'Return,' the information presented in Fig. 2 appears on the screen.

The cursor is highlighted at the column '$0.30.' The user keys in the new selling price and presses 'Return.' Once the new selling price appears on the screen, he presses 'Return' again. This updates the figures in the second table. Figures in the last 5 rows (i.e., Profit $/kg, Sales $/mo., Profit $/mo., Payback period /mos., and Breakeven acreage) are recalculated. The user can only

change the selling price, as the other cells are protected. The following instructions on the control panel reappear:

Price Acreage Graphs Quit

Projected profitability for a given price

To change the acreage available, the user moves the cursor to the 'Acreage' option and presses the 'Return' key. Fig. 3 shows what then appears on the screen.

The user enters the new acreage at the cell highlighted 'Your Acreage Please ?' He presses 'Return' to update the acreage. When the figure (new acreage) appears, he presses 'Return' again to update the table. The following is recalculated: 'Time req'd/mths,' 'Gross Profit $,' and 'Net Returns $.' The user can only change the 'Acreage' figure, as the rest of the cells are protected. The level 2 submenu reappears:

Price Acreage Graphs Quit

Projected profitability for a given price

The 'Graphs' option allows the user to view the following parameters graphically: Breakeven period, the Payback period, Monthly Profit, Investment required and the time required to complete the acreage available. He selects the 'Graphs' option and presses 'Return.' The level-3 submenu appears:

Breakeven Payback Monthly Profit Investment
    Time Save Quit

Breakeven acreage at given price of charcoal for
    various kilns options

The user selects any one of the options by highlighting that option and pressing 'Return.' Graphs then appear on the screen. To print a hard copy of a graph, the user selects the 'Save' option immediately after viewing the graph and the computer then prompts the user:

Save the Graph of Quit

Net Returns and Time required to clear Holdings
    (small model)

267

Cost and profit figures for 1, 2, and 3 kilns.

Selling price please ($0.00) ?: $0.30

	Ia	Ib	IIa	IIb	IIIa
No. of kilns	1	1	2	2	3
No. of burns/month	8	8	16	16	22
Acreage used/month	0.64	0.64	1.28	1.28	1.76
Charcoal pdn kg/month	4,000	4,000	8,000	8,000	11,000
No. of laborers	2	2	3	3	3
Fixed costs, $	2,930	930.00	4930.00	930.00	6,930.00
Capital depreciation, $	70.44	14.88	140.87	29.76	207.59
Total Cost, $/month	1,050.44	1,094.88	1,680.87	1,769.76	1,852.59
Cost of charcoal, $/kg	0.26	0.27	0.21	0.22	0.17
Profit $/kg	0.04	0.03	0.09	0.08	0.13
Sales, $/month	1,200.00	1,200.00	2,400.00	2,400.00	3,300.00
Profit, $/month	149.56	105.12	719.13	630.24	1,447.41
Payback period, months	20.29	9.00	6.95	1.48	4.83
Breakeven acreage	12.99	5.76	8.89	1.90	8.51

Fig. 2. Price option screen.

Investment, profit, and returns for 1, 2, or 3 kilns

Your acreage please ?: 5 acres
　　　　　Price/kg: $0.30

	Ia	Ib	IIa	IIb	IIIa
Profit, $/kg	0.04	0.03	0.09	0.08	0.13
Time required, months	7.81	7.81	3.91	3.91	2.84
Gross profit, $	1,168.47	821.25	2,461.87	2,461.87	4,111.97
Investment, $	2930.00	930.00	930.00	930.00	6,930.00
Net returns, $	-1,761.53	-108.75	-2120.90	1,531.87	-2818.03

Fig. 3. Acreage option screen.

Charcoal production parameters

Cost per kiln	$2,000.00
Other equipment	930.00
Cost of labor/day	15.00
Cost of bucking/burn	5.00
Packaging cost/kg	0.025
Rental of kiln/month	100.00
Depreciation of kiln/month	55.56
Depreciation of tools/burn	1.86
No. of rubber trees/burn	8.00
Acreage used per burn	0.08
Charcoal production (kg/burn)	500.00
Bank interest rate (p.a.)	4.00%
Bank interest rate (p.m.)	0.33%

Fig. 4. Basic data input table.

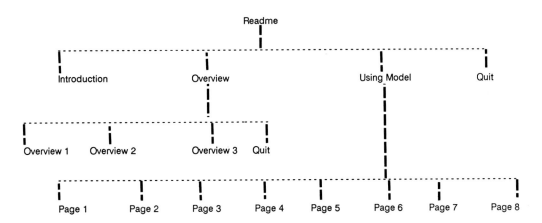

Fig. 5. Readme option screen.

He selects the option 'Save the Graph of' by pressing the 'Return' key. The computer prompts the user again:

Enter name of Graph file (8 chars max.):

The user types in a picture file name and presses 'Return.' The graph is stored with a <filename>.PIC extension. The Printgraph disk is used to print the desired graph. To return to level 2, the user selects the 'Quit' option and presses 'Return.' Similarly, to return to level 1, he selects the 'Quit' option and presses 'Return' again. The main menu reappears on the control panel:

Operating Schedule Input Readme End Quit

Operating schedule for 1, 2, and 3 kilns

The user moves the cursor to the 'Input' option and presses 'Return.' The basic data input appears in Fig. 4.

The user can change any or all of the data, and can use the up or down arrow key to move from one set of data to another. He presses 'Return' when all the changes have been made and updating will occur. This allows the model to be modified for other multipurpose wood species as input data, such as charcoal output, wages, and cost of investment, vary among species and location. The opportunity cost of not investing is also considered, and users are allowed to choose the interest rate. In this model, the current rate of 4% per annum is used. The level-1 menu instructions reappear:

Operating Schedule Input Readme End Quit

Operating schedule for 1, 2, and 3 kilns

The 'Readme' option provides useful information on how to use the model. Users are strongly advised to read this section before proceeding to the model proper. The layout of the 'Readme' option appears in Fig. 5.

The 'Introduction' option consists of 1 screen (page) of general information on rubberwood charcoal production using the TMK in Peninsular Malaysia. Three pages of information on the general layout of the model are given in the 'Overview' option, and 8 pages on the 'Using Model' option. To modify this model, the user

selects the 'End' option. This retrieves the Lotus worksheet environment where changes can be made. To end the modeling session, the user selects 'Quit.'

Conclusion

The TMK was found technically and economically feasible for rubberwood charcoal production in rubber smallholdings in Peninsular Malaysia. This model is used to determine the optimum operating schedule and only requires 2 inputs from the user (the selling price and acreage available). Four parameters are calculated automatically and can be viewed graphically. The parameters are breakeven acreage, the payback period in months, the monthly profits derived, and the time (in months) required to plant completely the total area of rubberwood available. The model is flexible and easily modified for other multipurpose tree species. Since the model is developed using Lotus 1-2-3 software, it is user-friendly and easily accessed by scientists using a microcomputer.

Anyone interested in obtaining a free copy of the program can write to the author.

Acknowledgment

The author is indebted to Mr. C.T. Wan for his invaluable assistance in writing the program (model) on the Lotus 1-2-3 software.

REFERENCES

Hoi, W.K., C.K. Low, and W.C. Wong. 1985. The Production of Charcoal by the Improved Transportable Metal Kiln Method. Paper presented at the 2nd Asian Conference on Technology For Rural Development, December 4-7, 1985, Kuala Lumpur.

Hoi, W.K., Mohd Ali Sujan, C.K. Low, and M. Megathevan. 1986. Small-scale Rubberwood Charcoal Production. Paper presented at the Grower's Conference, Oct. 1986, Ipoh, Malaysia.

Jalaluddin Harun, Abdul Rahman Md. Derus, and W.C. Wong. 1983. *Wood residue and its utilization in Peninsular Malaysia*. Serdang: Universiti Pertanian Malaysia.

Mohd., Rashid Ahmad. 1985. The Need To Modernize Smallholding Agriculture. Paper presented at the PETA'85 Conference on Current Agricultural Issues in Malaysia, May 14-15, 1985.

Whitehead, W.D.J. 1980. The construction of a transportable charcoal kiln. *Rural technology guide*. Tropical Products Institute, No.13.

Woon, W.C., W.K. Hoi, and C.K. Low. 1987. Economics of Rubberwood Charcoal Production Using the Transportable Metal Kiln. Paper submitted to The Planters, pending publication.

Fast-Growing Nitrogen Fixing Trees as MPTS for Fuelwood and Charcoal on Small Farms

Suttijed Chantrasiri

Thailand Institute of Scientific and Technological Research
Bangkok, Thailand

In Thailand, the ever-greater demand for fuelwood and charcoal is causing rapid degradation of natural forests and increased wood shortages. More than 90% of the fuelwood and 70% of the charcoal used by rural people come from the forest. Research and development on fast-growing, nitrogen-fixing trees is a feasible way to approach wood shortage, particularly as it affects small farmers.

Analysis and testing determined the following species suitable for potential use as fuelwood and charcoal: *Acacia mangium, A. auriculiformis, Leucaena leucocephala, L. diversifolia, Casuarina equisetifolia, C. junghuhniana, Calliandra calothyrsus, Gliricidia sepium*, and *Eucalyptus camaldulensis*. All have high calorific values, charcoal properties, and ability to coppice (Tables 1 and 2).

Table 1. Calorific values (kcal/kg) of five-year-old trees (dry basis).*

Species	Fuelwood	Charcoal (500 C)
Acacia auriculiformis	4690	7410
A. mangium	4640	7780
Calliandra calothyrsus	4610	7580
Eucalyptus camaldulensis	4550	7910
Gliricidia sepium	4550	7150
Casuarina equisetifolia	4550	7580
Albizia falcataria	4510	7780
Cassia siamea	4510	7680
Leucaena leucocephala	4450	7790

* Analysis by Energy Research Laboratory, TISTR.

Table 2. Regrowth data of trees one year after coppicing.

Species	Cutting at 15 mos.			Cutting at 24 mos.		
	Sprouting (%)	H (m)	Dbh (mm)	Sprouting (%)	H (m)	Dbh (mm)
Acacia auriculiformis	26	3.8	31	72	3.4	25
A. mangium	10	4.1	39	22	3.4	34
Albizia falcataria	56	4.5	33	76	4.5	31
Calliandra calothyrsus	90	2.9	20	52	2.2	15
Cassia siamea	100	3.1	21	97	2.4	18
Casuarina equisetifolia	2	3.5	27	1	0.6	4
Eucalyptus camaldulensis	90	4.7	29	88	4.8	29
Gliricidia sepium	99	2.2	16	73	1.4	7
Leucaena leucocephala	100	2.2	10	74	1.9	9

Introduction to the Acacia Hybrid

Jaffirin Lapongan

Forest Research Center, Forestry Department
Sabah, Malaysia

A hybrid of *Acacia auriculiformis* and *A. mangium* has been observed in Sabah since 1971, developing naturally where the 2 species are planted in close proximity. The hybrid has slender branchlets, large phyllodes approaching those of *A. mangium* in size, and white to cream-colored flowers, close in appearance to *A. mangium* flowers. It appears to grow better than *A. mangium* (Table 1), and adapts well to degraded soils. Potential uses for the hybrid include general construction, pulp and paper, fence posts, fuel, shade for cattle, and soil amelioration.

Characteristics

Acacia auriculiformis has slender branchlets, elongate phyllodes usually less than 30 mm wide, and lemon-yellow flowers (Fig. 1). *A. mangium* has thick branchlets that are acutely triangular. Its phyllodes are large and 2.5 to 5.3 times as long as their width (Fig. 2). Flower color is white to cream. The hybrid has slender branchlets, large phyllodes approaching those of *A. mangium* in size, and white to cream colored flowers, close in appearance to *A. mangium* flowers (Fig. 3).

The seeds of *Acacia mangium* are smaller and lighter than those of *A. auriculiformis*. Hybrid seeds most closely resemble seeds of *A. mangium*.

Adaptability

The hybrid is not demanding as far as soil is concerned. It grows well in degraded soil and competes successfully with *Eupatorium odoratum*, *Imperata cylindrica*, and other weeds.

Regeneration

Hybrid seedlings often grow naturally under parent trees (Fig. 4).

Timber Properties

Heartwood is brown, and sapwood is fairly narrow and pale cream to yellowish in color. Its grain is straight on the tangential face and slightly interlocked on the radial face. Its basic density is 0.53 g/cm^3.

Possible Uses

Potential uses include general construction, pulp and paper, fence posts, fuel, shade for cattle, and soil amelioration.

Table 1. Growth comparison of *Acacia mangium* and the Acacia hybrid at 2 sites.

Site and species	Age (yrs)	Height (m)	Dbh (cm)
Ulu Kukut			
Acacia hybrid	11	27.19	31.79
A. mangium	11	25.54	28.55
Bengkoka			
A. hybrid	7	19.71	24.44
A. mangium	7	16.83	18.08

Fig. 1. *Acacia auriculiformis*.

Fig. 2. *Acacia mangium*.

Fig. 3. The Acacia hybrid.

Fig. 4. Seedlings are often found growing naturally under the parent tree.

Appendices

Field Trip Summary

The workshop field trip, organized by the Kasetsart University Faculty of Forestry, provided the opportunity for participants to observe a variety of trees and agricultural crops grown in plantations and homesteads in Rayong Province near Pattaya.

Rayong Province has an area of 3,552 km². Its population is 416,894 and is increasing at an annual rate of 1.51%. The population density is 117 persons per km². The undulating land is suitable for many agricultural crops. The annual rainfall is about 1,324 mm with about 74 rainy days. The average temperature is 23.8°C and the average humidity 80.5%.

The forest area has been continuously converted for growing cash crops and fruit trees. About 7% of the land is covered by forest, most of which is located in the northern and eastern parts of the province. In the north, people still use fuelwood of mixed species. In the south, they use charcoal made from mangrove and rubber trees. No wood or charcoal scarcity exists in this province.

The workshop participants observed coconut plantations, mixed species in home gardens, plantations of *Casuarina junghuhniana, Eucalyptus camaldulensis, Bamboo* spp., and *Hevea braziliensis.*

Participants examine pineapple species.

The group visited Patanakaset Farm, Pluak Daeng District, which is owned by Mr. Sompong Kasetpibal, and observed intensive farming of *Ananas* spp., var. *battavia* and *Hevea braziliensis.* Another visit was made to the Rayong Agricultural Society, Pluak Daeng Branch, which was established 20 years ago. The society has about 1,500 members that farm a variety of agricultural crops, with sugarcane the primary crop.

Participants also visited a farm in Sri Racha District in Chol Buri. Its 6 hectares have various species grown in mixed plantings, including *Cocos nucifera, Tamarindus indica, Sandoricum indicum, Spondias pinnata, Artocarpus heterophyllus, A. lakoocha, Carica papaya, Nephelion lappaceum, N. longana, Annona squamosa, Salacia* spp., Ginger, *Azadirachta indica, Leucaena leucocephala, Areca catechu, Ceiba pentandra, Bamboo* spp., and *Mangifera indica.*

In some parts of the homestead, coconut is grown at a spacing of 10 x 10 m and pineapples are planted between the coconut rows. The group noted that mixed species plantings are suitable for homesteads. Old trees and branches are cut down when needed and are used as fuelwood.

Patrick Robinson (left) and Kovith Yantasath (right) discuss recent research efforts.

Participants

Australia

Brian Palmer
Commonwealth Scientific and Industrial Research
    Organization
Division of Tropical Crops and Pastures
Cunningham Laboratory
306 Carmody Road
St. Lucia, Queensland 4067

Bangladesh

Mohd. Zainul Abedin
Bangladesh Agricultural Research Institute
Joydebpur Gazipur 1701
Joydebpur

Kibriaul Khaleque
University of Dhaka
Department of Sociology
Dhaka 2

People's Republic of China

Lan Linfu
The Chinese Academy of Forestry
Subtropical Research Institute of Forestry
Fuyang, Zhejiang

Zheng Haishui
Research Institute of Tropical Forestry
Chinese Academy of Forestry
Longyandong
Guangzhou

Zhu Zhaohua
Chinese Academy of Forestry
Wan Shou Shan
Beijing

India

Amarnath Chaturvedi
TATA Energy Research Institute
7 Jor Bagh
New Delhi 110 003

Narayan Hegde
The Bharatiya Agro-Industries Foundation
Senapati Bapat Road
Pune 411 016

Karim Oka
International Development Research Centre
11, Jor Bagh
New Delhi 110003

A.N. Purohit
High Altitude Plant Physiology Research Centre
Garhwal University
Srinagar, Garhwal

Lal Relwani
The Bharatiya Agro-Industries Foundation
Uruli-Kanchan 412 202

Indonesia

Gunawan Sumadi
Ministry of Forestry
Manggala Wanabakti, Flor 13, Blok. I
Jl. Gatot Subroto, Senayan
Jakarta 10270

Malaysia

Jaffirin Lapongan
Forest Research Centre
P.O. Box 1407
Sandakan Sabah

Francis Ng
Forest Research Institute Malaysia
Kepong, Selangor
52109, Kuala Lumpur

Weng Chuen Woon
Forest Research Institute Malaysia
Kepong, Selangor
52109, Kuala Lumpur

Nepal

Pradeepmani Dixit
Farm Forestry Project
Jorepipal, Tangal Gairidhara
Kathmandu

Madhav Karki
Institute of Forestry
P.O. Box 43
Pokhara, Gandaki Zone
Pokhara

Yam B. Malla
Nepal-Australia Forestry Project
P.O. Box 208
Kathmandu

Patrick Robinson
Nepal-UK Forestry Research Project
P.O. Box 106
Kathmandu

Maheshwar Sapkota
Institute of Agriculture and Animal Science
Rampur, Chitwan

Ramesh Shakya
Forestry Research Project
Forest Survey and Research Office
Babar Mahal
Kathmandu

George Taylor
USAID/Office of Agriculture and Resource
    Conservation
Rabi Bhawan
Kathmandu

Khubchand G. Tejwani
International Centre for Integrated Mountain
    Development
P.O. Box 3226
Kathmandu

Kevin Joseph White
UNDP Asian Development Bank
Sagarnath Forest Project
P.O. Box 107
Kathmandu

Papua New Guinea

Kamane Saroa
Wau Ecology Institute
P.O. Box 77
Wau

Philippines

Rene Rafael Espino
University of the Philippines at Los Banos
Department of Horticulture
Laguna 3720

Dominador Gonzal
Visayas State College of Agriculture
No. 8 Lourdes Street
Pasay 3129

Saturnina Halos
Natural Sciences Research Institute
University of the Philippines, Diliman
Quezon 3004

Warlito Laquihon
Mindanao Baptist Rural Life Center
P.O. Box 94
Davao City 9501

Eduardo Mangaoang
Visayas State College of Agriculture
Department of Forestry
Baybay, Leyte 7127

Republic of Singapore

C. Devendra
International Development Research Centre
Tanglin P.O. Box 101
Singapore 9124

Cherla Sastry
International Development Research Centre
Tanglin P.O. Box 101
Singapore 9124

Sri Lanka

Vajira Liyanage
Coconut Research Institute
Lunuwila

Thailand

Jacques Amyot
Chulalongkorn University Social Research
  Institute
Bangkok 10500

Suree Bhumibhamon
Kasetsart University
Department of Silviculture
Faculty of Forestry
Bangkok 10903

Suttijed Chantrasiri
Thailand Institute of Scientific
  andTechnological Research
196 Phaholyothin Road
Bangkhen
Bangkok 10900

Kenneth G. MacDicken
F/FRED Coordinating Unit
c/o Faculty of Forestry
Kasetsart University
P.O. Box 1038
Kasetsart Post Office
Bangkok 10903

Charles B. Mehl
F/FRED Coordinating Unit
c/o Faculty of Forestry
Kasetsart University
P.O. Box 1038
Kasetsart Post Office
Bangkok 10903

G. Lamar Robert
Payap Research and Development Center
Payap University
Chiang Mai 50000

Narong Srisawas
Kasetsart University
Department of Sociology and Anthropology
Faculty of Social Sciences
Bangkhen
Bangkok 10900

Arjen Sterk
Food and Agriculture Organization
  of the United Nations
P.O. Box 13
Phu Wiang 40150

Cor Veer
c/o FAO/RAPA
GCP/RAS/111/NET
Maliwan Mansion
Phra Atit Road
Bangkok

Kovith Yantasath
Thailand Institute of Scientific and
  Technological Research
196 Phaholyothin Road
Bangkhen
Bangkok 10900

United Kingdom

Julian Evans
Forestry and Land Use Programme
International Institute for Environment
  and Development
London

United States

Dennis V. Johnson
USDA Forest Service/USAID Forestry
  Support Program
P.O. Box 96090
Washington, D.C. 20090

Workshop participants gather outside the Royal Cliff Beach Hotel conference center in Pattaya, Thailand.